AF566956

MTP International Review of Science

Radiochemistry

MTP International Review of Science

Publisher's Note

The MTP International Review of Science is an important new venture in scientific publishing, which we present in association with MTP Medical and Technical Publishing Co. Ltd. and University Park Press, Baltimore. The basic concept of the Review is to provide regular authoritative reviews of entire disciplines. We are starting with chemistry because the problems of literature survey are probably more acute in this subject than in any other. As a matter of policy, the authorship of the MTP Review of Chemistry is international and distinguished; the subject coverage is extensive, systematic and critical; and most important of all, new issues of the Review will be published every two years.

In the MTP Review of Chemistry (Series One), Inorganic, Physical and Organic Chemistry are comprehensively reviewed in 33 text volumes and 3 index volumes, details of which are shown opposite. In general, the reviews cover the period 1967 to 1971. In 1974, it is planned to issue the MTP Review of Chemistry (Series Two), consisting of a similar set of volumes covering the period 1971 to 1973. Series Three is planned for 1976, and so on.

The MTP Review of Chemistry has been conceived within a carefully organised editorial framework. The over-all plan was drawn up, and the volume editors were appointed, by three consultant editors. In turn, each volume editor planned the coverage of his field and appointed authors to write on subjects which were within the area of their own research experience. No geographical restriction was imposed. Hence, the 300 or so contributions to the MTP Review of Chemistry come from many countries of the world and provide an authoritative account of progress in chemistry.

To facilitate rapid production, individual volumes do not have an index. Instead, each chapter has been prefaced with a detailed list of contents, and an index to the 10 volumes of the MTP Review of Inorganic Chemistry (Series One) will appear, as a separate volume, after publication of the final volume. Similar arrangements will apply to the MTP Review of Physical Chemistry (Series One) and to subsequent series.

Butterworth & Co. (Publishers) Ltd.

Inorganic Chemistry
Series One
Consultant Editor
H. J. Emeléus, F.R.S.
Department of Chemistry
University of Cambridge

Volume titles and Editors

1 **MAIN GROUP ELEMENTS—HYDROGEN AND GROUPS I–IV**
Professor M. F. Lappert, *University of Sussex*

2 **MAIN GROUP ELEMENTS—GROUPS V AND VI**
Professor C. C. Addison, F.R.S. and Dr. D. B. Sowerby, *University of Nottingham*

3 **MAIN GROUP ELEMENTS—GROUP VII AND NOBLE GASES**
Professor Viktor Gutmann, *Technical University of Vienna*

4 **ORGANOMETALLIC DERIVATIVES OF THE MAIN GROUP ELEMENTS**
Dr. B. J. Aylett, *Westfield College, University of London*

5 **TRANSITION METALS—PART 1**
Professor D. W. A. Sharp, *University of Glasgow*

6 **TRANSITION METALS—PART 2**
Dr. M. J. Mays, *University of Cambridge*

7 **LANTHANIDES AND ACTINIDES**
Professor K. W. Bagnall, *University of Manchester*

8 **RADIOCHEMISTRY**
Dr. A. G. Maddock, *University of Cambridge*

9 **REACTION MECHANISMS IN INORGANIC CHEMISTRY**
Professor M. L. Tobe, *University College, University of London*

10 **SOLID STATE CHEMISTRY**
Dr. L. E. J. Roberts, *Atomic Energy Research Establishment, Harwell*

INDEX VOLUME

Physical Chemistry
Series One
Consultant Editor
A. D. Buckingham
Department of Chemistry
University of Cambridge

Volume titles and Editors

1 **THEORETICAL CHEMISTRY**
Professor W. Byers Brown, *University of Manchester*

2 **MOLECULAR STRUCTURE AND PROPERTIES**
Professor G. Allen, *University of Manchester*

3 **SPECTROSCOPY**
Dr. D. A. Ramsay, *National Research Council of Canada*

4 **MAGNETIC RESONANCE**
Professor C. A. McDowell, *University of British Columbia*

5 **MASS SPECTROMETRY**
Professor A. Maccoll, *University College, University of London*

6 **ELECTROCHEMISTRY**
Professor J. O'M Bockris, *University of Pennsylvania,*

7 **SURFACE CHEMISTRY AND COLLOIDS**
Professor M. Kerker, *Clarkson College of Technology, New York*

8 **MACROMOLECULAR SCIENCE**
Professor C. E. H. Bawn, F.R.S., *University of Liverpool*

9 **CHEMICAL KINETICS**
Professor J. C. Polanyi, F.R.S., *University of Toronto*

10 **THERMOCHEMISTRY AND THERMODYNAMICS**
Dr. H. A. Skinner, *University of Manchester*

11 **CHEMICAL CRYSTALLOGRAPHY**
Professor J. Monteath Robertson F.R.S., *University of Glasgow*

12 **ANALYTICAL CHEMISTRY – PART 1**
Professor T. S. West, *Imperial College, University of London*

13 **ANALYTICAL CHEMISTRY – PART 2**
Professor T. S. West, *Imperial College, University of London*

INDEX VOLUME

Organic Chemistry
Series One
Consultant Editor
D. H. Hey, F.R.S.
Department of Chemistry
King's College, University of London

Volume titles and Editors

1 **STRUCTURE DETERMINATION IN ORGANIC CHEMISTRY**
Professor W. D. Ollis, *University of Sheffield*

2 **ALIPHATIC COMPOUNDS**
Professor N. B. Chapman, *Duke University, North Carolina*

3 **AROMATIC COMPOUNDS**
Professor H. Zollinger, *Swiss Federal Institute of Technology*

4 **HETEROCYCLIC COMPOUNDS**
Dr. K. Schofield, *University of Exeter*

5 **ALICYCLIC COMPOUNDS**
Professor W. Parker, *University of Stirling*

6 **AMINO ACIDS AND PEPTIDES**
Professor D. H. Hey, F.R.S. and Dr. D. I. Johns, *King's College, University of London*

7 **CARBOHYDRATES**
Professor G. O. Aspinall, *University of Trent, Ontario*

8 **STEROIDS**
Dr. W. D. Johns, *G. D. Searle & Co., Chicago*

9 **ALKALOIDS**
Professor K. Wiesner, *University of New Brunswick*

10 **FREE RADICAL REACTIONS**
Professor W. A. Waters, F.R.S., *University of Oxford*

INDEX VOLUME

Inorganic Chemistry
Series One

Consultant Editor
H. J. Emeléus, F.R.S.

MTP International Review of Science

Volume 8
Radiochemistry

Edited by **A. G. Maddock**
University of Cambridge

Butterworths · London
University Park Press · Baltimore

THE BUTTERWORTH GROUP

ENGLAND
Butterworth & Co (Publishers) Ltd
London: 88 Kingsway, WC2B 6AB

AUSTRALIA
Butterworth & Co (Australia) Ltd
Sydney: 586 Pacific Highway 2067
Melbourne: 343 Little Collins Street, 3000
Brisbane: 240 Queen Street, 4000

NEW ZEALAND
Butterworth & Co (New Zealand) Ltd
Wellington: 26–28 Waring Taylor Street, 1

SOUTH AFRICA
Butterworth & Co (South Africa) (Pty) Ltd
Durban: 152–154 Gale Street

ISBN 0 408 70260 5

UNIVERSITY PARK PRESS

U.S.A. and CANADA
University Park Press Inc
Chamber of Commerce Building
Baltimore, Maryland, 21202

Library of Congress Cataloging in Publication Data

Maddock, A. G.
Radiochemistry

(Inorganic chemistry, series one, v. 8) (MTP international review of science)
Includes bibliographies
1. Radiochemistry. I. Title
QD151.2.15 vol. 8 [QD601.2] 541′.38 70–160330
ISBN 0–8391–1011–1

MTP MEDICAL AND TECHNICAL PUBLISHING CO. LTD.
Seacourt Tower
West Way
Oxford. OX2 OJW
and
BUTTERWORTH & CO. (PUBLISHERS) LTD.

Filmset by Photoprint Plates Ltd., Rayleigh, Essex
Printed in England by Redwood Press Ltd., Trowbridge, Wilts
and bound by R. J. Acford Ltd., Chichester, Sussex

Consultant Editor's Note

The problem of keeping abreast of research literature on as broad a front as possible is one that confronts all chemists. In the past this difficulty has been met, in the main, by literature surveys and by several uncorrelated reviews of progress in certain subject areas. There are obvious inadequacies in this approach, which have become increasingly apparent in recent years. I was, therefore, grateful for the opportunity of helping to plan this new series, which has been designed to provide a comprehensive, critical survey of each of the main branches of chemistry.

This section of the MTP International Review of Science deals with progress in Inorganic Chemistry. The subject is developing at an astonishing rate and in many directions. Fortunately, however, it lends itself to a systematic treatment. Ten volumes have been prepared, three dealing with the main group elements and two with the general chemistry of the transition metals. Organometallic derivatives of the main group elements and lanthanides and actinides are covered separately, as is the subject of reaction mechanisms. The two remaining volumes on radiochemistry and solid state chemistry have been planned to avoid, as far as possible, overlap with those that have gone before.

It is a pleasure to thank the many experts who have collaborated as authors and volume editors in making this publication possible. While working to a pre-arranged over-all plan, they have been able to assess and interpret the literature in terms of their own experience in specialised fields. I believe that in this way they will not only provide a record of what has been done, but will stimulate further exploration in this fascinating branch of chemistry.

Cambridge

H. J. Emeléus

Preface

The field of radiochemistry is, perhaps, not as clearly defined as that indicated by the titles of many of the other volumes in this series. It is certainly no less active, as the contributions to this volume show. Heavy element chemistry has received a considerable stimulus by the report earlier this year of the very economic discovery of isotopes in the region of comparative nuclear stability believed to exist round element 112.

The radiochemistry of fission, a subject curiously neglected for many years, has blossomed again. Some topics, such as positronium and mesonic chemistry and the use of angular correlations as a guide to chemical environment, have been recognised for several years, but it is only in the last few years that a sufficient body of results has been forthcoming to allow a reasonably critical assessment of their value.

In the area of hot atom chemistry the very successful theoretical model developed by Wolfgang, Estrup, Rowlands and collaborators has begun to break down under recent more searching investigations. How far the original ideas can be salvaged and the theory modified is a controversial issue at present. In the solid state considerable progress has been made towards an understanding of the effects of nuclear transformations, but it is very clear that even in the simplest systems the processes are rather complex. An interesting spin-off from these studies has been the data accumulated on isotopic exchange reactions in the solid state.

Cambridge A. G. Maddock

Contents

1
Nuclear Fission

G. N. WALTON
Imperial College of Science, University of London

1.1 INTRODUCTION

Since the discovery of nuclear fission in 1939[1], the experimental investigation of the process has never ceased to reveal strange and unexpected features. The theoretical structures that have been built on the experimental evidence are continually changing. For the most part the theory is extended rather than destroyed and re-built, so that the old concepts, as for example that of the liquid drop model, which was proposed by Bohr and Wheeler[2] within a few weeks of the original discovery, remain valid but greatly enriched.

The process of nuclear fission has a fascination rather like the pupation of an insect. On the one hand, the initial processes in which the compound nucleus is formed, and on the other, the final process in the emergence of fission products may be separately studied, but the changes that go on in between, like those that must go on in the chrysalis as the caterpillar changes to the butterfly, are less easily probed. The metaphor may be pursued a little; a liquid drop may lead to a multiplicity of products as a result of external forces, such as the proximity of other drops, pressure differences and so forth, but the more that is known about fission, the more it appears that the resulting products are dependent on the internal complexity of the nucleus. This is like the pupation of an insect, which depends mostly upon the internal molecular structure of the chromosomes and less on the soil and climate in which the process occurs.

The early work on fission has been extensively reviewed, notably by Earl Hyde[3] and by Halpern[4]. Work leading up to the first International Atomic Energy Agency Symposium on the Physics and Chemistry of Fission in Salzburg in 1965, was reviewed by Gindler and Huizenga[5] in 1968. Since that time a second I.A.E.A. Symposium has been held in Vienna in which a number of major new developments were reported and discussed[6]. This article describes the current position as seen at that Symposium. There has also been considerable renewed attention to those aspects of fission which have technological significance in the development of nuclear reactors, and this will also be discussed.

1.2 SPONTANEOUS FISSION ISOMERS

The most important development of the last few years is the recognition of a metastable state, as an intermediate structure, in the fission process. The evidence for this comes from the reconciliation of several widely-different phenomena:

(a) Spontaneous fission of very short half-life, observed in the nuclear reactions of fissile nuclides;

(b) the effect of the shell structure of nuclei on the asymmetry of low energy fission;

(c) narrow resonance effects in the fission cross-section at low energies;

(d) broad resonance in the fission cross-section at high energies.

1.2.1 Discovery of $^{242}Am_m$

Spontaneous fission isomers were first observed by Polikanov[7] in heavy ion experiments using the 300 cm cyclotron at the Joint Institute of Nuclear Research at Dubna. ^{16}O, ^{20}Ne, ^{22}Ne and ^{11}B ions were accelerated to bombard various targets including ^{235}U, ^{238}U, ^{232}Th. Recoil atoms, produced by 'knock-on' in the target, were estimated to have ranges of the order of 25 μg cm^{-2}, and the targets were made about 60 μg cm^{-2} thick so that a large fraction of the recoil atoms formed could escape. The recoil atoms were caught on an aluminium disc revolving at about 800 rev/min and carried by the disc to two solid-state detectors placed at intervals to scan the surface of the disc. Fission fragment pulses were observed in the detectors, and the difference in count rate in the two detectors suggested that the source of the fragments was decaying with a half-life of 12 to 14 ms. The amounts formed tended to show a maximum at different energies of the bombarding particles in the range 90 to 150 MeV. In the initial experiments, from the fact that $^{238}_{92}U$ and $^{11}_{5}B$ produced a positive source it was inferred that the atomic number must be $\leqslant 97$. The same source appeared to be formed in the reaction $^{238}_{92}U + ^{20}_{10}Ne$. From this it was inferred that the phenomenon was not associated with a heavy compound nucleus because there did not appear to be enough energy in the latter case to 'boil-off' sufficient neutrons and protons to give the same product as that which arose from the reaction with $^{11}_{5}B$. In this first paper it was, therefore, suggested that the fragments were being generated by the spontaneous fission of an isomeric state of a nucleus formed by fragmentation of the bombarding particle. It was noticed that the nature of the initial particle seemed to have little effect on the yield, whereas the nature of the target had a large effect.

In subsequent experiments[8], using α-particles and deuterium nuclei accelerated in a 150 cm cyclotron in an extracted beam, it was found that, while targets of Pu (both ^{239}Pu and a mixture of ^{240}Pu, ^{241}Pu and ^{242}Pu were tried) ^{241}Am, and ^{243}Am, gave the spontaneous fission source; ^{238}U did not. Furthermore, ^{239}Pu produced the product with α-particle bombardment but not with deuterons. This narrowed the possibilities for the atomic number of the spontaneously-fissile material to 95, i.e. americium. If neptunium isotopes were formed, they would have to arise by reactions in which charged particles were released, and these charged particles, such as α-particles and protons, were unlikely to be released on energy considerations.

The reactions in which the ^{242}Pu, ^{240}Pu mixture was bombarded with 19.8 MeV deuterons gave the highest yield, and it was considered that the product was most likely to be an isomer of $^{241}_{95}Am$ or $^{242}_{95}Am$. Subsequent work confirmed the reaction to be $^{242}Pu(d, 2n)\ ^{242}Am^{m}$ [9]. The spontaneous fission half-life of ^{242}Am is about 10^{12} years, and the isomeric state therefore has a half-life of the order 10^{19} times shorter. This was difficult to understand. Short half-lives for isomers might be associated with high angular momentum changes, but the angular momenta introduced by 16 MeV deuterons could not exceed 10 or 12 BM. It also appeared unlikely that cascade processes in γ-emission during formation could introduce very high spin.

The isomer was also formed by the neutron irradiation of ^{243}Am [10], and it was observed that the reaction had a very sharp threshold. The reaction rate rose from that corresponding to a cross-section of 0.024×10^{-28} cm^2 at 9.4 MeV to 0.20×10^{-28} cm^2 at 10.2 MeV, and a threshold energy of 9.2 ± 0.3 MeV was defined. In this reaction there is no coulomb barrier effect, and by subtracting the threshold energy for the formation of the ground state of ^{242}Am (6.37 MeV) a value of 2.9 MeV was assigned to the excitation energy of the fission isomer. Again, it was considered that an ordinary type of nucleus at this excitation, and of low spin state, should decay by γ-radiation faster than the observed half-life for spontaneous fission, and an unusual type of stable deformation was indicated.

1.2.2 Methods of detection

Since the discovery of $^{242}Am^{m}$, a search has been made for other isomers. Although a number have been found, $^{242}Am^{m}$ remains the isomer with the longest half-life so far to be identified. Two new techniques have been employed as described by Lark *et al.*[11]. In the first method, two large detectors

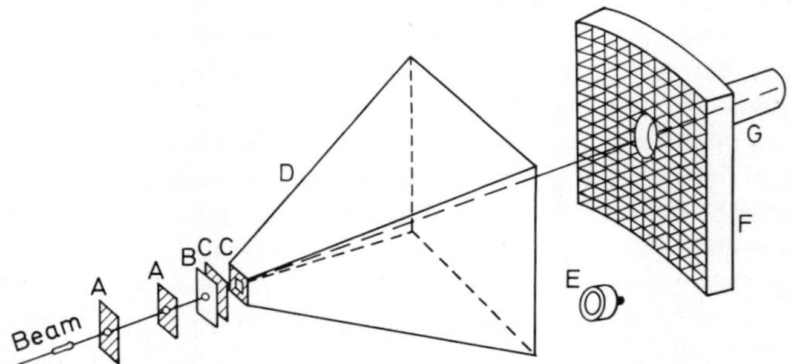

Figure 1.1 Recoil arrangement used for measuring half-lives of isomers. The parts indicated are: (A) beam collimator; (B) target; (C) fission isomer collimators; (D) detector cone; (E) monitor counter; (F) 'black body' isomer trap; (G) Faraday cup. (From Lark *et al.*[11], by courtesy of North-Holland Publishing Company)

are placed with their sensitive surfaces following the beam direction, behind a very thin target, which is bombarded with particles from an accelerator, as shown in Figure 1.1. The compound nucleus recoils from the target as a knock-on atom at a velocity of about 10^7 cm s^{-1}. If the nucleus undergoes spontaneous fission during the time that it is traversing the length of the detectors, the fission fragments are distributed along the detector according to the time between formation and decay. The assumption is made that the full momentum of the incident particle is transferred to the target nucleus. Alternative decay processes such as α- or γ-emission are not detected. The apparatus is calibrated by running a point source of ^{252}Cf, which is spontaneously fissile, at velocities varying exponentially between the detectors.

The method detects half-lives up to about 1 μs. If a catcher foil is placed between the detectors, a check can be made as to whether any longer-lived isomers are being formed. The detectors consist of polycarbonate foils (Makrofol) 0.7 mg cm^{-2} thick. These are etched for 25 min in 6 N NaOH

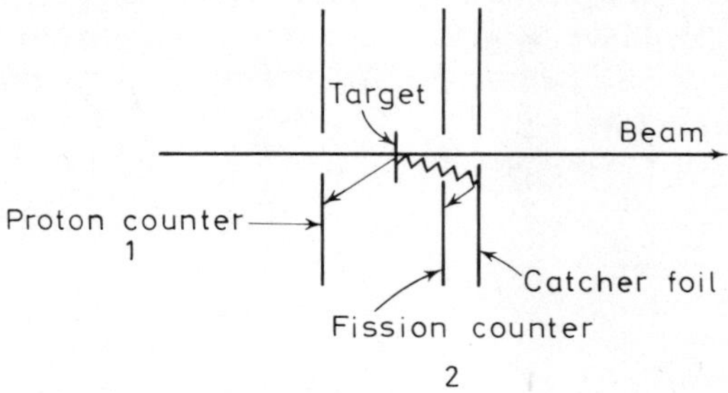

Figure 1.2 Apparatus for detection of spontaneous fission isomers by direct timing following proton emission. (From Lark *et al.*[11], by courtesy of North-Holland Publishing Company)

at 70 °C so as to enlarge the fission tracks and make them visible in an optical microscope. Under some circumstances the tracks appear in an electron microscope without etching. For automatic recording, the tracks may be filled by condensing metal vapour into them and recording the electrical conductivity[12].

The second method depends on delayed coincidences between proton emission and fission in the reaction (d, pf) following deuteron bombardment. This has a great advantage over the previous method in that assignment of the mass and charge of the product is more clearly defined. The detectors, consisting of annular shaped semiconductors, are placed with their sensitive surfaces looking down-stream along a proton beam, with a target and catcher foil placed centrally, as shown in Figure 1.2. Delayed coincidences are observed between protons detected in counter 1 and fission fragments detected in counter 2. A plot of the pulse rate against the coincidence delay time gives a half-life for spontaneously fissile products in the catcher foil.

These two methods gave results in good agreement with each other for many different targets. From the results of the ratio between the cross-section for formation of the spontaneously fissile isomer, and the prompt fission cross-section of the target nucleus, an absolute value for the former cross-section was obtained. Also, from the differences in the energy of the bombarding particle at the threshold of the reaction for forming the spontaneously fissile isomer, and the threshold energy for nuclear reactions in the ground state, estimates could be made of the excitation energy of the isomer.

Work, similar to the first method using large Makrofol detectors, was carried out by Metag *et al.* at the Max Planck Institute in Germany[13]. They used α-particle beams and obtained a number of new isomers, including some among the higher trans-uranic nuclides, but in these cases the attribution was rather more doubtful than the attribution of isomers obtained with lighter particle bombardments.

A third method for detecting spontaneously fissile isomers was reported by Vandenbosch and Wolf[14], in which a direct measurement was made between the time of a burst of particles from a pulsed cyclotron and the time distribution of the fission events. α-Particles were used, and the beam was pulsed in bursts of 2 ns with 264 ns between bursts. Prompt fissions were observed in various targets, and the identification of the spontaneously fissile isomers depended on an analysis of the 'tails' of the prompt fission curves. Results of the half-lives from these (α,xn) reactions were in the nanosecond range only.

1.2.3 Other isomers

In Table 1.1, a list of spontaneously fissile isomers is compiled for the results of the experiments described above up to the end of 1970. Figure 1.3 shows a plot of the half-lives of the isomers, compared with the well-known systematics of the half-lives of the ground states, as a function of the fission parameter Z^2/A. It does not, as yet, appear that the half-lives of the isomers show systematic variations comparable with those of the ground states. Large numbers of the isomers have half-lives in the 50 ns region. This may indicate that some type of highly distorted structure is being formed in all cases, or it may simply reflect the limitations of the experimental methods used for the detection of the isomers.

1.2.4 ^{241}Pu

In a measurement of the radioactivity of ^{241}Pu, Nisle and Stephan observed a short half-life component of 0.34 ± 0.11 years, in addition to the well-known half-life of 14 years for the β-decay of ^{241}Pu into ^{241}Am. It was suggested that this was a case of a very long-lived isomer[15]. A compilation by Cabell in 1953 of the measured half-life of ^{241}Pu also showed that the apparent value was 14.1 ± 0.2 years, while in 1968 the preferred value was 14.98 ± 0.33 years[16]. It was suggested that a discrepancy of this nature could be due to an unrecognised isomer. If such an isomer existed, it would be expected to have an excitation function for the fission cross-section different from that of the ground state. Careful measurements have now been completed of the fission cross-section of freshly-prepared ^{241}Pu, obtained by irradiating ^{240}Pu in a neutron flux, and of ^{241}Pu over 7 years old. Similar comparisons have been made of the γ-emission, and of the α-emission, of new and old ^{241}Pu, and in no case has evidence been found for an isomer[17].

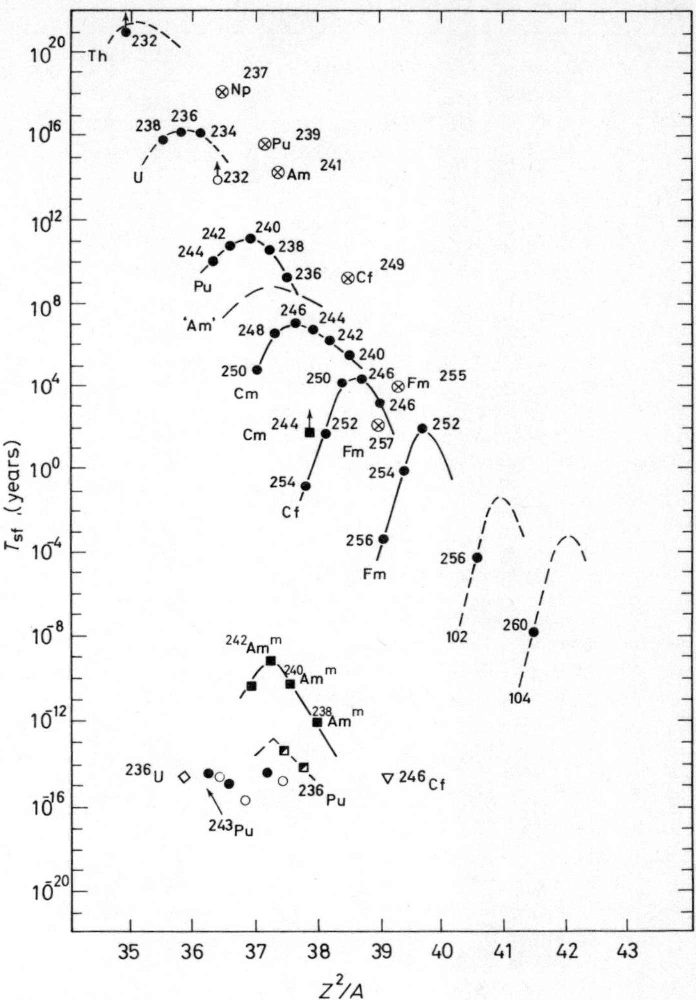

Figure 1.3 Spontaneous fission life-times as a function of Z^2/A. The short-lived isomers appear as a group in the lower left-hand corner of the figure. (From Strutinsky and Pauli[29], by courtesy of IAEA)

Table 1.1 Spontaneously-fissile isomers

Isomer	*Mode of formation*	*Half-life* µs	*Excitation energy* MeV	*Reference*
^{234}U	^{238}U (d, p)	0.030 ± 0.005		14c
	^{235}U (n, 2n)	0.020 ± 0.005	2.97	14d
^{235}U	^{234}U (n, γ)	0.020 ± 0.005	3.32	14c
^{236}U	^{235}U (d, p)	0.070 ± 0.020		14e
^{236}U	^{236}U (d, pn)	0.070 ± 0.020	2.6	14e
^{236}U	^{235}U (n, γ)	0.067 ± 0.009		14d
^{236}U	^{235}U (d, p)	0.110 ± 0.050		11
^{238}U	^{238}U (d, pn)	0.110 ± 0.030		14e
^{235}Pu	^{233}U (α, 2n)	0.020		13
^{236}Pu	^{237}Np (p, 2n)	0.034 ± 0.008		11
^{237}Pu^{m_i}	^{235}U (α, 2n)	0.082 ± 0.008		14c
^{237}Pu^{m_i}	^{236}U (α, n)	0.082 ± 0.008		14c
^{237}Pu^{m_i}	^{237}Np (d, 2n)	0.082 ± 0.008		14c
^{237}Pu^{m_i}	^{235}U (α, 2n)	1.120 ± 0.080		14c
^{237}Pu^{m_i}	^{236}U (α, n)	1.120 ± 0.080		14c
^{237}Pu^{m_i}	^{237}Np (d, 2n)	1.120 ± 0.080		14c
^{238}Pu	^{235}U (α, n)	< 0.002		11
^{239}Pu	^{236}U (α, n)	0.034	4.1	11
^{240}Pu	^{239}Pu (d, p)	0.009 ± 0.004	2.4	11
^{240}Pu	^{238}U (α, 2n)	0.004 ± 0.001		13
^{240}Pu	^{239}Pu (n, γ)	0.029 ± 0.004		14d
^{240}Pu	^{239}Pu (n, γ)	0.004		14d
^{241}Pu	^{240}Pu (d, p)	0.030 ± 0.005		11
^{242}Pu	^{241}Pu (d, p)	0.050 ± 0.030	< 3.3	11
^{243}Pu	^{242}Pu (n, γ)	0.060	3.4	11
^{243}Pu	^{242}Pu (d, p)	0.060	3.4	11
^{238}Am		60.0	2.9	33
^{239}Am	^{239}Pu (d, 2n)	0.240 ± 0.080		11
^{239}Am	^{240}Pu (p, 2n)	0.160 ± 0.040	3.2	11
^{239}Am	^{237}Np (α, 2n)	0.110		13
^{240}Am	^{241}Pu (p, 2n)	900 ± 70	3.15	80
^{241}Am	^{241}Pu (d, 2n)	1.5 ± 0.6	2.9	11
^{241}Am	^{242}Pu (p, 2n)	1.5 ± 0.6		11
^{242}Am	^{241}Am (n, γ)	14000	3.1	8
^{242}Am	^{243}Am (n, 2n)	14000		
^{242}Am	^{242}Pu (d, 2n)	14000		
^{241}Cm	^{239}Pu (α, 2n)	0.020		13
^{244}Cm	^{242}Pu (α, 2n)	0.012		13
^{245}Cm	^{242}Pu (α, n)	0.50		
^{243}Bk	^{241}Am (α, 2n)			13
^{244}Bk	^{241}Am (α, n)	0.100		

1.3 SHELL STRUCTURE EFFECTS AND FISSION THEORY

1.3.1 The statistical theory of fission

It has long been supposed that the asymmetry of low-energy fission is associated with the shell structure and stability of the fragments, but the precise nature of the association has evaded analysis. Figure 1.4 shows a number of fission yield curves superimposed. This shows that in all types of fission mass yields appear to predominate at A = 132, which would correspond

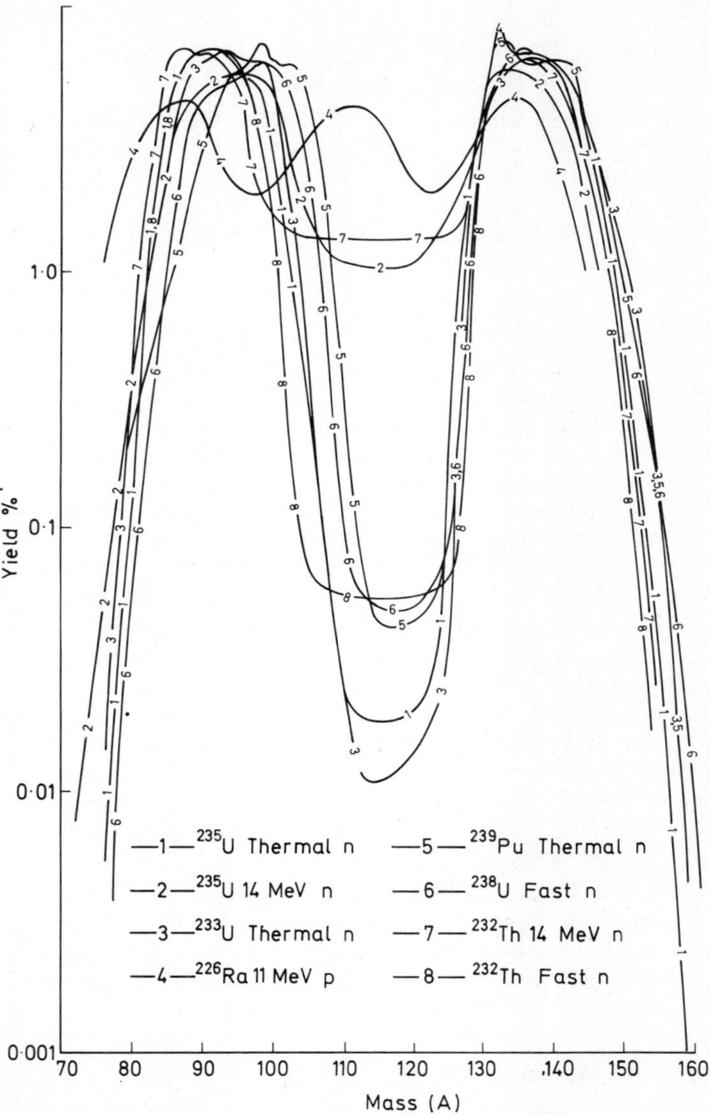

Figure 1.4 Fission yield curves superimposed to show the predominance of yields in the mass region 132. (Compiled by I. F. Croall)

to the sum of the magic numbers 50 protons and 82 neutrons. However, while in the systematics of the ground states of nuclei there are sharp discontinuities at the shells corresponding to magic numbers, in the mass yield, nuclear charge distribution, and neutron probabilities of fission, only trends and no very pronounced discontinuities can be discerned. In 1955 Fong put forward his statistical theory of fission[18]. In this, the separation of fragments during the fission process was assumed to be sufficiently slow, and the multiplicity of available energy levels sufficiently high, to allow thermodynamic equilibrium to be set up in the distribution of nuclei in the energy levels available to the compound nucleus as it separated into two fragments. The available excitation energy, E, was written

$$E = M_{AZ} - (M_{A_H Z_H} + M_{A_L Z_L}) - C - D \tag{1.1}$$

where M_{AZ} represents the masses of the compound nucleus and light (L) and heavy (H) fragments. C is the energy associated with coulomb repulsion (which appears as the kinetic energy of the fragments), and D is the energy of deformation (which eventually appears as prompt neutrons and γ-rays). The value of E was expected to control the density of energy levels, $\rho(E)$, by an equation of the type

$$\rho(E) = C_{\exp}\sqrt{aE} \tag{1.2}$$

where C and a are constants for a particular nuclear structure, and in turn the mass distribution for any particular energy would be determined by the density of energy levels for that energy. If the combination of masses had shell structure effects which would lower the values of $(M_{A_H Z_H} + M_{A_L Z_L})$, then the values of E, and the mass yields, would be high, and vice versa. Similarly, as the value of Z changed for constant A the value of E and the yields would change in a manner similar to the effect of changing A. These considerations provided a good qualitative explanation of the effect of the mass and charge of the products on fission yields. No good quantitative assessment could be made because masses of the highly deformed fragments initially formed in fission, suitable for inserting in equation (1.1), could not be evaluated. It was pointed out by Walton[19] that if the different known stable and near stable masses are extrapolated, the value of E becomes high for combinations of masses corresponding to those that show high yield in fission, and also for combinations of charges corresponding to the most probable charges observed in fission.

Further speculation and development of the statistical theory have been made for instance by Ramanna[20] and Brueckner[21], and the current position is shown in papers for instance by Nörenberg[22] and Schmitt[23].

This statistical theory of fission was relevant to considerations of the products of nuclear fission, but a very different theoretical picture, which seemed at first inconsistent with the statistical approach, was developed, originally by Bohr[24], as a result of evidence arising during the initiation of the fission process.

1.3.2 The channel theory of fission

The channel theory of fission, which has been very fully reviewed by Gindler and Huizenga[5], receives its main support from the observation that cross-sections for fission above the threshold often show marked steps at different energies of the bombarding particle[25], and the angular distributions of the products of fission with respect to the direction of the bombarding particle also show changes at the same energies of the particle. Figure 1.5 shows a recent measurement of the neutron cross-section of ^{232}Th, together with the

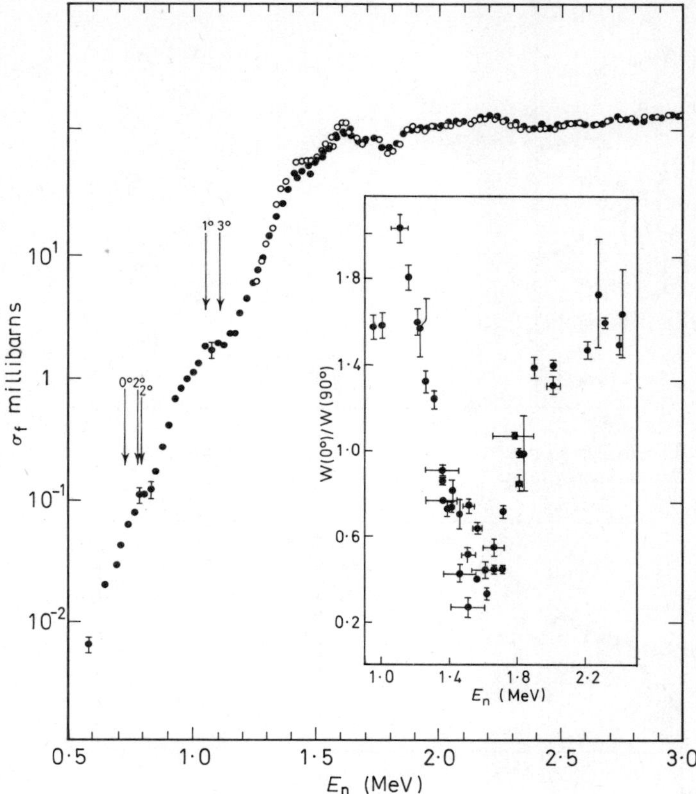

Figure 1.5 Fission cross-section of ^{232}Th for neutrons in the energy range 0.5–3.0 MeV. The inset shows changes in the angular distribution of the fission products over the corresponding energy range. σ_f millibarns; E_n MeV (From Kovacs *et al.*[26], by courtesy of IAEA)

corresponding measurements of the changes in angular distribution as a function of neutron energy[26]. $W_{0°}/W_{90°}$ is the ratio between the intensity of products emitted along the direction of the beam to the intensity measured at right angles to the beam. The inserted diagram in Figure 1.5 is taken from the data shown in Figure 1.6, where the distribution is measured over all angles. The effect appears to be most marked in ^{232}Th, and although it may be a general effect, recent developments, as shown below, suggest that this might not be the case. Figure 1.7 shows the effect in several other nuclides. Here A is given by the relationship:

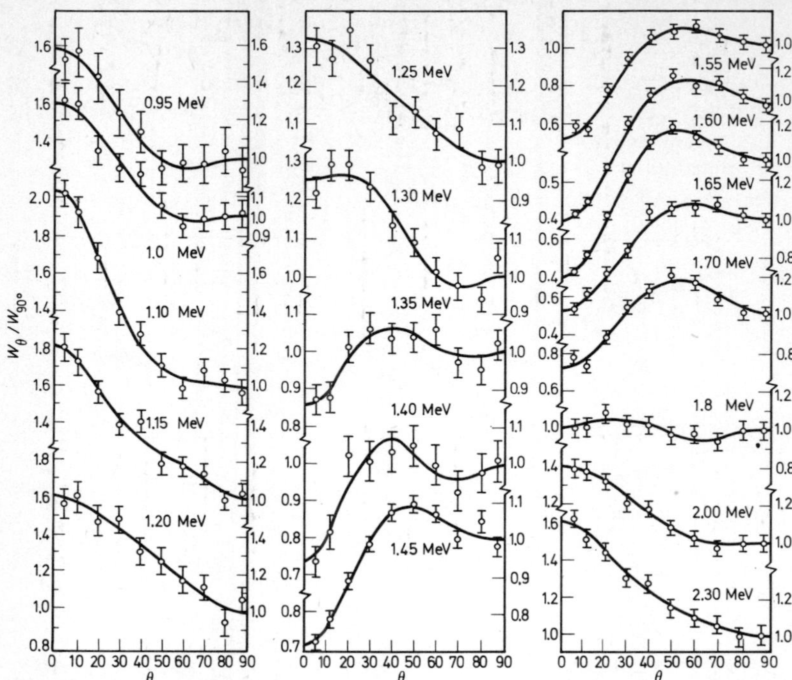

Figure 1.6 Changes in the angular distribution of fission products from ^{232}Th with neutrons over the energy range 0.95–2.30 MeV. (Ref. 26)

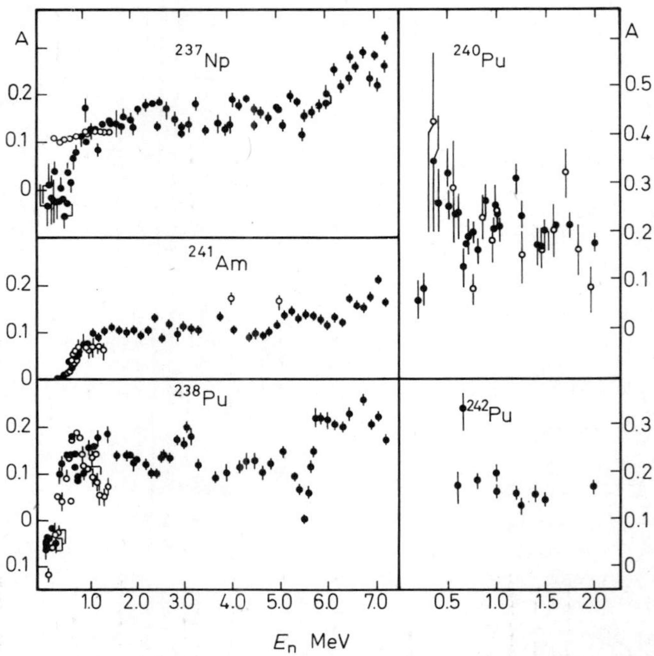

Figure 1.7 Changes in the angular distribution of fission products from various nuclides with neutrons over the energy range 0.5–7.0 MeV. (From Kovacs *et al.*[26], by courtesy of IAEA)

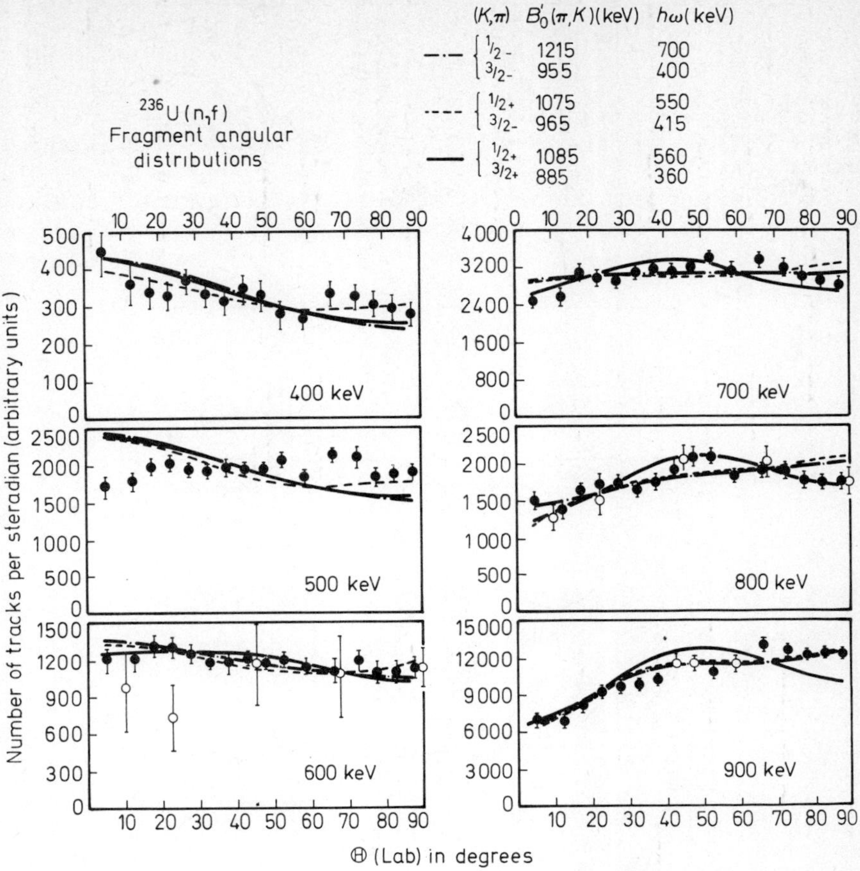

$^{236}U(n,f)$
Fragment angular distributions

	(K,π)	$B'_0(\pi,K)$(keV)	$\hbar\omega$(keV)
—·—	1/2−	1215	700
	3/2−	955	400
- - - -	1/2+	1075	550
	3/2−	965	415
——	1/2+	1085	560
	3/2+	885	360

Number of tracks per steradian (arbitrary units)

400 keV — 700 keV — 500 keV — 800 keV — 600 keV — 900 keV

Θ (Lab) in degrees

Figure 1.8 Changes in the angular distribution of fission products from ^{236}U with neutrons over the energy range 400 keV to 900 keV. The lines show predicted changes corresponding to transition states with data sets shown above. (From Huizenga *et al.*[27], by courtesy of IAEA)

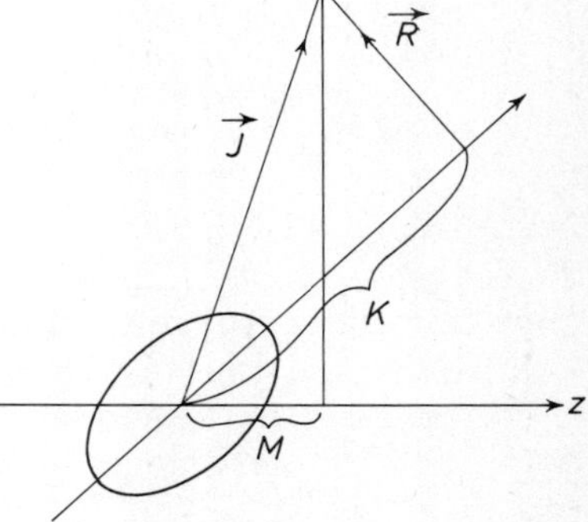

Figure 1.9 Angular momentum coupling scheme for a deformed nucleus. The vector J defines the total angular momentum. The quantity M is the component of the total angular momentum on the space-fixed z axis. This direction is defined as the beam direction. The quantity K is the component of the total angular momentum along the nuclear symmetry axis. The collective rotational angular momentum, R, is perpendicular to the nuclear symmetry axis; thus K is entirely a property of the intrinsic motion. (From Huizenga *et al.*[27], by courtesy of IAEA)

$$\frac{W_\theta}{W_{90°}} = 1 + A\cos^2\theta \qquad (1.3)$$

In Figure 1.8 the effect is shown for $^{236}U(n,f)$ together with some predicted values[27].

According to Bohr[24], these effects suggest that, in contrast to the statistical theory which pre-supposes a high density of energy levels available to the nucleus as it undergoes fission, there are only a very few widely-spaced

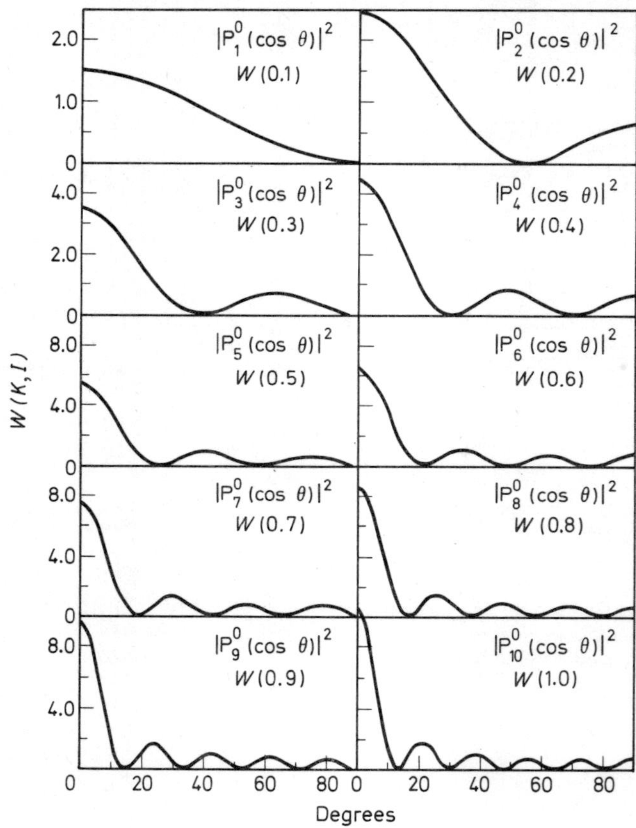

Figure 1.10 Theoretical values of the relative probability $W(K, I)$ of fission fragments being emitted in different directions relative to the beam axis from nuclei with different angular momenta, I, in cases where $K = 0$ and $M = 0$. (From Gindler and Huizenga[5], by courtesy of Academic Press)

levels. This could well be the case if, in the formation of the transition nucleus, the energy available, E, is nearly all absorbed in distorting the nucleus, so that very little is available as excitation energy and the nucleus is effectively 'cold' during the transition. Associated with the transition from each energy level, there is a particular set of quantum numbers J, K, M, π; where J represents the total angular momentum, π is the parity, K is the projection of J on the nuclear symmetry axis, and M is the projection of J on some experimentally fixed axis, such as the direction of the beam of particles causing fission. If it is assumed that the nucleus will split so that the fission fragments separate along the nuclear symmetry axis, then it will be expected that each energy transition will be associated with its particular angular distribution of fragments. Moreover, each energy level will have its own

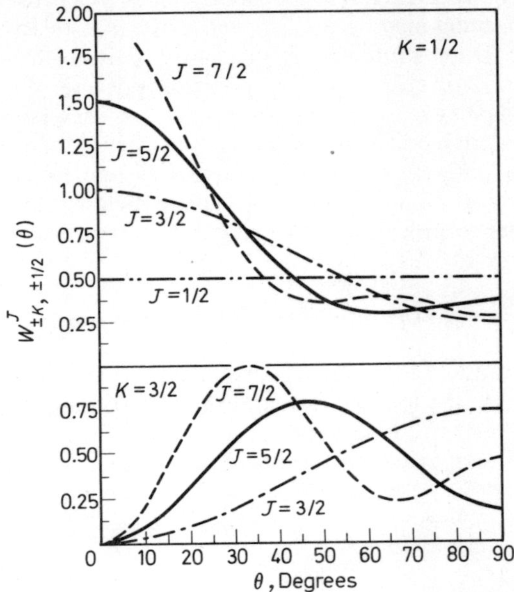

Figure 1.11 Theoretical values of the relative probability $W(K, I)$ of fission fragments from even–even nuclei being emitted in different directions relative to the beam axis from nuclei with different angular momenta vector J, and values of $K = \frac{1}{2}, \frac{3}{2}$.
Axes $W(K, I)$
θ, degrees
(From Huizenga *et al.*[27], by courtesy of IAEA)

transition width, and the cross-section for fission will be expected to change as the different levels are occupied with increasing energy of the bombarding particle.

This theory was amenable to detailed quantitative treatment. Figure 1.9 shows the different axes involved in the assessment of the manner in which the total angular momentum will affect the angular distribution of the products[27]. If K is zero the fragments will be distributed in the manner which will be expected if a glass ball is struck by a particle. Fragments will be emitted at right angles to the axis of the spin caused by the impact which in turn will be at right angles to the direction of the particle, and the most probable direction of the fragments will be backwards and forwards along the direction of the particle. The quantised estimate for the angular distribution if $K = 0$ is shown in Figure 1.10 which shows the preferential emission at $\theta = 0$, where θ is the angle between the fragment and the oncoming particle. For values of K above zero, the angular distribution will change for example, as shown in Figure 1.11. By fitting theoretical distributions to experimental points, it has been possible to allocate quantum states to the transition nuclei as shown in Figure 1.8. Here $B'_0(\pi K)$ is the fission barrier energy, and $h\omega$ (keV) depends on an arbitrary estimate of the barrier shape.

In fission which produces a charged particle, such as a proton, the direction of the charged particle may be used to define the axis relative to which the directions of the fission products are measured. This, for example, has been done for ^{234}U(d,pf), and greater precision appears possible than in irradiations where no charged particle is emitted. Although it has been possible to allocate quantum states and barrier shapes in a wide variety of fission reactions, induced by many different particles, no further systematic correlations have appeared. This may be because, as discussed below, the barrier shapes appear to be more complex than the original theory pre-supposed[28].

1.3.3 Shell structure effects in the liquid drop model

The liquid drop model of the nucleus provides an analogy of the manner in which nuclear stability follows the composition of neutrons and protons (i.e. mass and charge) from the helium atom to the trans-uranics. However, there are many irregularities in the properties of nuclei which are not pre-

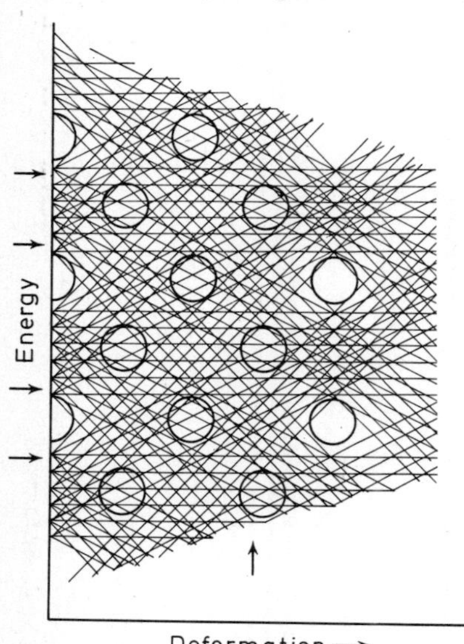

Figure 1.12 Qualitative picture of the distribution of single particle states in the deformed nucleus. The low density regions (shells) are shown by circles. Arrows show the places where transitions between sphericity and non-sphericity occur. (From Strutinsky[29], by courtesy of IAEA)

dicted by the liquid drop model, and many of these can be attributed to the extra binding energy associated with the magic numbers, or shells, 20, 50, 82 etc., protons or neutrons. This extra binding energy amounts to 5–10 MeV. In papers by Strutinsky, working at the Niels Bohr Institute, Copenhagen[29], it has been pointed out that this energy is comparable to the height of fission barriers, and that any shell formation occurring during fission should have a drastic effect on the process. Calculations suggest that, while 50 and 82 etc. protons or neutrons give stability to spherical nuclei, different numbers of protons and neutrons would give stability to deformed nuclei. Thus, for *any* particular number of protons or neutrons there would be certain deformations at which extra stability appeared. As shown in the Nilsson diagrams[30], the energy corresponding to any quantum state in the

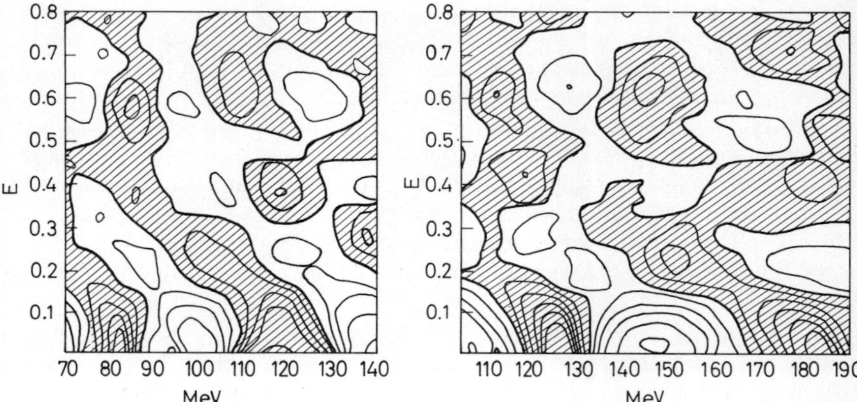

Figure 1.13 Contour diagrams of the proton (left) and neutron energy shell correction plotted as functions of the nucleon numbers and the ellipsoidal deformation parameter E. The increment between contour lines is 1 MeV. Regions of negative shell corrections ('the shells') are shaded. (From Strutinsky and Pauli[29b], by courtesy of IAEA)

single particle model of the nucleus may be calculated for different shapes of the nucleus. Just as the energy levels of electrons in atoms control the stability of ions and molecules, so, when the values of the energies for different quantum states of a particular nucleus are close to each other the nucleus shows instability because it may readily occupy different states. When the values are widely separated, the nucleus shows stability because large energy changes are required to alter the nucleus from one state to another. Figure 1.12, from Strutinsky, shows the manner in which more-stable and less-stable states of the nucleus might be expected to arise as the nucleus becomes more deformed. The theory is treated quantitatively by considering the density of energy levels, g, in an equation of the following type.

$$\delta g(E,\beta) = g_{\text{shell}}(E,\beta) - \tilde{g}(E,\beta) \tag{1.4}$$

$g_{\text{shell}}(E,\beta)$ is the local density of single particle states averaged over an energy interval of 1–2 MeV at energy E, at a deformation β, and $\tilde{g}(E,\beta)$ is the average density of energy levels over a wide energy range. Deformation may be described in many ways, one of which is, for example, the ratio between major and minor axes of an ellipsoid structure. $\delta g(E,\beta)$ thus represents the deviation from the average. The shell correction for the mass term in the liquid drop model is then related to the level density fluctuation by an expression of the form

$$\delta M(N,\beta) = \int_{\infty}^{\lambda} (E-\lambda)\delta g(E,\beta)\mathrm{d}E \tag{1.5}$$

where δM is the correction to the mass of a nucleus of N nucleons with deformation β, and λ is the effective energy for a distorted liquid drop without shell corrections (when $\delta M = 0$). At a shell $\delta g(E,\beta)$ is negative, and the mass M of the nucleus is lowered. On this basis, contour maps of deformed nuclei may be drawn, such as those shown in Figure 1.13 where the contours correspond to constant values of δM. Here the contour lines are 1 MeV apart. Negative values of δM, corresponding to the shell structures, are shaded. These diagrams show that for any given number of nucleons there is usually more than one minimum. These minima may give rise to metastable shapes of the nucleus and the assumption is made that the metastable shapes appear as spontaneously fissile isomers.

On this assumption, the potential energy of the nucleus as a function of distortion would, in general, be expected first to *decrease* as distortion increased, then to pass through a minimum and over a first fission barrier, and on into a second minimum and a second barrier. This is illustrated in Figure 1.14, taken from Strutinsky. The assumption of a single barrier as predicted by the liquid drop model without shell corrections is represented by the line

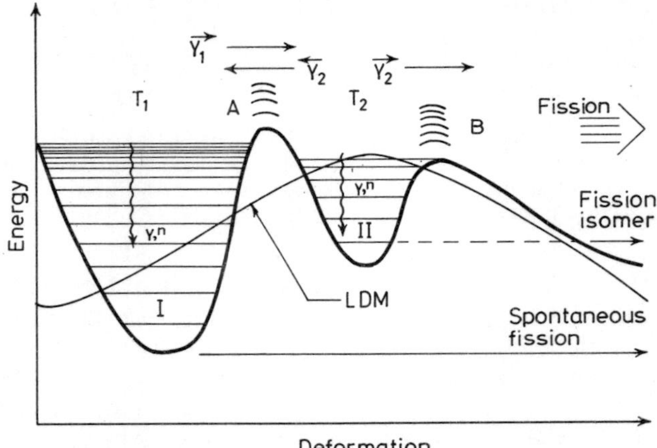

Figure 1.14 Two-humped barrier in the deformation energy of a heavy nucleus. Transitions between the two families of stationary states, class I and class II, are indicated. The line LDM corresponds to the barrier predicted by the unmodified liquid drop model. (From Strutinsky and Pauli[29b], by courtesy of IAEA)

LDM. The horizontal lines represent vibrational levels of a nucleus in the two different states of deformation. Following the nomenclature introduced by Lynn[31, 32], who was amongst the first to analyse the process on this basis, the nuclei in the first deformation are called class I states and those in the second, class II states.

1.3.4 Fission cross-sections and the shell structure effect

The assumption of metastable intermediate states helps to resolve a number of experimental observations that the unmodified liquid drop model failed to resolve. As the mass and charge of a nucleus increases the coulomb repulsion term, Z^2/A, in the mass equation of the liquid drop model increases, and the nucleus becomes less stable. The fission threshold energy accordingly is high for low masses, reaches values between 5 and 10 MeV for nuclides which are fissile to thermal neutron excitation, and would be expected to decrease for still higher masses when spontaneous fission half-lives become

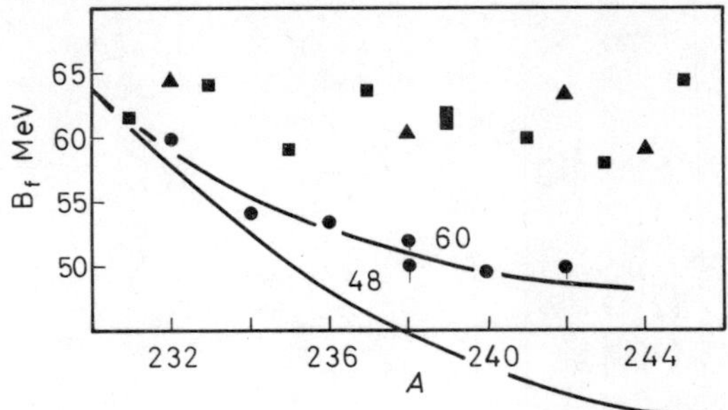

Figure 1.15 Fission barrier heights obtained from the threshold of measured fission cross-sections. The two lines correspond to predicted values of the barrier height for the unmodified liquid drop model with different values of the critical value of Z^2/A for fissionability. (From Strutinsky and Pauli[29b], by courtesy of IAEA)

small. The general trend is illustrated by the full line in Figure 1.15. However, experimental values of the fission barrier, as inferred from the threshold energy for fission cross-sections, do not decrease in the manner predicted by the liquid drop model. Figure 1.15 illustrates the discrepancy in question. Referring back to Figure 1.14, it is seen that the initial deformation will be expected to heighten the fission barrier because the initial energy falls below the ground state predicted by the liquid drop model, while the first barrier is above that predicted. As a result, the fission barriers will be larger than those predicted by the liquid drop model. They will be dependent on the manner in which the shell structure changes with deformation, and not necessarily dependent on the parameters of the liquid drop model, as indicated in Figure 1.15.

1.3.5 Resonances and the shell structure effect

Another feature of fission cross-sections is that resonance effects occur in the excitation functions. On the channel theory of fission, as the excitation energy of the compound nucleus increases, more fission channels are opened, and the cross-section for fission would be expected always to increase with energy. As shown in Figure 1.16, the cross-section, although it shows stepwise increases as predicted, sometimes decreases with increasing energy. From Figure 1.14, showing the double fission barrier, it would be expected that the vibrational energy levels of the class II states would be more widely spaced than those in class I states where the potential well is deeper. Consequently, class I states will more readily undergo loss of energy by γ-emission to lower states, and the vibrations will, in effect, be more readily damped than class II states. However, when the energy level in a class I state matches that of a class II state, the class II vibrational condition can occur, and fission will be more probable. This provides a possible explanation of the fall of

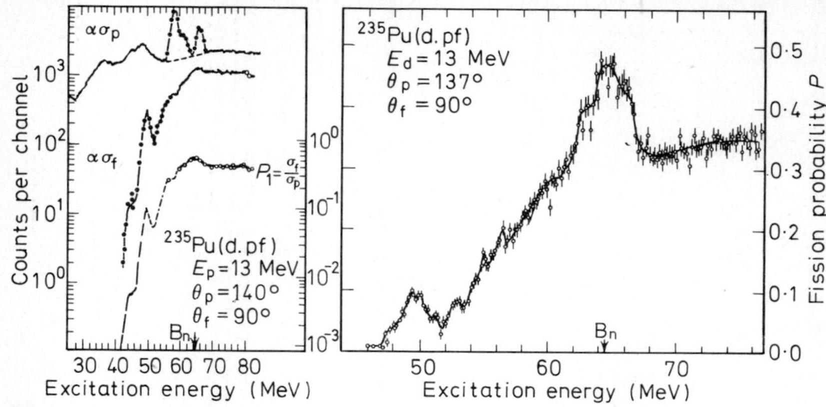

Figure 1.16 Broad resonances in the deuteron fission of ^{239}Pu (d, pf). To the left the single proton spectrum (σ_p) and the fission spectrum (σ_f) are shown. To the right the fission probability σ_f/σ_p is shown in higher resolution. (From Back *et al.*[55], by courtesy of IAEA)

the fission cross-section when the excitation of the compound nucleus fails to match the class II level structure.

The effect has been treated quantitatively by Lynn, who first analysed this link between spontaneous fission isomers and the excitation functions, and a number of papers have been published[33].

Dynamic equations of the following type can be written for transitions into, and out of, the two potential wells and also for transitions between the wells:

$$\frac{dn_1}{dt} = p_{2\gamma}n_2 - (p_{1\gamma} + p_{1n} + p_{1f})n_1 \tag{1.6}$$

$$\frac{dn_2}{dt} = p_{1\gamma}n_1 - (p_{2\gamma} + p_{2n} + p_{2f})n_2 \tag{1.7}$$

where n_1, n_2 represent the population (density of energy levels at a given excitation) in the two potential wells and $p_{1\gamma}$ etc. represent the probabilities of various kinds of decay by γ-emission, neutron emission, or fission. For the neutron capture case, the initial conditions will be:

$$n_1 = 1, n_2 = 0, \text{ when } t = 0$$

and for the cases in which the compound nucleus is formed by de-excitation of highly excited nuclei formed, for instance, in charged particle reactions, the initial condition may be:

$$n_1 = 0, n_2 = 1, \text{ when } t = 0$$

As discussed above, the probability of fission in the second well may be comparable with the probability of γ-emission, whereas in the first well the probability of γ-emission is much greater.

The spontaneous fission isomers will consist of pure class II states, i.e. all the nuclei in question will be in vibrational levels in the class II potential well. These will have a higher probability of fission not only because the levels are well separated, so that there is low probability of damping by γ-emission to lower states, but also because the amplitude of the vibration

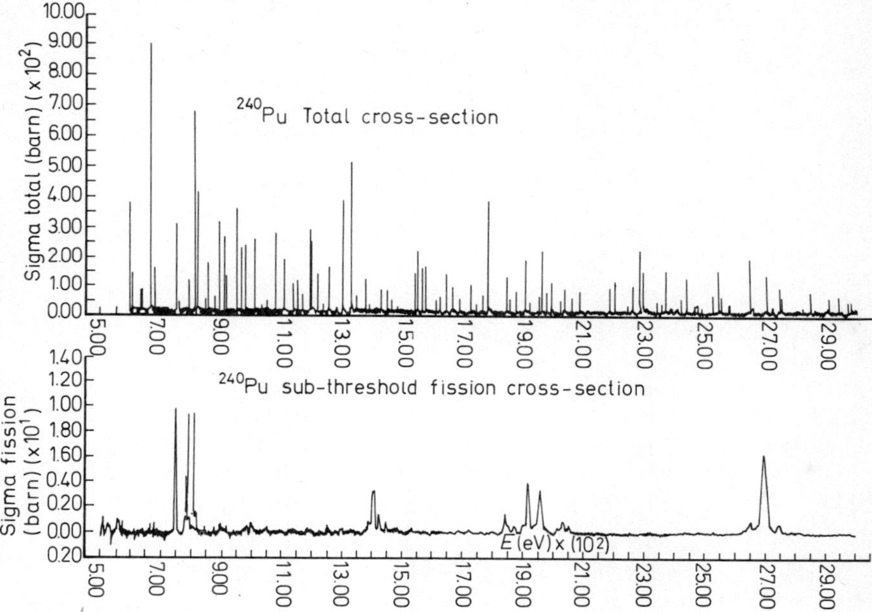

Figure 1.17 Narrow resonances. The fission cross-section of ^{240}Pu is compared to the total cross-section to show the grouping of the resonances. (From Migneco and Theobald[35], by courtesy of North-Holland Publishing Company)

is high, the restoring forces are small and the coulomb repulsion is high, the centres of charged particle distribution being already well separated in the distorted shape.

Calculations on these lines show that it is possible to account for the factor of 10^{19} between the half-lives for spontaneous fission of the isomers and their ground states.

1.3.5.1 Narrow resonances

In 1968, at the first Vienna Symposium on the physics and chemistry of fission, it was pointed out that when the resonance fission cross-section for ^{237}Np was compared with the total cross-section for neutron absorption it was seen that the fission resonances were very widely spaced compared with the neutron absorption resonances, and that the fission resonances corresponded to peaks in the neutron absorption resonances[34]. Neutron irradiation of ^{240}Pu below the fission threshold showed the same effect[35]. Figure 1.17 shows this remarkable correspondence. The explanation readily appears on the assumption of the double fission potential well. Neutron absorption occurs to excite the multiplicity of levels in the first well (class I states).

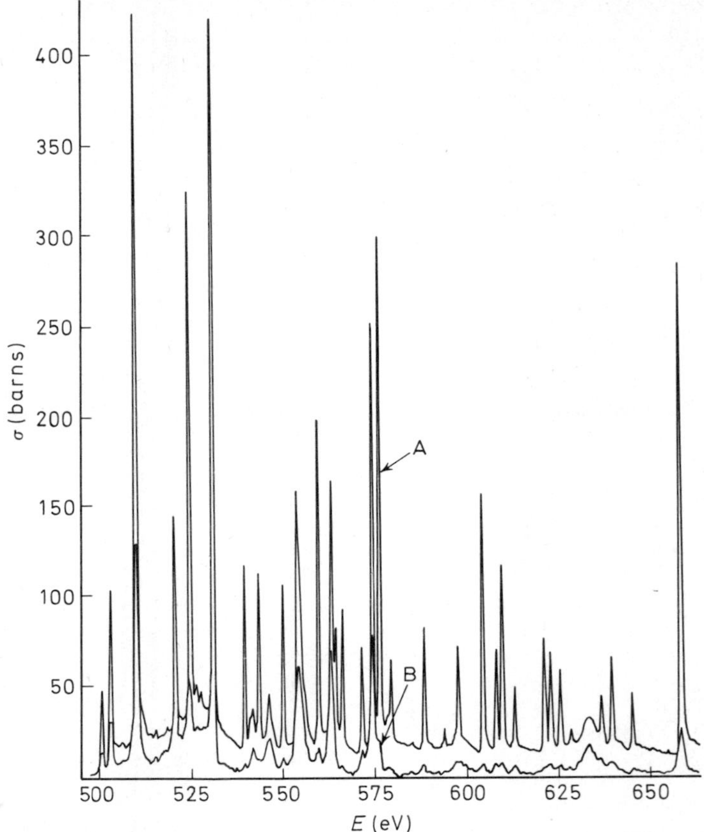

Figure 1.18 Narrow resonances. The fission cross-section of ^{239}Pu (B) is compared to the total cross-section (A) to show the grouping of the resonances. (From Paya *et al.*[36], by courtesy of IAEA)

When, and only when, the energy corresponds to an energy in the second well, where the levels are more widely spaced, the resonance vibration in the second well is excited, and fission becomes more probable. The effect appears to be general, and further examples are being found. The comparison of the fission and total neutron cross-section for ^{239}Pu is shown in Figure 1.18[36].

1.3.5.2 Broad resonances

Broad resonances have been most clearly seen in fission by deuteron bombardment, i.e. (d,pf) reactions. Here the neutron can be introduced at energies

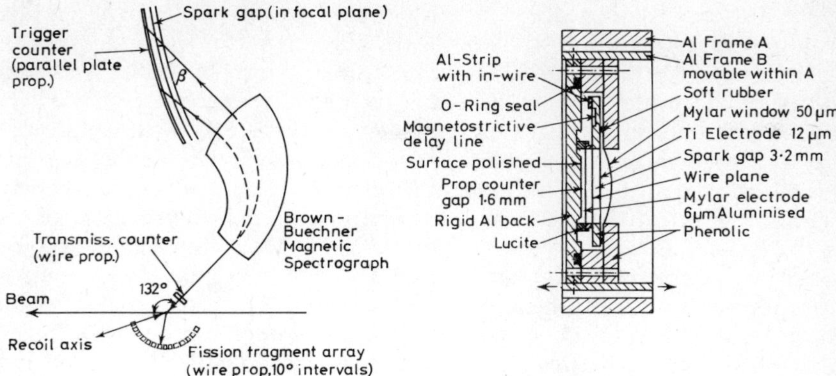

Figure 1.19 Diagram of the apparatus for a high-resolution study of the ^{239}Pu (d,pf) reaction. (From Specht *et al.*[37], by courtesy of IAEA)

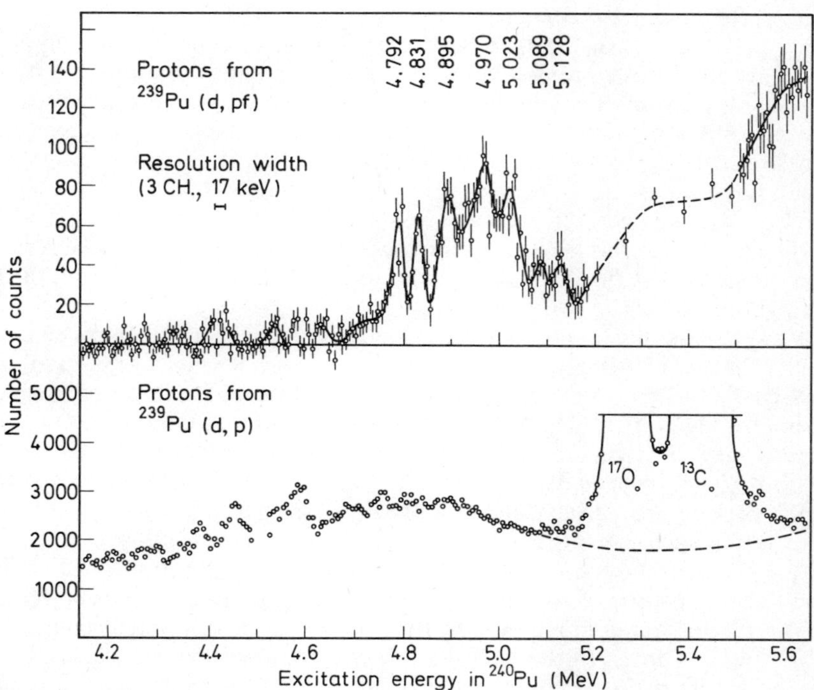

Figure 1.20 Resonances in the ^{239}Pu (d,pf) reaction. The upper curve shows the spectrum of protons in coincidence with fission, and the lower curve shows the total proton emission for different excitation energies. Protons from ^{17}O (d,p) and ^{13}C (d,p) reactions in the targets have not been subtracted. (From Specht *et al.*[37], by courtesy of IAEA)

which are clearly defined by the energy and distribution of the emitted proton. A high-resolution study of this reaction in ^{239}Pu has been reported from Canada[37]. A diagram of the apparatus is shown in Figure 1.19. The energy of the proton was measured in a magnetic spectrograph and the angular distribution of the fission fragments in coincidence with the protons was measured in an array of proportional counters. The deuteron beam was taken from an 11.5 MeV tandem accelerator. The results of this experiment are shown in Figure 1.20, and were interpreted on the basis of the double potential well. Over the range 4.6–5.0 MeV the 'singles' proton spectrum (i.e. protons not in coincidence with fission events), shows neutron absorption attributed to the levels in the first well. The broad peak over the range 4.8–5.2 MeV in the spectrum of protons in coincidence with fission indicates resonance between the first and second potential wells, and fine structure observed in this peak is attributed to the overlapping of individual levels.

1.4 COMPARISON OF THE FISSION THEORIES

Although the three theoretical approaches described, i.e. the statistical theory, the channel theory, and the double-barrier theory, all derive from the liquid drop model and all have a great deal of experimental support, they are not entirely consistent with each other and numerous questions remain unresolved. It is not clear whether the remarkable anisotropy which occurs in fission should be attributed to low-lying levels in the nucleus as it passes from the class I states, or the class II states. It has been suggested[38] that if the nucleus passes from class I states the changes in anisotropy, which can be correlated with changes in cross-section, will be observed, and when the nucleus passes via class II states changes in anisotropy will not be observed. This suggests that the occurrence of anisotropy will be an indication of the relative heights of the two barriers.

A more fundamental problem relates to energy-level density. In the channel theory, fission is thought to take place through a *few* widely-separated and well-characterised levels, and again the second potential well, through which fission passes, is attributed to regions of distortion corresponding to *low-level* density. On the other hand, high fission yields in the statistical theory are attributed to *high-level* density. Calculations show that the second potential well is produced by shell structure in the distorted nucleus, but the distortion envisaged is such as to produce major and minor axes of an ellipsoid structure with a ratio of about two, and does not correspond to a dumb-bell shape. The shell structure of the distorted nucleus in the double-barrier theory does not correspond in any way to the shell structure that might be formed in the two products. In the statistical theory, the shell structure corrections that enter into the mass equations, that enable the asymmetry of fission to be predicted, do not appear to be related to the shell structure corrections that enable the spontaneous fissile isomers to be predicted in the double-barrier theory. The processes that lead nuclei in a few different quantum states to a wide multiplicity of different products remain unresolved.

1.5 TERNARY FISSION

1.5.1 Types of ternary fission

Although nuclear fission is normally regarded as a binary process, there are always other particles emitted. Renewed attention is being given to these

Table 1.2 Ternary fission. Long range particles in ^{252}Cf spontaneous fission

Particle	*Yield* *relative to* ^{4}He	*Energy range* MeV
^{1}H	1.10 $\pm$0.15	7.3–18.8
^{2}H	0.63 $\pm$0.03	5.3–21.5
^{3}H	6.42 $\pm$0.20	6.5–24.3
^{3}He	$<7.5\times10^{-3}$	14.2–21.3
^{4}He	100	8.3–37.7
^{6}He	1.95 $\pm$0.15	10.0–33.3
^{8}He	6.2×10^{-2}	9.3–27.7
Li isotopes	0.126$\pm$0.015	15.2–37.3
Be isotopes	0.156$\pm$0.016	23.0–49.1
'True' ternary products	0.048	~50

in the hope that clues may be found particularly to the nature of the final stages of fission. A recent review has been written by Feather[39]. Table 1.2 shows the list of particles which have been detected in the spontaneous fission of ^{252}Cf which, as discussed below, is probably representative in this instance of most types of fission.

1.5.2 α-Particle emission

The most notable features of α-particle emission in fission are first, that the yield is very invariant for different nuclei and different fission excitations, second, that the α-particles tend to be emitted at right angles to the direction of the fragments, and third, that the energy is high and shows a broad distribution.

1.5.2.1 Origin of α-particles

The invariant character of α-particle emission suggests that the 'scission' configuration, i.e. the configuration at the instant when the two fragments separate, is very nearly the same for all excitations, and that the α-particle emission occurs at the same instant. The angular distribution suggests that the α-particle is emitted from the 'neck' of the 'scission' state, where the coulomb repulsion would be expected to project the particle at right angles to the axis of the two main fragments. It also shows that the α-particle is not emitted by one of the fragments after formation. Theoretical studies of this process, notably by Boneh[40], have been made. For instance, it may be supposed that the three particles act as point charges, and that the α-particle originates isotropically with a velocity distribution corresponding to the nuclear 'temperature'. The point of origin is assumed to be the point of minimum electrostatic potential between the two fragments when they have effectively zero velocity. Computation of the α-particle emission based on this model shows many of the features observed experimentally, except that the energy distribution of the α-particle is considerably lower than the observed distribution. The distance between the centres on the instant of fission to produce the observed α-particle energy is 26×10^{-15} m (which may be compared to the compound nucleus diameter of 18.6×10^{-15} m and the distance between the centres of two spheres in contact 14.7×10^{-15} m).

1.5.2.2 *Angular distribution*

Detailed studies have been made of energy and mass correlations in α-particle emission during fission. Work at Saclay[41,44] has shown that the energy of the α-particles appears to be independent of the mass ratio of the fragments, but the angle between the α-particle and the light fragment increases with the

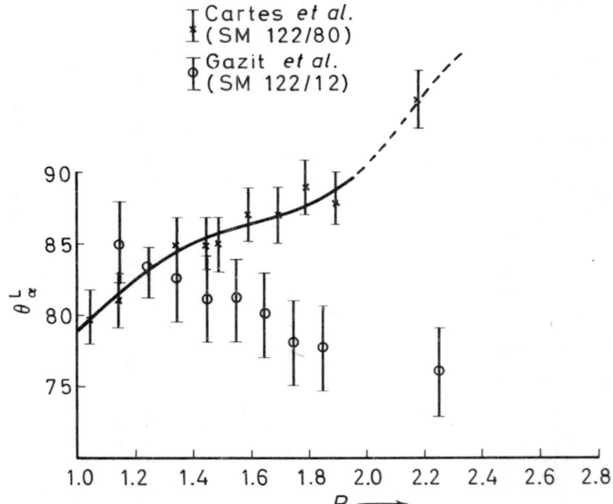

Figure 1.21 Comparison between two different sets of experimental values of the angle between the trajectory of long-range α-particles and the trajectory of the light fragment (θ_α^L) in fission with different degrees of asymmetry (R = ratio of mass of heavy to light fragments). (From Gazit *et al.*[43], by courtesy of IAEA)

ratio of the masses, i.e., the α-particle veers towards the heavy fragment. It has also been shown that the energy of the α-particle is lowest when it is emitted at right angles, and increases when it is emitted in directions veering towards both light and heavy fragments. Fong[42] has pointed out that the statistical theory enables predictions to be made about the velocity distribution, the positions, and times, of the α-particles and the fragments at their formation, and that the subsequent behaviour can be computed. This he has done and claims, in contrast to the Saclay findings, that the α-particle should veer towards the *light* fragment[42]. However, the experimental results are by no means agreed. Figure 1.21, put forward by Ben-David[43], shows two conflicting sets of results on the question as to whether

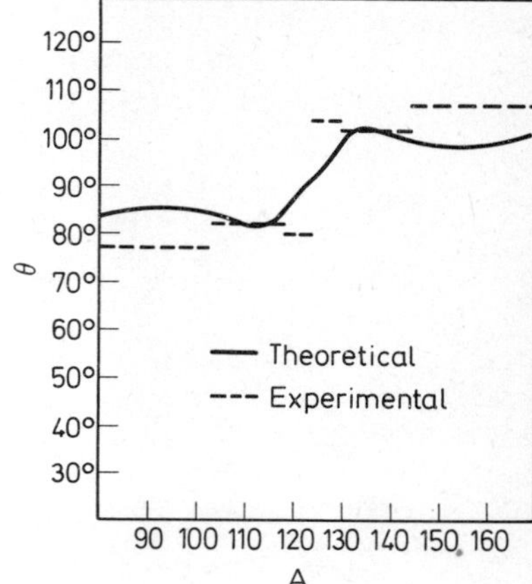

Figure 1.22 Theoretical values of the angle between the trajectory of long-range α-particles and trajectory of fission fragment of mass A as calculated from the statistical theory of fission. (From Fong[42], by courtesy of IAEA)

the α-particle veers towards the heavy or light fragment. In Figure 1.21, θ_α^L is the angle between the α-particle and light fragment trajectories, and R is the mass ratio in the fission. Fong's calculations are shown in Figure 1.22 and appear to be in good agreement with the results of Gazit *et al.*[43]. Predictions of the final trajectories are very sensitive to the starting conditions, and further experimental studies will be interesting.

1.5.3 Prompt neutron emission

Prompt neutron emission in fission is not usually regarded as a ternary process. However, as with α-particle measurements, detailed observations are proving very significant. Current evidence appears to suggest that there are three distinct sources of prompt neutrons in various types of fission:

(a) 'Evaporation neutrons' are emitted in competition with fission, during de-excitation of the compound nucleus. These 'pre-fission neutrons' have been most clearly distinguished in, for instance, proton-induced fission where the products and cross-sections for (p,nf), (p,2nf), and (p,3nf) may be observed. The ratios of the fission products' yields change sharply as '1st chance', '2nd chance', '3rd chance' fission succeed each other as one, two, or three successive neutrons are 'boiled-off' from the compound nucleus[45];

(b) scission neutrons;

(c) fragment neutrons.

Studies of the angular distribution of prompt neutrons with respect to the directions of the heavy and light fragments show that most of the neutrons originate from the moving fragments, and are called fragment neutrons. This has led to the well-known analysis of the 'saw tooth' plot, which shows that in near-symmetric fission most of the neutrons are given off by the lighter fragment, whereas in highly-unsymmetric fission the neutrons are emitted from the heavy fragment; an effect that is attributed to the shell structure of the products[46]. However, it has been shown by a number of authors[47] that not all the neutrons can be attributed to the motion of the fragments, but that some of them are isotropically distributed from the original compound nucleus. These latter neutrons, amounting to 10–15% in thermal fission, have been called 'scission' or 'central' neutrons.

It is by no means clear how 'scission' neutrons can be distinguished

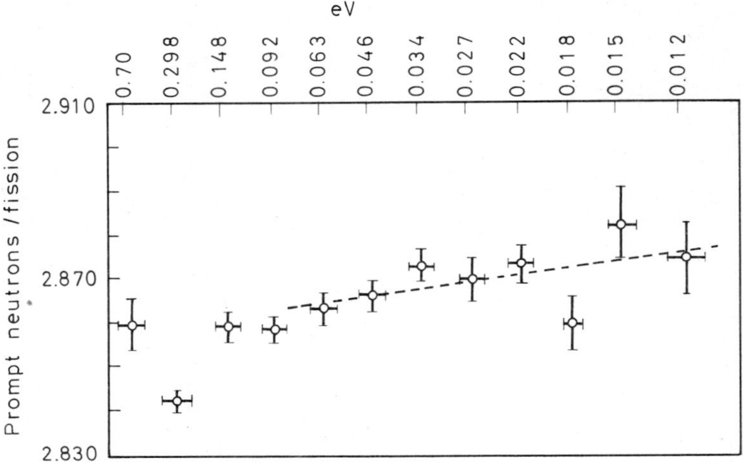

Figure 1.23 Apparent correlation of prompt neutrons emitted in neutron fission of ^{239}Pu with resonances in the neutron cross-section. (From Weinstein, *et al.*[48], by courtesy of IAEA)

experimentally from 'evaporation' neutrons. In recent work[48], an attempt has been made to see if neutron emission is correlated with resonance fission at different energies of neutron absorption. Correlations have been claimed, as shown for instance in Figure 1.23, by several authors[49], but the claims as yet are not consistent with each other. It is open to question whether any correlation is to be expected between the number of neutrons per fission and resonance effects in the fission cross-section. The latter is determined by the energy levels of the compound nucleus and the 'channels' through which fission occurs, but the former may be associated with the energy levels and shape of the scission nucleus when the products are formed, and as discussed above, the system of energy levels at the two stages of fission may be quite different. On the other hand, it would be expected that 'evaporation neutrons' would be closely associated with fission cross-sections.

1.5.4 Tritium emission

In the emission of other types of particles in fission no very clear theoretical conclusions appear to emerge. However, the processes are important in technology. For instance, the α-particle emission in thermal fission of uranium is about 2.1×10^{-3} particles per fission, and the emission of tritium atoms is about one in every 4700 fissions. This has radiological implications in the operation of reactors and the chemical processing of nuclear fuel, because tritium can readily escape into the atmosphere and emission should eventually require some control.

1.5.5 'True' ternary fission

'True' ternary fission may be said to occur when the nucleus divides into three fragments of about equal mass. This is of considerable interest because nuclear binding energies are such that ternary fission would be expected to release more energy than binary fission. However, although evidence has been claimed in triple coincidence experiments to give the yield figure shown in Table 1.2[50], the observed kinetic energy released was lower than that in binary fission. Also, radiochemical methods for the detection of likely products have failed to show any evidence for ternary processes in low-energy fission[51].

1.6 POST-FISSION PHENOMENA

1.6.1 Kinetic energy and angular momenta

A new line of investigation was opened up in experiments performed by Unik, Cuninghame and Croall on the Variable Energy Cyclotron at Harwell[52]. They measured the mass and kinetic energy of fission fragments emitted from the same compound nucleus formed in different nuclear reactions.

Table 1.3 Kinetic energy and angular momentum in the fission of ^{210}Po

Reaction	*Excitation energy* MeV	*Angular momentum* I^2	*Total kinetic energy* MeV
$^{209}Bi + p \rightarrow {}^{210}Po$	31.8	53	142.9 ± 0.2
	44.1	75	143.7 ± 0.2
	57.6	94	145.0 ± 0.2
$^{206}Pb + \alpha \rightarrow {}^{210}Po$	30.7	199	143.8 ± 0.5
	44.5	308	145.1 ± 0.3
	57.2	435	147.0 ± 0.3
$^{198}Pt + {}^{12}C \rightarrow {}^{210}Po$	58.5	1240	151.6 ± 2.0

The compound nucleus chosen was ^{210}Po, and three modes of formation were involved:

$$\left.\begin{array}{l} ^{209}Bi + p \\ ^{206}Pb + \alpha \\ ^{198}Pb + {}^{12}C \end{array}\right\} \rightarrow {}^{210}Po \rightarrow \text{fission}$$

The protons, α-particles, and carbon ions, were accelerated with energy E_p, so as to give the same excitation E^* to the compound nucleus. The mass and energy of the fission fragments were measured by the pulses in gold surface barrier detectors in coincidence on opposite sides of the targets in a scattering chamber. Table 1.3 shows the results obtained for the total kinetic energy of the fragments. I^2 is the angular momentum calculated to be introduced by the bombarding particles at different energies. This varies over a wide range as a result of the different masses of the bombarding particles. The most interesting result of this study was that the mean kinetic energy, TKE, was dependent on the excitation energy, E^*, and also, for a given excitation energy, the kinetic energy increased with the angular momentum introduced by the bombarding particle. The results fitted an equation of the type

$$E_R = AI^2 + BT \quad (1.8)$$

where E_R is the extra kinetic energy introduced by the angular momentum of the spinning nucleus, T is the estimated nuclear temperature corresponding to the excitation E^*, and A and B are arbitrary constants. Although the effect of nuclear temperature was higher than theoretical estimates would suggest, the value of A came out at 6 keV/(BM)^2, which is close to the expected value.

1.6.2 Prompt γ-rays

Another experiment that probes closely the final states of the fission process was carried out at Karlsruhe in the apparatus shown in Figure 1.24 [53]. Here, a neutron beam from a reactor strikes a thin fissile target placed between two fission fragment detectors. A lithium-drifted γ-ray detector is placed behind shielding in such a way that it can only 'see' the γ-ray emission from the fission fragments as they fly between the target and the detectors, and it does not 'see' the γ-emission from the fragments after they are stopped in the detectors or elsewhere. The γ- spectra obtained show clearly the Doppler shift between fragments flying away from the detector and those flying towards the detector. Figure 1.25 shows typical spectra for specific masses. The masses are derived from the ratios of the energy levels in the fragment detectors with which the γ-rays are in coincidence. These spectra represent the γ-emission from fragments within 10^{-9}s of their formation, while they are probably still in a highly distorted shape, and therefore may be giving information on the level structure of distorted nuclei.

Similar work is being done at Julich[54]. Here, the fragment energy and masses are measured by 'time-of-flight' methods with silicon surface barrier detectors which are more definitive than measurements on the energy ratios of fragments in coincidence with each other. The initiation of a fission track is observed by the scintillation produced by the complementary fragment in the backing of the source. Measurements have been made on the energy and nuclear anisotropy of γ-rays emitted in time intervals of 3×10^{-12} to 2×10^{-10}s after fission. It has been concluded that the initial spin of the fragments formed in fission is very high and has the value of about 15 BM. In 1955, Fong predicted in his statistical theory of fission that fragments would be formed in very high spin states[18]. Very high spin states of the initial fragments have also been inferred from studies of the relative yields

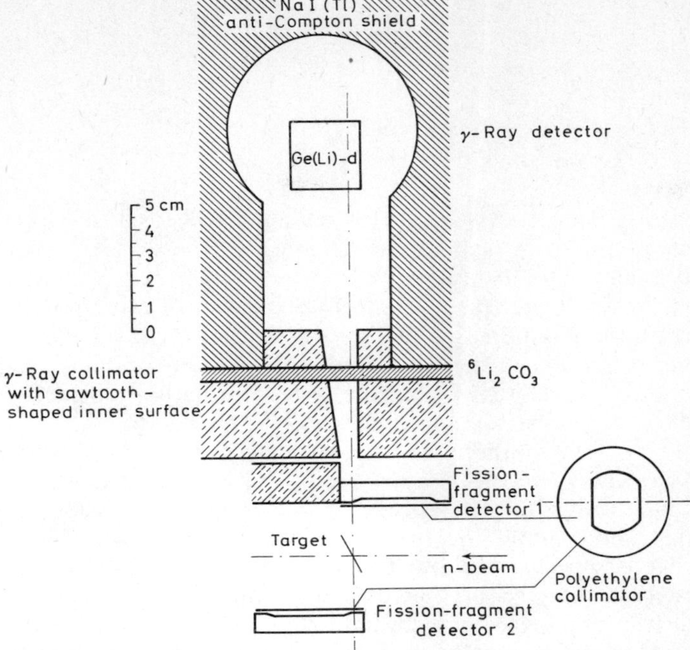

Figure 1.24 Apparatus for the detection of γ-rays emitted from fission products as they recoil from the target. (From Horsch *et al.*[53], by courtesy of IAEA)

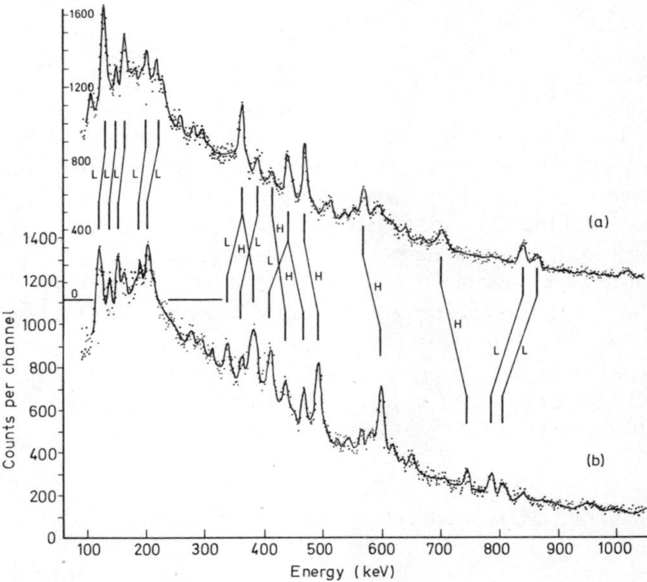

Figure 1.25 Observed prompt γ-ray spectra demonstrating the dependence of the γ-ray energy on the velocity and direction of the fragment motion. The two spectra represent the cases:
(a) Light fragments moving towards γ-ray detector,
(b) Heavy fragments moving towards γ-ray detector.
The letters L and H indicate some assignments to the light and heavy fragments respectively. (From Horsch *et al.*[53], by courtesy of IAEA)

of isomeric states when two or more in the same nucleus are formed independently in fission[56, 80].

1.6.3 K x-ray emission

During the fission process, x-rays are emitted as a result of electronic rearrangements in the freshly-formed fragments. These rearrangements are probably mainly due to internal conversion of the prompt γ-rays emitted in fission. In particular, the hardest of the x-rays will correspond to the K shell electron-binding energy and this energy, in turn, will be dependent on the nuclear charge of the initial fragments. A study of the K x-ray distribution should, therefore, in principle, give information on nuclear charge distribution. A number of studies have been reported for the spontaneous fission of ^{252}Cf, and for thermal fission, of which a recent one is by Glendenin *et al.*[57]. The energy of the x-rays from fission are detected in coincidence with two detectors for the energy of complementary fission fragments. From this information, the masses of the fragments and their nuclear charges over the period of about 10^{-9}s after fission are inferred. Figure 1.26 shows the type of analysis required. Results from this work suggest that the charge distribution in the initial formation of a given set of isobars is narrower than the value found by radiochemical analysis of products after the initial formation period. It is inferred that prompt neutron emission occurs so as to widen the charge distribution for a given mass.

1.6.4 Delayed neutrons

There is renewed interest in delayed neutrons emitted in fission, not only because they provide information on the condition of nuclei far from stability, but also because delayed neutrons are of technical importance. In particular, the dynamic response of reactor systems is dependent on the half-life, energy, and intensity, of delayed neutrons. As discussed, for instance, by Keepin[58] and by Yiftah and Saphier[59], the delayed neutron contribution in fast reactors may be different in the core from that in the blanket, and the stability of reactivity as the reactor starts up or shuts down may be affected. Furthermore, delayed neutrons may give a faster signal in the detection of burst fuel in reactors than the conventional signal from rare gas fission products, and there has also been renewed interest for this reason.

1.6.4.1 Sources of delayed neutrons

A summary of the current knowledge of delayed neutron emitters has been compiled by Amiel[60]. Some 60% of all delayed neutrons are emitted from nuclides in the light mass peak in the thermal fission of ^{235}U, and the remainder from the heavy mass peak. According to Amiel, about 12% of the total delayed neutron emission in ^{235}U is still unaccounted, i.e. is not attributed to some particular precursor.

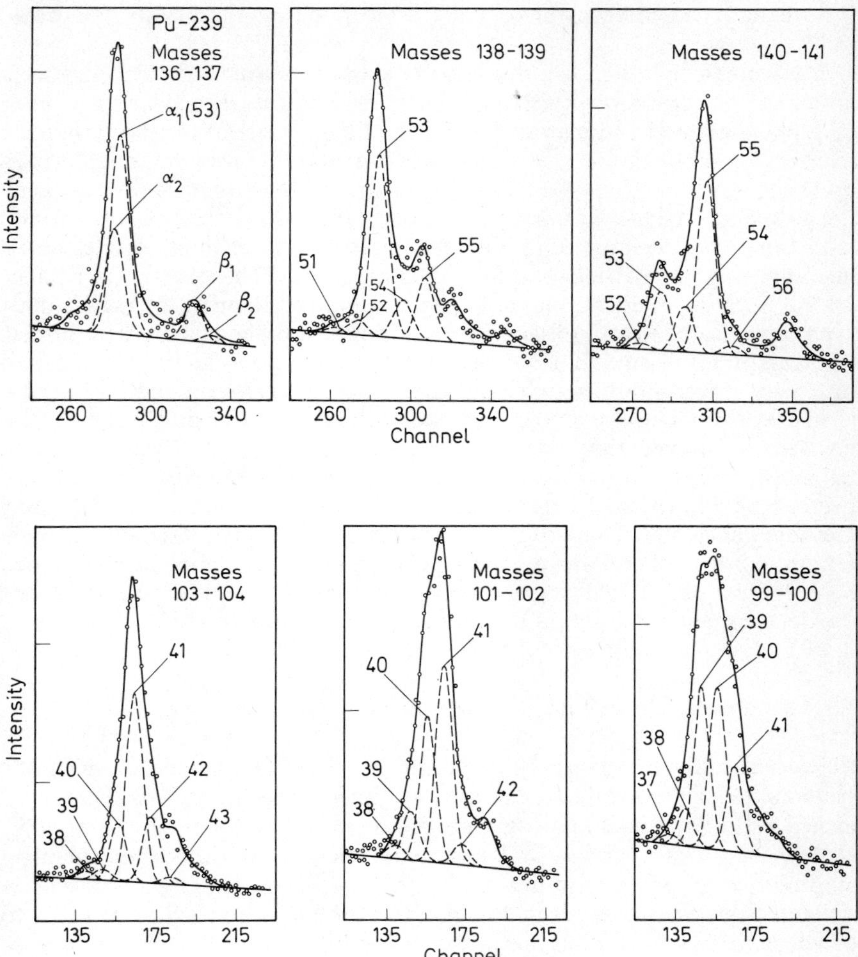

Figure 1.26 Mass-sorted K x-ray spectra for some adjacent complementary fragment mass intervals in thermal neutron fission of ^{239}Pu. The relative contributions from the various elements were determined by computer fit (solid curves) to the observed data based on the known relative intensities of the K x-ray components (α_1, α_2, β_1, β_2) for each element – as shown for $Z = 53$ (dashed curves) in the mass interval 136–137. For visual clarity, only the α_1 component for each element is shown in the other mass intervals. (From Glendenin *et al.*[57], by courtesy of IAEA)

Although the sources of delayed neutrons in fission are becoming well established, information on the energy spectra of the neutrons is almost non-existent. Very early measurements[61], in which energy resolution was low, were made on the energy spectrum of the total delayed neutron emission, but virtually no measurements have been made on the energy spectra in individual emitters.

The longest half-life for a delayed neutron precursor is 55 s, and identification of precursors can only be made for elements which are amenable to fast radiochemical separation from the fissile material in which they are formed. Bromine, iodine, rubidium, caesium, and to a lesser extent, krypton and xenon, have been identified as precursors by direct volatilisation of the element. Arsenic, antimony, selenium, and to a lesser extent, tin and tellurium, have been measured by forming the hydrides in a gas stream and decomposing the hydrides at different temperatures. These methods produce all the isotopes of a given element. By passing the element through a mass separator the neutron emission from specific isotopes can be determined. The caesium and rubidium isotopes readily undergo differential ionisation in a mass spectrograph source, and have been fully investigated[62].

In the investigation of the chemical behaviour of fission fragments, it has been observed that isobars formed independently in fission tend to behave differently from those formed by the decay of precursors[63]. This has been applied in delayed neutron studies, where it has been possible to label methane and methyl iodide with halogens produced independently in fission, whereas there was discrimination against the same species formed by precursor decay. By fast gas chromatography, it was possible to correlate the delayed neutron fraction with independent fission yields[64].

1.6.4.2 Probability of neutron emission

The occurrence of delayed neutron emitters is in fair agreement with the systematics of the available energy, and many predictions have been confirmed in the experimental programme. Delayed neutron emission is possible when β-decay of a precursor leads to an energy level above the neutron binding energy of the product, i.e. when $Q_\beta - B_n > 0$ where Q_β is the mass difference between the precursor and the ground state of the product to which it decays, and B_n is the neutron binding energy in the product, as shown in Figure 1.27. For instance, in the decay chain.

$$^{87}_{35}\text{Br} \xrightarrow{\beta} {}^{86}_{36}\text{Kr} \xrightarrow{n} {}^{85}_{36}\text{Kr}$$

the mass difference between ^{87}Br and ^{86}Kr is 1.0114 amu, whereas that between ^{86}Kr and ^{85}Kr is 0.9982 amu. With increasing distance from stability along the isobaric chains, Q_β values increase, neutron binding energies decrease, and neutron emission becomes more likely. However, selection rules may hinder the transition, in which case de-excitation will occur by γ-emission. The probability of neutron emission relative to the total decay of the precursor is called the 'neutron branching ratio', P_n. This may be determined experimentally as the ratio of the yield of delayed neutrons per fission to the cumulative yield of the precursor, after making allowances

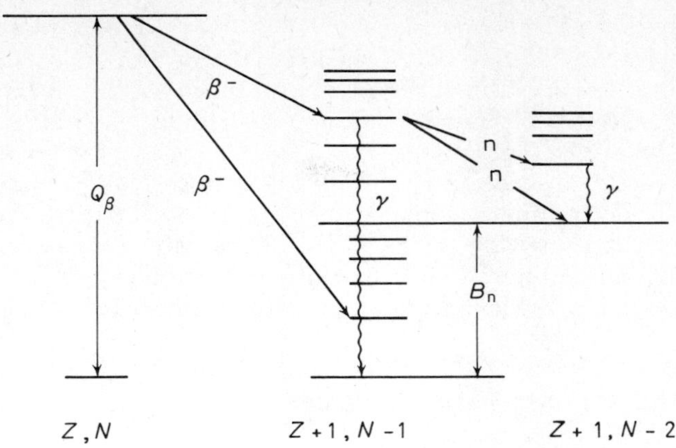

Figure 1.27 Energy level diagram for delayed neutron emission.

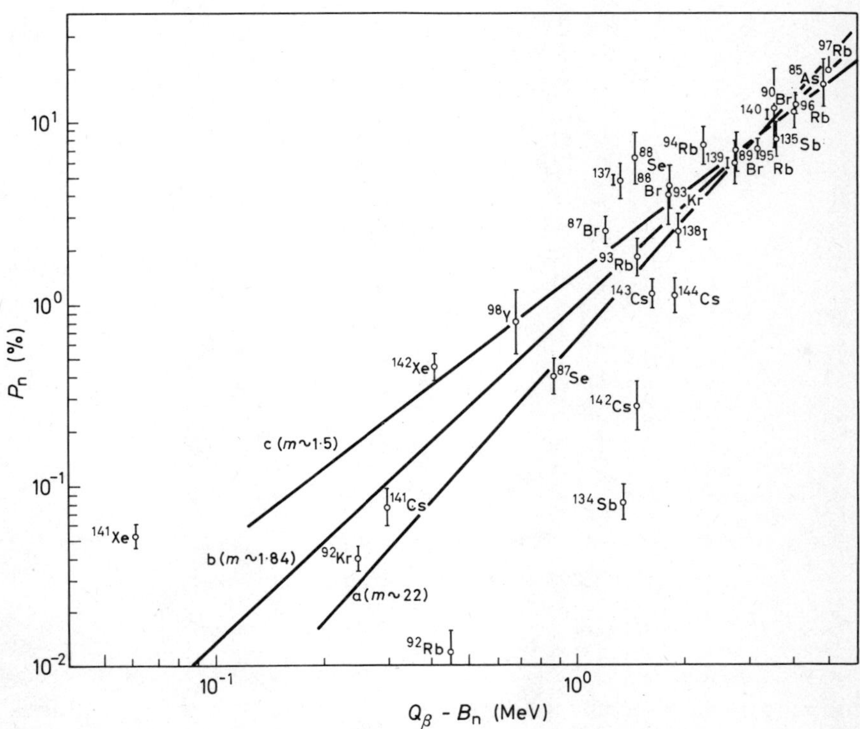

Figure 1.28 Correlation of experimental neutron emission probabilities, P_n, with the parameter $(Q_\beta - B_n)$. (From Amiel[60], by courtesy of IAEA)

for the formation of any isomers of the precursor. At high values of E_n, where $E_n = Q_\beta - B_n$, it may be anticipated that P_n will be a function of the energy level density of the emitter, which in turn will be an exponential function of E_n, say $(E_n)^m$. Figure 1.28 shows a logarithmic plot of the observed values of P_n against E_n. The letters a, b, c, show the values, *m*, of the exponent. At low values of E_n, a wide scatter is observed which may be attributed to nuclear shell effects. For instance, $^{134}_{51}Sb$ decays to a closed shell of 82 neutrons and has a low value of P_n [65]. This is the only example discovered, so far, of a closed shell nuclide which is also a neutron emitter. This type of information provides evidence for nuclear shell structure far from stability.

1.6.4.3 Delayed photo-fission neutrons

In using delayed neutron emission for the detection of burst fuel in reactors, it has become important to determine if there are any delayed neutron precursors with long half-lives, i.e. longer than 55 s, which is the longest

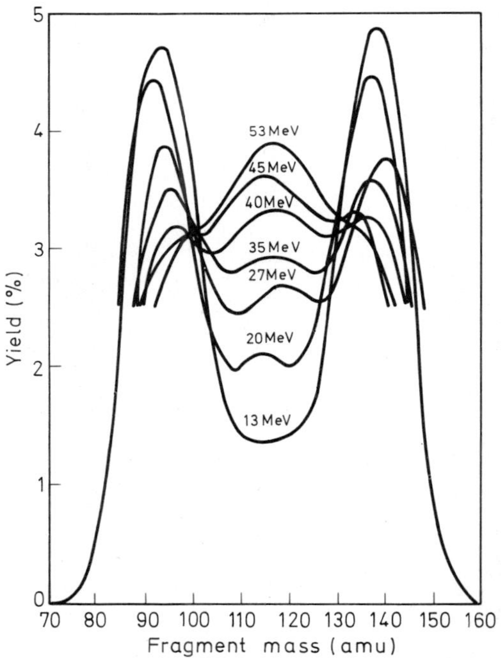

Figure 1.29 'Initial' mass distribution curves for the fission of ^{238}U bombarded with 13 to 53 MeV protons. (From Croall and Cuninghame[68], by courtesy of North-Holland Publishing Company)

hitherto known. A search for delayed neutron precursors with long half-lives has been made [66]. No true delayed neutron precursors with half-lives longer than 55 s were detected (yields were shown to be $<10^{-12}$ neutrons per fission) but apparent delayed neutron groups of half-lives 3, 17, and 111 min were found. By varying the amount of uranium and plutonium involved in the experiments, it was found that these apparent delayed neutrons could be attributed to prompt neutrons from photo-emission caused by high-energy γ-rays emitted from short-lived fission products. The photo-fission cross-section for ^{239}Pu has a lower energy threshold than that for ^{238}U, and the effect was more marked in the case of plutonium. The photo-fission

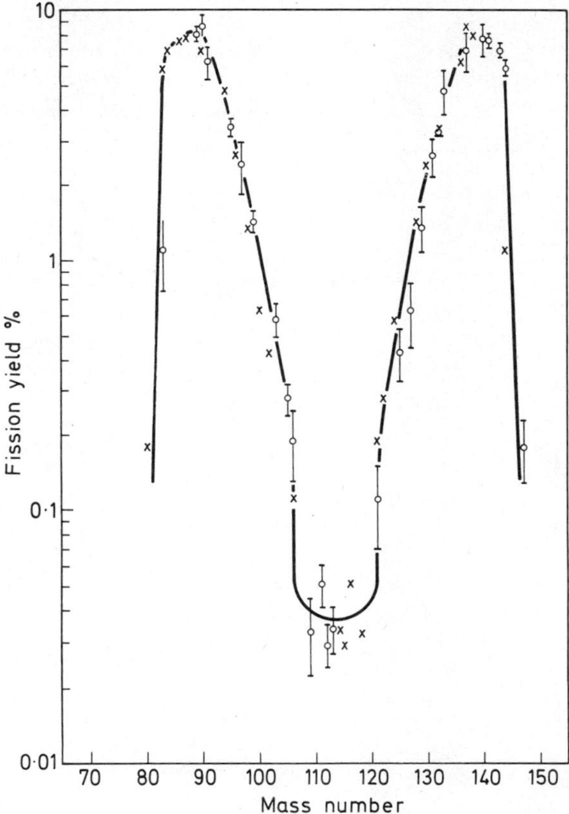

Figure 1.30 Mass yield curve for thermal neutron fission of ^{227}Th. o measured points, x reflected points. (From Flynn and Von Guten[69], by courtesy of IAEA)

effect is not readily observed in small samples of fissile material extracted from the core of a reactor, but will be expected to occur inside the core of reactors where the bulk of the fissile material can absorb γ-radiation. High-energy γ-rays are more prolific amongst short-lived fission products, and also in the activation products of the constructional materials in reactor cores, and it may be anticipated that these apparent delayed neutrons from photo-emission constitute an appreciable fraction of the effective delayed neutron component of reactor fluxes.

1.6.5 Mass and charge distribution

As reviewed by Pappas *et al.*[67], no very striking new developments have taken place in studies of the mass and charge distribution in fission although there has been much work confirming and amplifying the subject.

1.6.5.1 Mass yield curves

Figure 1.29 shows a series of yield curves obtained in the bombardment of ^{238}U with protons at varying energies from 13 MeV to 53 MeV [68]. As with other yield curves, the changes in shape shown in Figure 1.29 can be roughly

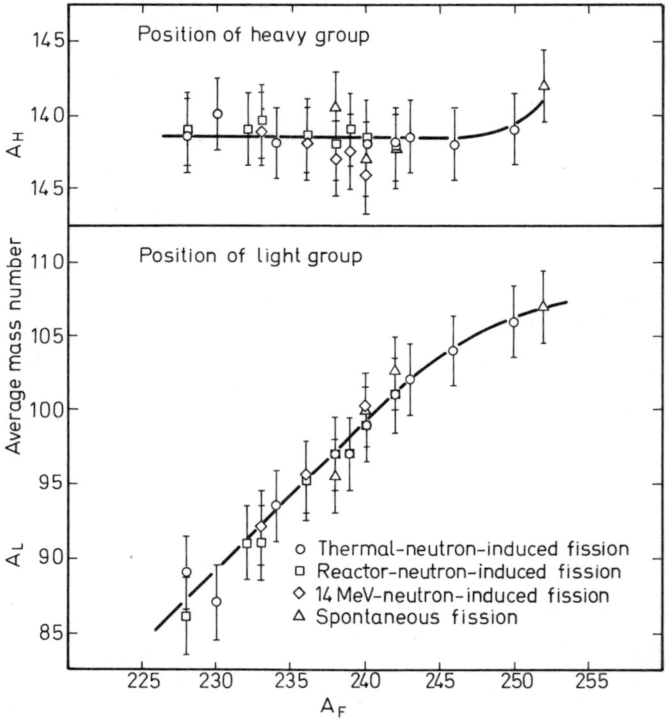

Figure 1.31 Positions of light, A_L, and heavy, A_H, groups in the fission of various nuclides of mass A_F. The average mass number at half-maximum of the peaks is plotted. (From Flynn and Von Guten[69], by courtesy of IAEA)

attributed to the sum of asymmetric and symmetric components in different proportions, but the fact that the composite curves do not intersect at single points shows that no precise analysis of this type can be made.

Figure 1.30 shows the mass yield curve for ^{227}Th which is the lowest mass nuclide to be fissionable by thermal neutrons[69]. This again illustrates the constancy of the mass of the heavy fragment for all types of fission.

Figure 1.31 shows a plot of the maximum of the light, A_L, and heavy,

A_H, groups, illustrating how the mass range of the heavy group remains constant while that of the light group changes with the mass, A_F, of the fissioning nucleus.

Figure 1.32 shows the single-humped fission yield curves obtained in

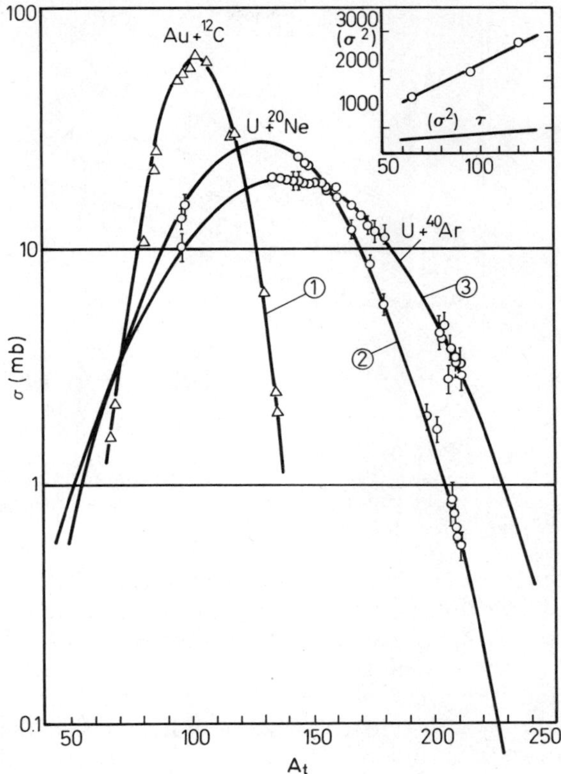

Figure 1.32 Mass yield curves for the fission of uranium nuclei with heavy ions, ^{12}C, ^{20}Ne, and ^{40}Ar. (From Karamyan *et al.*[70], by courtesy of IAEA)

the bombardment of ^{238}U with heavy ions[70]. Here there is a highly-charged compound nucleus (85–110 protons) and no asymmetric fission appears.

1.6.5.2 Charge distribution

New nuclear charge distribution studies in proton fission have been reported by Yaffé[71]. As the excitation energy of fission increases, it is observed that the most probable charge on the fragments, Z_P, moves closer to the most stable charge, Z_A, for any given mass, and that the same effect is observed as the neutron to proton ratio, N/Z, of the target nucleus decreases. Figure 1.33, obtained from studies of products in the mass range 129–139, shows this effect for a number of fissile nuclides.

As discussed by Wahl *et al.*[72], the cumulative yield of fission products as a function of their nuclear charge may be plotted on a probability scale to give linear plots for each mass. This method of representation, as shown in Figure 1.34, reveals that independent yields of nuclides with even Z are slightly higher than those with odd Z[73]. Although the mass equation and

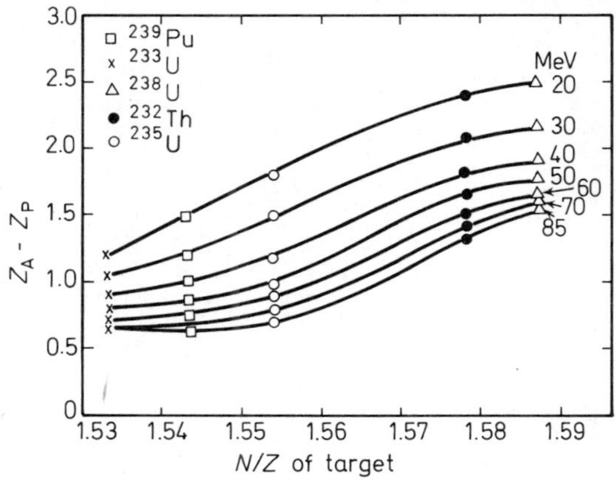

Figure 1.33 Displacement of most probable charge from stability as a function of the neutron–proton ratio of the target for a variety of bombarding energies. (From Yaffe[71], by courtesy of IAEA)

the statistical theory of fission would predict this effect, it has not previously been observed experimentally in a clear manner. Figure 1.34 also shows clearly the irregularities that occur in the region of closed shell nuclides. Close inspection shows that irregularities are not limited to closed shells of 82 neutrons and 50 protons only, which suggests, as has been discussed in connection with the double-barrier theory of fission, that the closed shells relevant in fission may not be the same as those relevant in undistorted nuclei.

Recent work has been reported by Sistemich *et al.*[74] on the primary charge of fission products as measured by a new technique. Fission fragments, as they were formed, were passed into a mass separator, on line, and the number of β-decays of each separate isotope was measured as it became stable. The results obtained by this method were essentially in agreement with those obtained in the past by radiochemical methods.

1.6.6 Mass yields in fission

For technological reasons, renewed attention has been given in recent years to the precise determination of fission yields. In the past, fission yields have often been determined by radiochemical methods, on fissile material irradiated for short periods, in which the accuracy is normally only 5–10%. Mass spectrometric methods of measuring fission yields give greater accuracy, but these can normally only be applied to highly-irradiated fissile material where the quantities of fission products are sufficient for mounting as sources in a mass spectrometer. It has been difficult to reconcile the results obtained from the radiochemical and mass spectrometric methods and the literature contains many inconsistent results.

1.6.6.1 Mass yields and 'burn-up'

In the determination of the amount of fissile material consumed, i.e. the 'burn-up', in a given fuel element in a nuclear reactor, one of the most direct methods is to measure the amounts of particular fission products

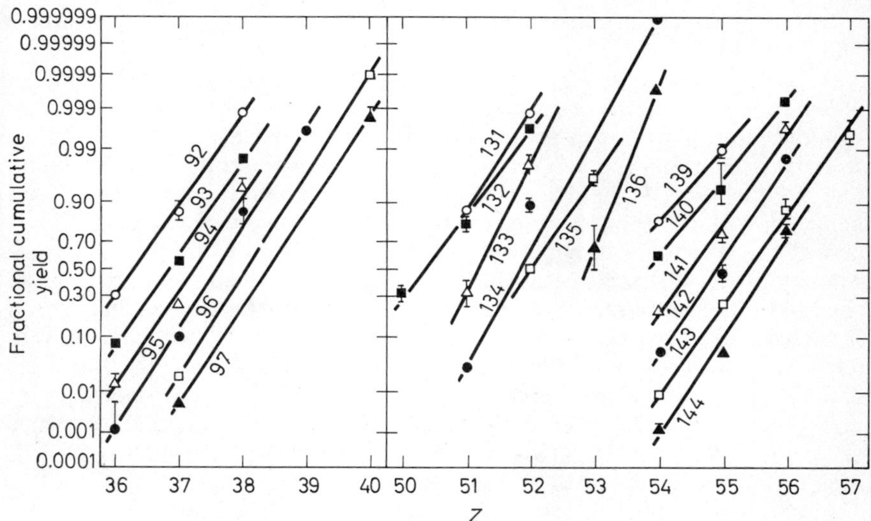

Figure 1.34 Probability scale plots of measured fractional cumulative yields against atomic number. Mass numbers are given near lines fitted visually to the data. (From Runnalls *et al.*[73], by courtesy of The American Institute of Physics[ε]

formed. If the fission yields are known accurately, then the burn-up can also be determined accurately. Heat output from fuel is determined by the fission rate, i.e. by the burn-up in a given time. It is required to optimise as precisely as possible the heat output from nuclear fuel with respect to fuel and coolant temperatures. Also the power from nuclear reactors represents the return from large capital investment, and accurate information is required for commercial purposes. Methods for the determination of burn-up have been reviewed by Rider and Ruiz[75], and a symposium has recently been held on the subject[76]. Current developments in this field include the observation that many fission yield curves overlap in the region of the stable neodymium isotopes ($^{142-146,148,150}$Nd) (see for instance, Figures 1.4 and 1.32). These nuclides may be determined by isotope dilution and mass spectrometric methods with good accuracy in highly-irradiated fuel, and therefore provide good standards for burn-up determinations. The problem of estimating to what extent the fission rate should be attributed to different fissile nuclides, i.e. ^{239}Pu or ^{235}U, in fuel is minimised. The effect of varying neutron spectra in thermal reactors is also minimised.

1.6.6.2 Variation with neutron energy

In fast reactor fuels the determination of burn-up is dependent on the manner in which fission yields vary with neutron energy. The neutron energy spectra change from a nearly pure fission spectrum in the core of an unmoderated assembly to various mixtures of 'fast', 'resonance', 'epithermal' and 'thermal' spectra, in the 'blanket' and intermediate regions of reactors where the materials of construction and coolants introduce varying amounts of moderation.

A study has been continued over a number of years at Los Alamos, in which underground nuclear explosions have been used to produce an instantaneous high flux of neutrons, and a further paper on this subject has recently been published[77]. Neutrons are passed down a tube 240 m in length and, by means of a rapidly rotating disc, samples may be irradiated with neutrons within a specific range. In the latest reference, neutrons are intercepted in the energy range 20 eV – 10^4 eV by rotating the disc to give a velocity of 1.5×10^4 cm s^{-1} at the rim. A film of ^{235}U was mounted on the rim, and radio autography showed the positions at which resonance fission occurred. Strips of film corresponding to the different resonances were removed from the rim of the disc and analysed for 2.3 day ^{115}Cd and 2.7 day ^{99}Mo. These nuclides represent the symmetric and asymmetric components of fission. The ratios of the nuclides vary sharply from one resonance to another, and evidence of this nature should assist in predicting the average fission yields to be expected in fast reactor spectra.

As this Review has tried to indicate, the effect of energy changes on cross-sections, fission yields, and other phenomena of fission, provide a challenge both in understanding the nature of nuclear processes and in making industrial use of them.

References

1. Hahn, O. and Strassmann, F. (1939). *Naturwissenschaften,* **27,** 11
2. Bohr, N. and Wheeler, J. A. (1939). *Phys. Rev.,* **56,** 426
3. Hyde, E. K. (1964). *The Nuclear Properties of the Heavy Elements, Vol. III, Fission Phenomena* (Englewood Cliffs, New Jersey: Prentice-Hall)
4. Halpern, I. (1959). *Ann. Rev. Nucl. Sci.,* **9,** 245
5. Gindler, J. E. and Huizenga, J. R. (1968). *Nuclear Chemistry II,* 1. Ed. L. Yaffe. (New York: Academic Press)
6. *Proc. Symp. Phys. Chem. Fission, 2nd, Vienna* (1969). (Vienna: International Atomic Energy Agency)
7. Polikanov, S. M., Druin, V. A., Karnaukhov, V. A., Mikheev, V. L., Pleve, AK. A., Skobelev, N. K., Subbotin, V. G., Ter-Akop'yan, G. M. and Formichev, V. A. (1962). *Soviet Phys. JETP (English Transl.),* **15,** 1016
8. Flerov, G. N., Polikanov, S. M., Gavrilov, K. A., Mikheev, V. I., Pevelygin, V. P. and Pleve, A. A. (1964). *Soviet Physics JETP (English Transl.),* **18,** 964, see opposite, *Soviet Physics JETP (English Transl.),* **17,** 544
9. Brenner, D. S., Westgaard, L. and Bjornholm, S. (1966). *Nucl. Phys.,* **89,** 267
10. Flerov, G. N., Pleve, A. A., Polikanov, S. M., Tretyakova, S. P., and Martalogu, M., Poenaru, D., Sezon, M., Vilcov, I. and Vilcov, N. (1967). *Nucl. Phys.,* **A97,** 444

11. Lark, N. L., Sletten, G., Pedersen, J. and Bjornholm, S. (1969). *Nucl. Phys.*, **A139,** 481
12. Lark, N. L. (1969). *Nucl. Instr.*, **67,** 137
13. Metag, V., Repnow, R., von Brentano, P. and Fox, J. D. (1969). *See* Reference 6, 449, IAEA–SM–122/29
14a. Vandenbosch, R. and Wolf, K. L. (1969). *See* Reference 6, 439, IAEA–SM–122/110
14b. Nix, J. R. and Walker, G. E. (1969). *Nucl. Phys.*, **A132,** 60
14c. Russo, P. A., Vandenbosch, R., Mehta, M., Termer, J. R. and Wolf, K. L. (1971). *Phys. Rev.*, **C3,** 1595
14d. Elwyn, A. J. and Ferguson, A. T. G. (1970). *Nucl. Phys.*, **A148,** 337
14e. Repnow, R., Metag, V., Fox, J. D. and von Brentano, P. (1970). *Nucl. Phys.*, **A147,** 183
15. Nisle, R. G. and Stephan, J. E. (1969). *Idaho Nucl. Corp. Report,* IN-1317, 149
16. Cabell, M. J. (1968). *J. Inorg. Nucl. Chem.*, **30,** 2583
17. James, G. D. (1971). *Atomic Energy Research Establishment, Harwell, Report R.6676*
18. Fong, P. (1955). *M.S. Thesis, University of Washington,* Published in Fong, P., *Statistical Theory of Fission* (1969). (New York: Gordon and Breach)
19. Walton, G. N. (1961). *Nature (London),* **192,** 1062
20. Ramanna, R., Subramanian, R. and Aiyer, N. (1965). *Nucl. Phys.*, **67,** 529
21. Brueckner, K. A., Buchler, J. R., Clark, R. C. and Lombard, R. J. (1969). *Phys. Rev.*, **181,** 1543
22. Nörenburg, W. (1969). *See* Reference 6, 51, IAEA–SM–122/30
23. Schmitt, H. W. (1969). *See* Reference 6, 67, IAEA–SM–122/122
24. Bohr, A. (1965). *Proc. Intern. Conf. Peaceful Uses At. Energy, Geneva,* **2,** 151. (New York: Columbia University Press)
25. Hughes, D. J. and Schwartz, R. B. (1958). *BNL,* **325,** *2nd Ed.* (New York: Brookhaven Nat. Lab.)
26. Kovacs, I., *et al.* (1969). *See* Reference 6, 419, IAEA–SM–122/134
27. Huizenga, J. R., Behkami, A. N. and Roberts, J. H. (1969). *See* Reference 6, 403 IAEA–SM–122/118
28. Otozai, K., Meadows, J. W., Behkami, A. N. and Huizenga, J. R. (1970). *Nucl. Phys.*, **A144,** 502
29. Strutinsky, V. (1968). *Nucl. Phys.*, **A122,** 1
29a. Strutinsky, V. and Bjornholm, S. (1968). *Proc. Int. Symp. Dubna,* IAEA, Vienna (1968) 431
29b. Strutinsky, V. and Pauli, H. C. (1969). *See* Reference 6, 165, IAEA–SM–122/203
30. Nilsson, S. G. (1955). *Dan, Mat. Fys. Medd,* **29,** 16. See also Nathan, O. and Nilsson, S. G. (1965). *Alpha-, Beta- and Gamma-Ray Spectroscopy,* Chap. **10,** Ed. K. Siegbahn. (Amsterdam: North Holland Publ. Co.)
31. Lynn, J. E. (1968). *Atomic Energy Research Establishment, Harwell, Report* **R5891**
32. Lynn, J. E. (1969). *See Reference 6,* 249, IAEA–SM–122/204
33. Lynn, J. E. (1968). (Oxford: Clarendon Press)
34. Fubini, A., Blons, J., Michaudon, A. and Paya, D. (1968). *Phys. Rev. Lett.*, **20,** 1373
35. Migneco, E. and Theobald, J. P. (1968). *Nucl. Phys.*, **A112,** 603
36. Paya, D., Blons, J., Derrien, H. and Michaudon, A. (1969). *See Reference 6,* 307, IAEA–SM–122/90
37. Specht, H. J., Fraser, J. S., Milton, J. C. D. and Davies, W. G. (1969). *See* Reference 6, 363, IAEA–SM–122/128
38. Vandenbosch, R. (1969). *Discussion. See Reference 6,* 418
39. Feather, N. (1969). *See Reference 6,* 83, IAEA–SM–122/201
Cosper, S. W., Cerny, J. and Gatti, R. C. (1967). *Phys. Rev.*, **154,** 1193
Nobles, R. A. (1962). *Phys. Rev.*, **126,** 1508
40. Nardi, E., Boneh, Y. and Frenkel, Z., (1969). *See* Reference 6, 143, IAEA–SM–122/11 See also *Phys. Rev.*, (1967). **156,** 1305
41. Carles, C., Asghar, M., Doan, T. P., Chastel, R. and Signarbieux, C. (1969). *See Reference 6,* 119, IAEA–SM–122/80
42. Fong, P. (1969). *See Reference 6,* 133, IAEA–SM–122/108
43. Gazit, Y., Katase, A., Ben-David, G. and Moreh, R. (1969). *See Reference 6*, 891, IAEA–SM–122/12
44. Asghar, M., Carles, C., Chastel, R., Doan, T. P., Ribrag, M. and Signarbieux, C. (1970). *Nucl. Phys.*, **A145,** 657. Nardi, E. and Frankel, Z. (1970). *Phys. Rev.*, **C2,** 1156

45. Bowles, B. J. and Becket, N. (1966). *Phys. Rev.*, **147,** 852
46. Apalin, V. F., Gritsyuk, Y. N., Kutikov, I. E., Lebedev, V. I. and Mikaelyan, L. A. (1965). *Proc. Symp. Phys. Chem. Fission, Salzburg, Austria, 1965,* IAEA–SM–160/92
47. Skarsvag, K. and Bergheim, K. (1963). *Nucl. Phys.*, **45,** 72
48. Weinstein, S., Reed, R. and Block, R. C. (1969). *See Reference 6,* 477, IAEA–SM–122/113
49. Asghar, M. (1967). *Nucl. Phys.*, **A98,** 33. See also Dermendzhiev, E. (1969). *See Reference 6*, 487, IAEA–SM–122/113
50. Muga, M. L. and Rice, C. R. (1969). *See Reference 6*, 107 IAEA–SM–122/99
51. Stoenner, R. W. and Hillman, M. (1966). *Phys. Rev.*, **142,** 716. See also Iyer, R. H. and Cohble, J. W. (1968). *Phys. Rev.*, **172,** 1186
52. Unik, J. P., Cuninghame, J. G. and Croall, I. F. (1969). *See Reference 6*, 717, IAEA–SM–122/58
53. Horsch, F. and Michaelis, W. (1969). *See Reference 6,* 527, IAEA–SM–122/44
54. Armbruster, P., Hossfeld, F., Labus, H. and Reichelt, K. (1969). *See Reference 6,* 545, IAEA–SM–122/23
55. Back, B. B., Bondorf, J. P., Otroshenko, G. A., Pedersen, J. and Rasmussen, B. (1969). *See Reference 6*, 351, IAEA–SM–122/74
56. Croall, I. F. and Willis, H. H. (1963). *J. Inorg. Nucl. Chem.*, **25,** 1213
57. Glendenin, L. E., Griffin, H. C., Reisdorf, W. and Unik, J. P. (1969). *See Reference 6,* 781, IAEA–SM–122/114
58. Keepin, G. R. (1968). *Proceedings of a Panel on Delayed Fission Neutrons, Vienna 1967,* IAEA, Vienna, 1968
59. Yiftah, S. and Saphier, D. (1968). *Proceedings of a Panel on Delayed Fission Neutrons, Vienna, 1967.* IAEA, Vienna 1968
60. Amiel, S. (1969). *See Reference 6,* 569, IAEA–SM–122/205
61. Batchelor, R. and McHyder, H. R. (1956). *J. Nucl. Energy,* **3,** 7; See also Talbert, W. L. and McConnel, J. R. (1967). *Arkiv für Fysik,* **36,** 99
62. Amarel, I., Bernas, R., Foucher, R., Jastrzebski, J., Johnson, A., Teillac, J. and Gauvin, H. (1967). *Phys. Lett.*, **24B,** 402; See also Amarel, I., Gauvin, H. and Johnson, A. (1969). *J. Inorg. Nucl. Chem.*, **31,** 577
63. Hall, D. and Walton, G. N. (1960). *J. Inorg. Nucl. Chem.*, **19,** 16
64. Amiel, S. and Paiss, Y. (1965). *Radiochimica Acta,* **4,** 157
65. Tomlinson, L. and Hurdus, M. H. (1967). *Phys. Lett.*, **25B,** 545
66. Tomlinson, L. and Hurdus, M. H. (1969). *See Reference 6,* 605, IAEA–SM–122/60
67. Pappas, A. C., Alstad, J. and Hagebø, E. (1969). *See Reference 6*, 669, IAEA–SM–122/206
68. Croall, I. F. and Cuninghame, J. G. (1969). *Nucl. Phys.*, **A125,** 402
69. Flynn, K. F. and Von Gunten, H. R. (1969). *See Reference 6*, 731, IAEA–SM–122/1
70. Karamyan, S. A., Oganesyan, T., Poostilnik, B. I. and Flerov, G. N. (1969). *See* Reference 6, 759, IAEA–SM–122/130
71. Yaffe, L. (1969). *See* Reference 6, 701, IAEA–SM–122/3
72. Wahl, A. C., Norris, A. E., Rouse, R. A. and Williams, J. C. (1969). *See* Reference 6, 813 IAEA–SM–122/116
73. Runnalls, N. G., Troutner, D. E. and Ferguson, R. L. (1969). *Phys. Rev.*, **179,** 1188
74. Sistemich, K., Armbruster, P., Eidens, J. and Roeckl, E. (1969). *Nucl. Phys.*, **A139,** 289
75. Rider, B. F. and Ruiz, C. P. (1962). *Progress in Nuclear Energy,* **3,** 25
76. I.A.E.A. (1967). "Fuel Burn-up Predictions in Thermal Reactors", *Panel Proceedings Series,* STI/PUB/172, Vienna, 10–14 April, 1967
77. Cowan, G. A., Bayhurst, B. P., Prestwood, R. J., Gilmore, J. S. and Knobeloch, G. W. (1970). *Phys. Rev.*, **C2,** 615
78. Fong, P. (1971). *Phys. Rev.*, **C3,**
79. Nardi, E. and Fraenkel, Z. (1970). *Phys. Rev.*, **C2,** 1156
80. Loveland, W., Shum, Y. S. and Hill, D. W. (1971). *Nucl Phys. Ann. Rep.*, 191 University of Washington

2
The Chemical Applications of Angular Correlations and Half-life Measurements

J. I. VARGAS

Nuclear Chemistry Laboratory, Grenoble

2.1 INTRODUCTION

The purpose of this brief review is to summarise, by choosing selected examples, some recent progress obtained in the application of γ–γ angular correlations and lifetime measurements to chemical problems. The applications of these techniques to chemistry are a consequence of the fact that, under certain circumstances, the directional correlation of two radiations emitted successively by an excited nucleus, and the transition rates of isotopes decaying by electron capture and/or by isomeric transition, can be markedly influenced by interactions between the nucleus and the orbital electrons.

The first technique has been extensively treated, from the point of view of its application to solid-state and nuclear physics problems, in the articles of Steffen and Frauenfelder[1,2]. The specifically chemical influences on angular correlations have received much less attention. In fact, this topic was only discussed in a review by De Benedetti *et al.*[3]. This article has not, however, included the chemically interesting hot atom effects, which, later were touched upon briefly as a part of the 'after effects' phenomena by Bäverstam *et al.*[4] and more recently by Thun[5], and by Gerholm and Holmberg[6]. Recent developments in the perturbed angular correlations (PAC) field are found in *Hyperfine Structure and Nuclear Radiations* edited by Matthias and Shirley[7]. In this book, the reader will find a compilation of all PAC measurements carried out until 1967. The utilisation of new powerful on-line techniques, closely related to the classical γ–γ angular correlations in radioactive sources, have been reviewed by Grodzins[8,9], by Deutch[10] and by Murnick *et al.*[11]. A wealth of information on PAC will shortly be available in the *Proceedings of an International Conference on Hyperfine Interactions Detected by Nuclear Radiations* (1970), Rebovot, Israel.

The paper by Cohen[12] as well as the earlier elementary treatment by Steffen[13] and the review by Heer and Novey[14] still serve as excellent introductions to the subject.

In spite of the scarcity of investigations dealing with chemically interesting systems, the potentialities of the PAC techniques are considerable.

As perturbed angular correlation (PAC) studies, with all their technical variants and refinements, constitute an enormous and expanding field of activity, complete coverage of all phenomena and technical developments, that might have a bearing on chemistry, is out of the question. In fact, it might be argued that every experiment conducted with insulator sources contains some kind of chemical information. Thus, we shall omit all the results which were not obtained either to elucidate a chemical process or to illustrate some principle important for the comprehension of the subject under review. Such criteria are, of course, arbitrary and we beg the pardon of those authors whose papers will not be included in the list of references.

In this review, emphasis will be given to experimental results. Nevertheless, references will be given, in passing, to recent theoretical work which, the author's opinion, may be relevant to chemical research.

In the first Section, a brief presentation of the unperturbed angular correlations of γ-rays will be given; in the second, the different perturbation mechanisms will be reviewed in a qualitative way to introduce the vocabulary and formulae which will be useful in later discussions. A third Section will report on results obtained in measurements carried out in gaseous, liquid and solid sources. The experimental information on solids will be limited mostly to quadrupole coupling in insulators, as the large majority of results on magnetic perturbations refer to metallic matrices.

Hot atom effects will next be presented and discussed; finally, the chemical influences on decay rates will be considered.

2.2 THE UNPERTURBED ANGULAR CORRELATION

The probability of emission of a radiation by an excited nucleus depends on the orientation of the nuclear spin axis relative to the direction of emission. In an ordinary radioactive source, the nuclei have their spins distributed at random in space and therefore the radiation emitted is isotropic. To produce an anisotropic intensity pattern, the nuclear spins must be oriented. This may be achieved by several more or less sophisticated methods[15].

In angular correlation work, the nuclear orientation is obtained simply by selecting, in a system composed of many radioactive nuclei, which decay to the ground state with the emission of two γ-rays in cascade, only those nuclei which have emitted the first γ-ray, γ_1, in a fixed direction $\vec{k}_1$. The intermediate nuclear states will have their spins aligned relative to $\vec{k}_1$, and therefore, the second γ-ray, γ_2, detected in coincidence with the first one, is not isotropic. This explains why the emission probability of γ_2 is a function of the angle θ between γ_1 and γ_2, that is to say, why the γ_1 and γ_2 are correlated. This angular dependence is described by the correlation function, $W(\theta)$, which measures the relative probability that the two γ-rays be emitted at an angle θ from each other.

$$W(\theta) = \sum_{K_{\text{even}} = 0}^{K_{\max}} A_K P_K(\cos\theta) \tag{2.1}$$

In most cases, the angular correlation function is suitably described by only two parameters A_2 and A_4, and so:

$$W(\theta) = A_2 P_2(\cos\theta) + A_4 P_4(\cos\theta) \tag{2.2}$$

The A_2 and A_4 are fixed by a large number of nuclear parameters (especially in the case of mixed multipoles) and cannot, at present, be calculated from first principles[3]. P_K (cos θ) are the Legendre polynomials.

From the experimental point of view, the angular correlation is observed by measuring γ_1 and γ_2 in coincidence, as a function of the angle between the counters, a fixed one which detects γ_1 and a mobile one which measures γ_2. Fast-slow circuits are commonly used. Two types of measurements are performed, depending on the mean life, τ, of the intermediate level and on the resolution time t_r of the coincidence circuit. The most valuable information is obtained from the time differential angular correlation (TDAC) measurements. For such measurements to be possible, the resolution time must be much smaller than the mean life of the intermediate level. In any actual experiment, the counter may detect the second radiation within a finite time interval t_1 to t_2 after the first one, and the measured correlation will be:

$$W(\theta, t_1 t_2) = \frac{\int_{t_1}^{t_2} W(\theta, t) \,.\, \exp - \left(\frac{t}{\tau}\right) . \,dt}{\int_{t_1}^{t_2} \exp - \left(\frac{t}{\tau}\right) . \,dt} \tag{2.3}$$

if account is taken of the exponential decay of the intermediate level with the time t.

The smallest time interval (t_1, t_2) which may be used is, of course, the resolving time t_r of the experimental set-up. If the resolving time t_r is much larger than the mean life τ, as frequently occurs, then one can only measure the integral correlation. On the contrary, if $t_r \ll \tau$, one can measure the coincidence counting rates as a function of the time spent by nuclei in the intermediate level before decaying, for a given angular setting between the counters. An unperturbed angular correlation will not depend on the time spent by nuclei in the intermediate state, and therefore $W(\theta, t) = W(\theta, 0)$.

A large effort has been made by nuclear spectroscopists to find perturbation-free media, so as to extract the nuclear information (spins, etc. . . .) contained in the A_K coefficients.

2.3 THE PERTURBED ANGULAR CORRELATION

An angular correlation may be perturbed when a coupling is established between the momenta of the intermediate nuclear state and extranuclear electromagnetic fields. In PAC, two cases merit attention: the interaction between the nuclear dipole moment and a magnetic field, and the interaction between the nuclear quadrupole moment and an electric field gradient. These fields may be static or dynamic, classic or hyperfine. The classical static fields are represented by stationary externally applied magnetic fields and, to a good approximation, by crystalline electric fields. The hyperfine fields result from the coupling of the nuclear angular momentum I, with some quantum mechanical system surrounding it, such as the electronic shell of the atoms or molecules. Both classical and hyperfine fields may be, of course, time-dependent[2]. The perturbation of the angular correlation of the radiations may be imagined classically by supposing that the fields exert a torque on the nuclear dipole or on the electric quadrupole moments, tending to align the nuclear magnetic dipole or electric quadrupole axis; the nuclear angular momentum I will respond to this torque by a precession around the magnetic field or around the electric field gradient axis. Consequently, it can easily be imagined that the correlation will change in the time interval elapsed between the emission of the first and second γ-rays. In other words, it will be reduced, and the formula (2.3) must now contain time-dependent $G_K(t)$ functions describing the attenuation of the correlation coefficients, A_K.

$$W(\theta, t) = \sum_K A_K G_K(t) P_K(\cos\theta) \tag{2.4}$$

All the physical information is contained in the $G_K(t)$ functions and most of the theoretical work had to deal with the derivation of appropriate formulae of $G_K(t)$ for each type of interaction. Theoretical expressions for all kinds of static interactions are given by Steffen and Frauenfelder[2].

2.3.1 Static magnetic interaction

In the common case of a magnetic field perpendicular to the plane formed by the two counters, the perturbation factor takes the simple form:

$$W(\theta, t) = \sum_K A_K P_K [\cos(\theta - \omega_L t)] \tag{2.5}$$

which confirms classical expectations, that is to say, the angular correlation precesses with the Larmor frequency, ω_L, around the field direction, so that, in a time t, the angular correlation pattern has rotated by an angle $\omega_L t$. In those cases where the nuclear lifetime is too short for a time dependent measurement to be made, the time integral correlation technique has to be used, and an integral rotation of the correlation by an angle $\Delta\theta \approx \omega_L \tau$ can be observed[1, 2].

If the static magnetic field is parallel to the direction of emission $\vec{k}_1$, chosen as the quantisation axis, then the influence of this field can be interpreted as a precession of the intermediate nuclear state around $\vec{k}_1$. The projections of $\vec{I}$ on $\vec{k}_1$ do not change, and the population of the $(2I+1)$ states with respect to $\vec{k}_1$ remains the same. Thus, the angular distribution of the second radiation with respect to $\vec{k}_1$ is not disturbed[1, 2].

2.3.2 Static electric field interaction

As already mentioned, detailed forms of the perturbation factor have been calculated for single crystals and polycrystalline powders[2]. The interaction Hamiltonian between the nuclear quadrupole moment and the electric field gradient is the product of two second-order tensors:

$$H = \tfrac{4}{5}\pi T^{(2)} V^{(2)} \tag{2.6}$$

The field gradient at the nucleus is described by the V_{ij} symmetrical tensor, which has six components. In the orthogonal (principal axis) system, this symmetrical tensor may be reduced to its diagonal form, that is to say, the number of components is reduced to V_{xx}, V_{yy}, V_{zz}. Since these components must obey Laplace's equation $\nabla^2 V = 0$, only two of them are independent. Usually, the principal axes are chosen such that:

$$|V_{xx}| \lesssim |V_{yy}| \lesssim |V_{zz}| \tag{2.7}$$

and one defines

$$eq = V_{zz} = \left(\frac{\mathrm{d}^2 V}{\mathrm{d}z^2}\right) \tag{2.8}$$

The field gradient asymmetry is measured by the usual asymmetry parameter

$$\eta = \frac{V_{xx} - V_{yy}}{V_{zz}} \tag{2.9}$$

From Laplace's equation and (2.9), it follows that V_{xx} and V_{yy} have the same sign and therefore $0 \lesssim \eta \lesssim 1$. For a cubic symmetry: $V_{xx} = V_{yy} = V_{zz}$ and $\nabla^2 V$ and $V_{ij} = 0$; therefore, the correlation is not perturbed.

In the case of a static quadrupole interaction, the precession frequencies correspond to the energy differences between neighbouring levels.

$$\omega_Q(m, m') = E_Q(m) - E_Q(m') \tag{2.10}$$

where

$$E(m) = \frac{3m^2 - I(I+1)}{4I(2I-1)} \cdot \Delta\nu_Q \tag{2.11}$$

and where $\Delta\nu_Q = eQV_{zz} = e^2Qq$ = quadrupole coupling strength. Q = quadrupole moment of the nuclear excited state, and m is the z component of I.

These frequencies can be expressed as multiples of the fundamental frequencies. For integral I:

$$\omega_Q = 3\frac{2\pi}{4I(2I-1)}\Delta\nu_Q \tag{2.12}$$

For a polycrystalline sample, the interactions must be averaged over all crystal directions and the perturbation factor is then given by:

$$G_K(t) = \sum_n S_{Kn} \cos(\eta\omega_0 t) \tag{2.13}$$

$$\omega_0 = 6\,\omega_Q \text{ and } m = |M^2 - M'^2|, \text{ for half integral } I,$$

and

$$\omega_0 = 3\,\omega_Q \text{ and } m = |M^2 - M'^2|, \text{ for integral } I.$$

The S_{Km} are given by Alder *et al.*[16] and Abragam and Pound[17] for an axial field; the corresponding values for non-axial fields are given by Béraud *et al.*, for $\frac{5}{2}$ spin[18]. The attenuation factors are in fact extremely sensitive to the asymmetry parameter, as shown in Figure 2.1, taken from the last reference.

The perturbation factor is thus a sum of periodic functions and, therefore, the static interaction in crystal powders cannot destroy the angular correlation: a mere rotation is produced provided the field strength is constant in time. With this proviso, the attenuation factor is:

$$\overline{G_K^{(\infty)}} = \sum_n S_{K_n} \frac{1}{1+(n\omega_0\tau)^2} \tag{2.14}$$

and so, even for an infinitely strong interaction, the perturbation factor is not cancelled. Therefore, a correlation always survives:

$$\overline{G_K{}^{(\infty)}_{\min}} = \frac{1}{2K+1} \tag{2.15}$$

This important result frequently serves to decide whether a static or dynamic interaction prevails in the sample: the latter may effectively lead to attentuation factors smaller than the 'hard core' values given by (2.15).

This discussion has until now assumed that the electric field gradients acting on the nuclear quadrupole moments are the same at every nuclear site. This is an idealisation: slight variations of the crystalline fields are expected, because of lattice vibrations, imperfections and impurity centres. It is reasonable to suppose that the variations of the electric field gradient from nucleus to nucleus results in a probability distribution[19] of the quadrupole interaction frequency ω_Q Gaussian and block distributions have frequently been considered as necessary to fit the experimental results[20, 21]; others assume a Lorentzian distribution as more appropriate to account for their observations[22, 23]. These distributions are characterised by a parameter

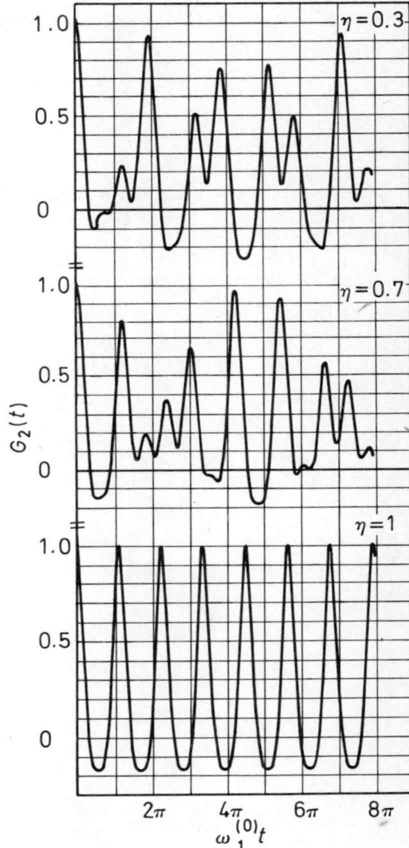

Figure 2.1 Time-dependent perturbation factor $G_2(t)$ for $I = 5/2$ and for different values of asymmetry parameter (From Beraud *et al.*[18], by courtesy of North-Holland Publishing Company)

$\delta = \sigma/\omega_Q$, where ω_Q is the centroid of the quadrupole coupling frequency and σ is the width of the distribution[24].

2.3.3 Static hyperfine interactions

Even when the classical static fields are eliminated—by turning off the external magnetic fields, by placing the nucleus in a crystal of cubic symmetry or by actually isolating the atoms (gaseous sources)—the intermediate state, as remarked above, can still be perturbed by coupling with its immediate surroundings, such as, for example, the electronic shells of atoms or molecules[1, 15, 25].

For free atoms, the magnetic hyperfine interaction is isotropic and hence, the attenuation factor of the A_K correlation coefficients is pseudo-periodic: the angular correlation pattern rotates with frequencies identical to those of the hyperfine interaction:

$$\hbar\omega = E_m - E_{m'}$$

Again, a 'hard core' is observed for the integral attenuation factor, which is, for the special case of $J = \frac{1}{2}$:

$$\overline{G_K{}^{(\infty)}} = \frac{1-K(K+1)}{(2I+1)^2} \tag{2.16}$$

The quadrupole coupling in free atoms has also been considered theoretically[1] and, similarly to crystalline powders, a 'hard core' is obtained[26].

Even if the electronic states are considered as stationary, the isotropic hyperfine coupling as it exists in free atoms, and was discussed above, is not a general enough description for atoms or ions in solids. In this case, the hyperfine interaction may not be isotropic and the coupling between the nuclear spin and the electronic states can be represented by the 'Spin Hamiltonian':

$$H = \vec{I}\,.\,\mathrm{T}\,.\,\vec{S} \tag{2.17}$$

where $\vec{S}$ is the spin of the electronic shell and T is the symmetric coupling tensor[17]. For a strongly anisotropic interaction: $H \approx A\,.\,I_z\,.\,S_z$, where A is the z component of T, previously reduced to the principal axis system. The integral perturbation factors are given by:

$$\overline{G_{KK}[^{\infty}]} = \frac{1}{2K+1} \sum_{N=-K}^{+K} \frac{1}{1+(AN\tau/2\hbar)^2} \tag{2.18}$$

which, for very large values of A, has the usual lower limit: $1/(2K+1)$.

2.3.4 Combined static interactions

A recent example dealing with combined static magnetic and axially symmetric electric interactions is the work of Lieder *et al.*[21], in which a hafnium single-crystal containing the ^{181}Ta probe was examined. The zero-field resonance behaviour of the integral correlation function was studied in a series of papers by Leisi[27]. The influence of a static magnetic field on the correlation perturbed by an isotropic hyperfine coupling ($\vec{I}\,.\,\vec{J}\,.$), corresponds to the Zeeman effect hyperfine structure in optical spectroscopy. The strong field case is analogous to the Paschen–Back effect: if the field is strong enough, it causes a decoupling of $\vec{I}$ and $\vec{J}$, and so the unperturbed correlation is re-established.

2.3.5 Time-dependent interactions

Time-dependent interactions occur when the fields acting on the intermediate nucleus fluctuate randomly with time, as for example happens in a liquid. The effect of these fields on the angular correlation will depend on a correlation time τ_c characteristic of the fluctuations, and on the mean life of the nuclear excited state τ. Roughly speaking, τ_c is the time required for the liquid to change its local configuration. If τ_c is much smaller than the mean life of the nuclear excited level, the fluctuating fields average to zero, and the angular correlation is not perturbed. If τ_c is increased, for instance by increasing the viscosity of the liquid, τ is approached and the correlation becomes perturbed.

For non-viscous liquids (short correlation times) and small perturbing molecular fields. Abragam and Pound[17] have shown that the time differential perturbation factor of the angular correlation is given by

$$G_K(t) = \exp(-\lambda_K t) \tag{2.19}$$

In the case of perturbations produced by random electric field gradients, the relaxation constants λ_K take the form:

$$\lambda_K = \frac{3}{80}\,4\pi^2\left(\frac{\tau_c}{\hbar^2}\right)(eQ)^2 < V_{zz}{}^2 > \frac{K(K+1)\left[4I(I+1)-K(K+1)-1\right]}{I^2(2I-1)^2} \tag{2.20}$$

where the symbols have their usual meaning.

As mentioned above, the theory should be valid only in the limit given by $<\omega_Q^2>^{\frac{1}{2}}.\ \tau_c \ll 1$ and $\tau_c \ll \tau$. The time integral perturbation coefficient is obviously:

$$\overline{G_K{}^{(\infty)}} = \frac{1}{1+\lambda_K\tau} \tag{2.21}$$

Although many time differential investigations have confirmed the exponential behaviour of the perturbation factors, additional information may be obtained if a model is adopted for the relaxation mechanism[2]. In the simple Debye model for polar liquids, the correlation time τ_c is:

$$\tau_c = \left(\frac{4\pi a^3}{3kT}\right)\eta \tag{2.22}$$

where η is the viscosity, T the temperature and a is a characteristic length (effective molecular diameter). Tests of the model were carried out by measuring the DPAC in solutions whose viscosities were varied by adding glycerine[28].

Abragam and Pound have also obtained an expression for the magnetic relaxation constant λ_K^M in liquids:

$$\lambda_K^M = \tfrac{1}{3}\omega_L^2\tau_S K(K+1) \tag{2.23}$$

which, together with equation (2.20), permits interesting conclusions to be drawn about the nature of the species interacting in solutions. For example, pure quadrupole interactions, $\lambda_4/\lambda_2 = 0.59$, while for the magnetic case such as occurs in solutions of paramagnetic ions, $\lambda_4/\lambda_2 = 3.33$.

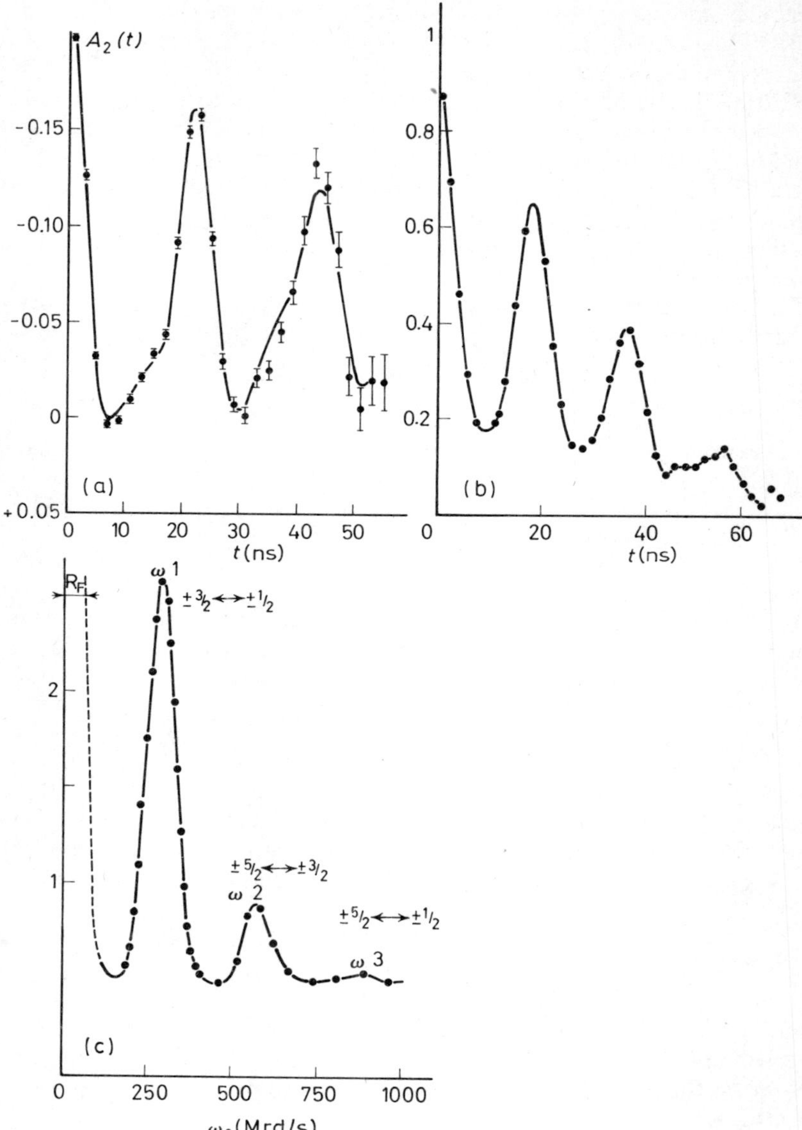

Figure 2.2 (a) Time-dependent perturbation coefficient $A_2(t)$ of ^{181}Ta in Hf metal. For the unperturbed correlation of the 133–482 keV cascade $A_4 = -0.071 \pm 0.007$ and $A_2 = -0.297 \pm 0.006$. The full line gives the result of a theoretical fit with the following values of quadrupole interaction parameters: $\omega_0 = 292$ Mrd/s, $\eta = 0.12$, $\delta = 4\%$. (b) Autocorrelated spectrum. (c) Fourier transform of (b).

The λ_4/λ_2 values describing the quadrupole interaction obviously depend on the spin of the intermediate level, as shown by formula (2.20). The values $\lambda_4/\lambda_2 = 0.59$ and $\lambda_4/\lambda_2 = 1.7$ refer to spins 2 and 5/2, respectively.

The electric and magnetic perturbations mentioned above are both isotropic, and therefore, may present analogies with the coupling of electronic spins and electric field gradients with lattice vibrations. The isotropic paramagnetic spin-lattice relaxation is thus given also by equation (2.23), provided the spin relaxation time τ_S is defined analogously to the correlation time τ_c, and that $\omega_S\tau_S \ll 1$; $\omega_S\hbar = a_S$, where a_S is a coupling constant proportional to the magnetic moment of the nuclear state.

Theoretical expressions for the angular correlation perturbed by isotropic and anisotropic mixed fluctuating and static electric field gradients (in Abragam and Pound's limit) were obtained by Andrade *et al.*[29] and applied to explain the interesting relaxation spectrum of the quadrupole interaction in $(NH_4)_3HfF_7$, which was detected by using the well known ^{181}Ta probe.

For longer correlation times and/or strong uncorrelated interactions, the relaxation cannot in general be described by formula (2.19), and several theoretical models were proposed which try to cover the whole range of correlation times[30–33]. An experimental approach to these questions was made by Cameron *et al.* who studied the quadrupole coupling of ^{181}Ta in frozen aqueous solutions[34].

In spite of the fact that in real substances all these interactions may take place simultaneously, the theoretical treatment sketched above may often help to decide on the nature of the interaction, even in the less favourable cases where only time integral measurements are performed. In fact, for a given nucleus (fixed I and τ), the time integral attenuation factors $\overline{G_2^{(\infty)}}$ and $\overline{G_4^{(\infty)}}$ assume well-defined values which characterise uniquely each type of interaction. An example of such a parametric graph is given by Avida *et. al.*[35].

The most powerful technique remains however the DPAC, particularly when the measurements are carried out at different temperatures and/or external fields. The information contained in the anisotropy time spectrum is obtained by adjusting the observed experimental points to a theoretical function in which account has been taken of corrections for the time resolution, solid angle of the counters, etc[2]. In the case of solids where a distribution of frequencies or of sites is possible, it is most convenient to submit the experimental time spectrum to a Fourier transformation, thus directly obtaining a frequency spectrum of the coupling. This Fourier spectrum is now completely equivalent to a nuclear quadrupole or nuclear magnetic resonance spectrum[7]. To this end, a preliminary smoothing of the data is obtained by performing an autocorrelation analysis of the raw coincidence-time numbers. The advantage of such procedures is shown in Figure 2.2 for a specially pure polycrystalline sample of metallic hafnium. The correlation spectrum was obtained with the 133–482 keV cascade of ^{181}Ta. The three peaks correspond to the three transitions of the 5/2 nucleus. A distribution of frequencies would be revealed—as in n.q.r. and n.m.r.—by the 'resonance' line widths, while the presence of distinctive crystal sites would be revealed by other peaks having appropriate intensities.

2.4 EXPERIMENTAL

Although a large number of nuclei may in principle be used in PAC investigations, only isotopes having half-lives longer than about 1 h can, because of counting statistics, be investigated in the conventional γ–γ experiments. Of these, only a few permit DPAC studies, and therefore, much of the new chemically interesting information was obtained with the well-known ^{111}Cd, ^{181}Ta and ^{204}Pb^{m} probes. The increased efficiency obtained by the use of four-counter systems[36] has, however, permitted the examination of relatively short-lived rare earth isotopes, where these present appropriate intermediate levels[37]. Long-lived intermediate levels may also cause difficulties, because of chance coincidence problems, and because of the need for long-term stability of apparatus[38]. In addition to the above mentioned probes, DeGrazia[39] has started using the extremely interesting 480 72 keV cascade in ^{181}Re for investigations on rhenium chemistry.

In using angular correlation, one must always keep in mind that, excepting the case of isomeric transition, the γ–γ cascade is fed by a prior energetic nuclear disintegration, which may have two consequences: firstly, hot atom effects may result from recoil and/or from rearrangements in the electronic shells, creating exceptional static and dynamic fields. Secondly, even when the hot atom effects are eliminated by a judicious choice of the sample and/or nuclear probe, the atomic number of the isotope tracer will be changed; consequently, the interaction parameters refer solely to an entirely new chemical entity.

2.4.1 Gases

The free hyperfine interaction has been studied in ^{127}I and ^{125}I atoms, produced by electron capture in ^{127}Xe and ^{125}Xe, respectively[40, 41]. Integral PAC measurements using the 172–203 keV and 55–188 keV cascades, indicated strong static and dynamic attenuations, attributed to the electron capture, which were gradually reduced by increasing the pressure of the buffer Xe gas. From the collected data, information on cross-sections for charge-exchange between the buffer and the highly-charged iodine atoms was obtained.

In a new experiment using iodine vapour mixed with the xenon, Gygax and Leisi[42] found the anisotropy of the correlation to exhibit a characteristic drop as the gas pressure is increased. The effect is due to the change of the hyperfine coupling during the lifetime of the intermediate state, either by change of atomic state, e.g. by charge exchange, or by electronic spin depolarisation. From the experiments with the 172–203 keV γ–γ correlation in the decay of the ^{127}Xe the geometrical cross-section for these processes was found to be 11.10^{-15} cm^{2}.

2.4.2 Liquids

The largest number of investigations on liquids by far has been carried out on aqueous solutions.

New results[37] on aqueous acid solutions of ^{154}Gd and ^{156}Gd, of ^{160}Dy and ^{172}Yb and of ^{152}Sm, ^{154}Gd, ^{156}Gd, ^{160}Dy, ^{166}Er and ^{176}Yb[43] are in agreement with Abragam and Pound's theory. Considerable chemical influences on the angular correlations are found in all cases. Thus, in ^{154}Gd 1.0 N $HClO_4$ solutions $\lambda_4/\lambda_2 = 2.99 \pm 0.52$, while in 2.6 N H_2SO_4 $\lambda_4/\lambda_2 = 0.73 \pm 0.17$.

An important contribution by Popp and Wagner[43] is the demonstration that PAC extends considerably the range of measurement of nuclear spin relaxation times, to about 10^{-9} s.

Perhaps the most interesting recent application of PAC is the use of probe nuclei to measure the 'tumbling time' of biologically important labelled macromolecules in solution[44, 45].

The 173–247 keV cascade in ^{111}Cd was used as a 'rotational tracer' for the relaxation. Changes in conformation were observed, for example, by altering the pH of a solution of bovine serum albumin (BSA): as the correlation time τ_c changed, a variation in the effective length or volume given by formula (2.22) was observed. τ_c was obtained by supposing the fluctuating electric field gradient in the liquid to be approximately the same as the static one characteristic of the highly viscous, practically solid solutions (hard-core values). Measurements were also performed in adenosine triphosphate (ATP), apo-carbonic anhydrase and lighter polydentate ^{111}Cd complexes. The range of the corresponding correlation times for different temperatures was from 10^{-11} s for the Cd^{II} ethylenediamine complex, to 10^{-6} s for BSA.

This work suggests many other biologically interesting experiments; similar work carried out *in vivo* should be particularly rewarding.

2.4.3 Solids

The single-crystal work of Lieder *et al.*[21] and that of Leisi[46], using respectively ^{181}Ta in metallic hafnium and ^{166}Ho in rare earth ethylsulphates, further confirm the calculations of Alder *et al.*[16].

Most of the PAC work was however carried out with polycrystalline materials. Measurements on hafnium powders confirm the single-crystal

Table 2.1 Field gradients in hafnium compounds: Comparison between PAC and Mössbauer values ($\times 10^{17}$ V/cm ^{2})

Sample	*Temperature* (K)	^{181}Ta	^{178}Hf
HfO_2	77	13.2 ± 0.6 [49]	12.3 ± 0.5 [22]
$(NH_4)_2HfF_6$	290	14.3 ± 0.8 [22]	13.6 ± 0.5 [22]
K_2HfF_6	290	20.0 ± 0.8 [52, 101]	23.9 ± 0.6 [52, 53, 101]
$(NH_4)_3HfF_7$	4.2	3.7 ± 0.3 [52, 101]	7.2 ± 0.3
K_3HfF_7	77	7.0 ± 0.3 [52, 101]	6.0 ± 0.3
Na_3HfF_7	290	10 [52, 101]	7.0 ± 0.3

work: the electric field gradient at the ^{181}Ta site possesses axial symmetry, as expected from the h.c.p. structure of the matrix[23]. Measurements on HfO_2 [22, 47, 49] are all in agreement with the monoclinic structure of the compound. In the cubic HfC, Oliveira[49] detects non-stoichiometry-induced perturbations. No after-effect of the K-capture was found in In_2O_3, In_2S_3 and $In(C_9H_7NO)_3$, in the work of Falk *et al.*[50].

The fluorine complexes of hafnium received considerable attention from Andrade[51], Gerdau *et al.*[22] and from Berthier *et al.* at Grenoble[52]. Table 2.1 includes for comparison Mössbauer results on the 2+ level of ^{178}Hf by Gerdau *et al.*[53] and observations made at Grenoble. These results confirm the striking similarity of the electric field gradient values found at the two

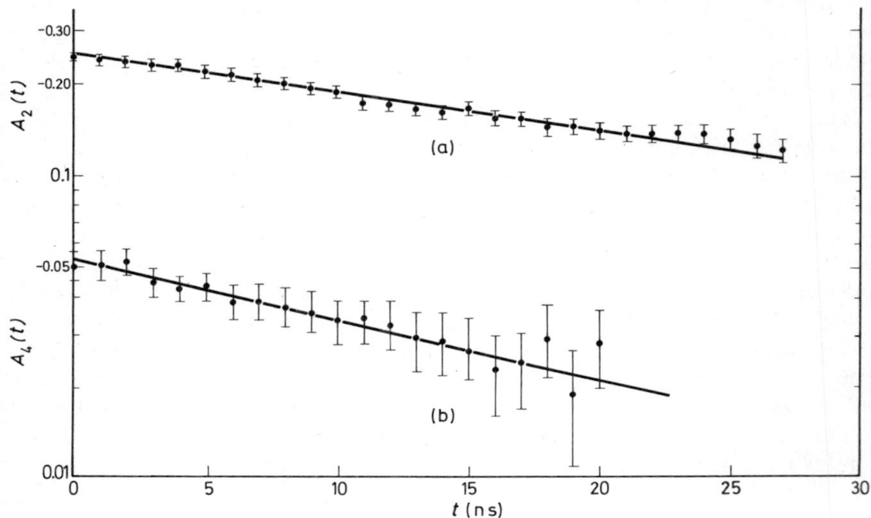

Figure 2.3 Time-dependent perturbed angular correlation coefficients for ^{181}Ta in solid $(NH_4)_3$ HfF_7. (a) = A_2 (*t*). (b) = A_4 (*t*) at 37 °C

nuclei, and indicate firstly, that β^- decay produces no effect, as expected from the long half-life (17 μs) of the primary 615 keV level of the ^{181}Ta 133–482 keV cascade and secondly, that the vacancy created by the change in atomic number resulting from the disintegration is not filled: otherwise, strong magnetic perturbation would result from the $5d^1$ structure of a Ta^{IV} ion. A fast relaxation of this interaction is not thought to be responsible for the observations: results of DPAC measurements in $(NH_4)_3HfF_7$, at liquid helium temperature, can be fitted with pure electric quadrupole interaction parameters, and, in addition, the relaxation in the room temperature cubic disordered phase of the material gives $\lambda_4/\lambda_2 = 1.7$, as it would for the postulated interaction[52]. This is shown in Figure 2.3. One further interesting aspect of the results is the observation of β^--synthesis of tantalum complexes in new configurations: indeed, the nascent $^{181}TaF_7$ ion is forced to adopt the pentagonal bipyramidal structure (D_{5h}) of HfF_7^{3-}, rather than the capped trigonal-prismatic one (C_{2v}) typical of normal TaF_7^{2-}.

The natural width of the 15.3 ns intermediate level in ^{181}Ta also allows the study of interesting conformational changes, as demonstrated by the data obtained on the quadrupole coupling ω_0, the asymmetry parameter η and the frequency distribution δ of ^{18}Ta, in the series of octacoordinate hafnium complexes assembled in Table 2.2. The presence of two sites in Hf(A.A)$_4$, Hf(NBPHA)$_4$ and Hf(Cup)$_4$ (made evident by the Fourier analysis of the correlation-time spectra), is interpreted as due to soft mode interconversions among possible conformational isomers. A typical example of these DPAC measurements is shown in Figure 2.4, for the tetrakis hafnium(IV) *N*-benzoyl-*N*-phenyl-hydroxylamine complex.

The Grenoble group also observes that the z component of the field gradient is related (for a given coordination number) to the electronic density at the ligands, so that, for the more abundant sites listed in Table 2.2 (which may represent the most stable molecular configurations), the quadrupole coupling increases in the order: $HfO_2 > Hf(A.A)_4 > Hf(NBPHA)_4 > Hf(T)_4$. A similar correlation is found for quadrupole coupling in the analogous high-spin Fe^{III} complexes[54].

Conclusions on the different geometries of the compounds are drawn from the asymmetry parameter: antiprismatic forms are improbable, because in all cases $\eta > 0$. However, care must be exercised in the interpretation, because of the extreme sensitivity of η to lattice perturbations. This is revealed

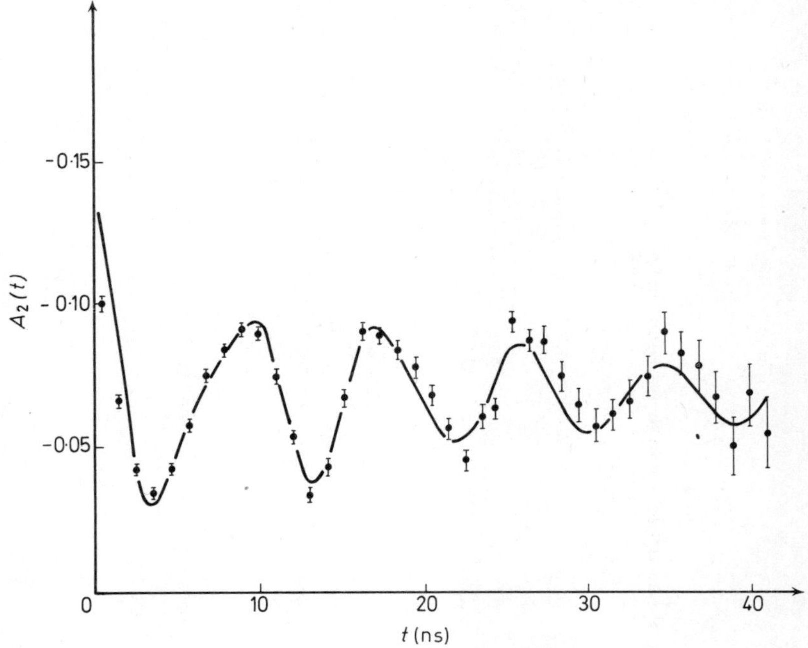

Figure 2.4 Time spectrum of ^{181}Ta in Hf(NBPHA)$_4$

by the large increase observed for this parameter when chloroform dopant was introduced in the Hf(T)$_4$ lattice[54]. Another example of the application of the DPAC technique to physico-chemical problems was the measurement of the relaxation rates of the ^{181}Ta nucleus in frozen aqueous solutions by Cameron *et al.*[34]. A close agreement between the PAC and the Mössbauer studies[55, 56] was observed. Analogous observations were made with ^{111}Cd by Barrett *et al.*[57].

Table 2.2 ^{181}Ta differential perturbed angular correlation in

Sample	T(K)	*Site abundance* (%)	ω_0 (Mrd/s)	η	δ	$V_{zz}(10^{18}$ V/cm$^2)$
$Hf(AA)_4$ [60]	293	site 1 40%	710 ± 60	0,1	8 at 12%	1.23
		site 2 60%	710 ± 60	1		1.23
$Hf(NBPHA)_4$ [51, 61, 62, 102]	293	site 1 74%	605 ± 10	0.42 ± 0.02	6%	1.05
		site 2 26%	760 ± 20	0.80 ± 0.05	6%	1.31
$Hf(Cup)_4$ [52, 62, 102]	293	site 1 74%	315 ± 5	0.32 ± 0.02	9%	0.54
		site 2 26%	720 ± 20	0.68 ± 0.04	9%	1.25
	77	site 1 74%	360 ± 5	0.21 ± 0.02	9%	0.62
		site 2 26%	640 ± 10	0.60 ± 0.04	9%	1.11
$Hf(T)_4$ [52, 54, 102]	293	site 1 100%	138 ± 5	0.40 ± 0.05	relaxes	0.24
	77	site 1 100%	157 ± 5	0.23 ± 0.05	27%	0.27
$K_4Hf(C_2O_4)_4 \cdot 5H_2O$ [23]	293	site 1 100%	310 ± 20	0.80 ± 0.10	relaxes	0.54
$Hf(T)_4 + CH \cdot Cl_3$ [52, 54, 102]	293	site 1 100%	145 ± 5	0.75 ± 0.05	relaxes	0.25

$Hf(AA)_4$ = Tetrakis-Hf^{IV} acetylacetone
$Hf(NBPHA)_4$ = Tetrakis-Hf^{IV} *N*-benzoyl-*N*-phenyl-hydroxylamine
$Hf(Cup)_4$ = Tetrakis-Hf^{IV} *N*-nitroso-*N*-phenyl-hydroxylamine
$Hf(T)_4$ = Tetrakis-Hf^{IV} tropolone

2.5 HOT ATOM CHEMISTRY

In the PAC studies – as in all other techniques employing radioactive tracers – the activity may be introduced in the sample by doping, or by means of a prior nuclear reaction. In the latter case, the probe nucleus may supply local information on the physico-chemical consequences of the nuclear transformation.

Among the large class of reactions which are in principle amenable to PAC studies[7,9], the simplest is the capture of a thermal neutron by some suitable target nucleus. Recoil effects may then be studied at leisure, if the induced activity is sufficiently long lived.

The first indication that PAC could aid in the solution of hot atom chemical problems, was the work of Sato *et al.*[58] on neutron-irradiated $KReO_4$. This substance was chosen by the Japanese workers because, among the high valence oxyanions, it was the only one not presenting a Szilard–Chalmers effect. A measurement on the angular correlation of ^{188}Os in the neutron-

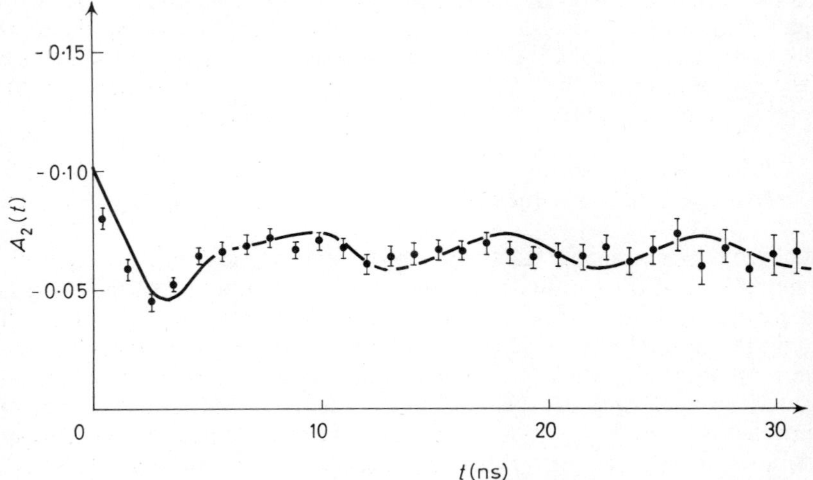

Figure 2.5 Time spectrum of irradiated Hf(NBPHA)$_4$. Full curve corresponds to calculated sum of three tantalum sites

irradiated and recrystallised samples, showed however that, in spite of the chemical measurement, the ^{188}Re recoil atoms were not in normal lattice sites, and probably also not in the same chemical form as the target molecules; therefore, the apparent failure of recoil bond-breaking should be sought in dissolution (oxidation) reactions which hampered the separation of the lower valence ^{188}Re recoil species.

More recent radiochemical and PAC investigations on several Re^{IV} and Re^{VII} compounds by Bădică *et al.*[59], indicate that analogies exist between the chemical behaviour of the ^{188}Re recoils and the coupling parameters of ^{188}Os, measured by the integral angular correlation of the 155–931 keV γ–γ cascade. Differential measurements using ^{181}Ta as a probe in neutron-irradiated Hf(A.A)$_4$ by Béraud *et al.*[60] and in Hf(NBPHA)$_4$ by Vargas *et al.*[61], led to more precise information. These investigations revealed that distinct chemical forms, rather than broad non-specific defects, were indeed produced in the solid targets. More detailed radiochemical and DPAC investigations on the second compound, demonstrated a close parallelism to exist between results obtained by both techniques[62,63]. Thus, radiochromatographic analysis of the irradiated samples detected three distinct ^{181}Hf fractions, which accounted for 37.0, 13.0 and 50.0%, respectively, of the total activity. The first fraction is the retention value for ^{181}Hf. The spectrum of the irradiated material, shown in Figure 2.5, was obtained with a computer, by taking into account only the value of the initial retention. The results of computer calculations agreed within 2–3% with the chemical measurements. The frequencies, ω, for each recoil fragment, were determined, and also shown to be correlated with the chemical behaviour of the recoil species. For instance, the more labile fragment had the smaller quadrupole coupling.

The mechanism of thermal annealing of the (n,γ) damage in $Hf(T)_4$ was thoroughly studied by radiochemical methods[54]. Parallel DPAC measurements on the low temperature irradiated solid and on subsequently annealed samples demonstrated that, in spite of the information given by the chemical methods, no really complete re-formation of the lattice takes place in this case.

The continuation of this type of work will help to throw light on the extremely complex problems related to hot atom chemistry.

2.6 HALF-LIFE MEASUREMENTS

Bouchez and Depommier[64] start their review on electron capture by writing: 'the simple fact of the capture of an electron by a proton inside the nucleus points to the influence of the atomic electrons on nuclear properties' . . . If this remark is generalised to all processes involving direct or indirect electron-nucleus interactions one would expect that most nuclear rate processes, measured by the simplest experimental parameter which takes account of the interaction – the decay rate 'constant' would be affected by the changes in electronic structure produced by chemical combination. Although nothing can be said against such a sweeping statement, the fact remains that the anticipated small chemical influences on transition rates were actually observed only in a limited number of cases, and, until recently, solely for the 'direct' electron capture and internal conversion processes. The earlier literature on the subject is presented in a review article of De Benedetti *et al.*[3], and by Perlman and Emery[65].

2.6.1 Electron capture

The possibility of altering the rate of decay by electron capture of ^{7}Be was suggested in 1947 by Segré[66], Daudel[67] and by Bouchez *et al.*[68,69]. Measurements carried out in France and at the Brookhaven National Laboratory confirmed these suggestions: the decay rate in metallic Be was effectively larger than in BeO and BeF_2. The maximum observed difference was $\lambda(\text{Be}) - \lambda(\text{BeF}_2) = (7.41 \pm 0.47) \times 10^{-4}$ relative to λ(Be)[70]. Johlige *et al.*

confirmed the earlier observations and extended the measurements to six new compounds[71]. The variations are distributed in a scale of electronic density at the ^{7}Be nucleus, relative to metallic beryllium, as shown in Figure 2.6. The precision of the observations was such that a distinction between the

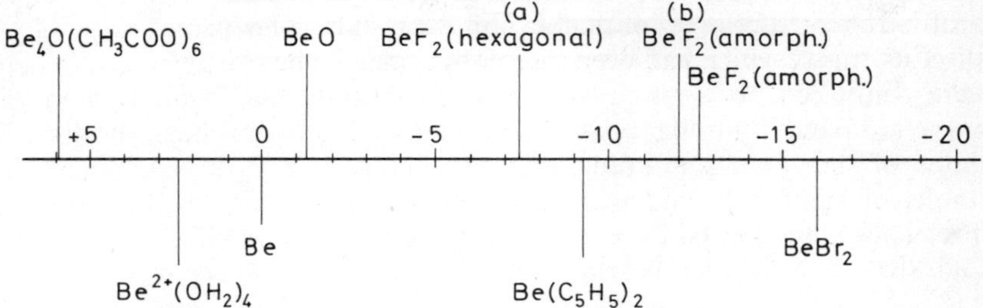

Figure 2.6 Differences of electron densities at the ^{7}Be nucleus. (a) and (b) Ref. 69 (From Johlige *et al.*[71], by courtesy of The American Institute of Physics)

hexagonal and amorphous forms of BeF_2 is clearly detected. No correlation was, however, found between the observed 'shifts' and any general physico-chemical parameter defining the compounds, For instance, in contrast with the electronegativity of the atoms, and common chemical feeling, $\lambda_{BeF_2} > \lambda_{BeBr_2}$. Equally intriguing is to find $\lambda_{Be^{2+}(H_2O)_4} < \lambda_{Be}$, when $\lambda_{BeO} > \lambda_{Be}$.

These stimulating and important results constitute the first clear-cut example of the discovery of new chemical facts through half-life measurements. It is most important that photoelectron spectroscopic measurements (ESCA) be performed on the compounds studied by the Munich group. Judging by the resolution obtained for the couple Be–BeO by Siegbahn[72], the chemical shifts of the other compounds should be easily detected.

Apart from ^{7}Be, all other chemically interesting isotopes decaying by E.C. seemed excluded for such studies, firstly, because of the considerable experimental difficulties involved in these delicate measurements, and secondly, because of theoretical arguments which indicated the chemically-induced variation of s-electron density to be negligible for the heavier elements. For instance, for chemically 'bound' 4s electrons, it would amount to about 10^{-5}–10^{-4} of the total electron density at the nucleus. Results were, however, reported by Kemeny[73], indicating important chemical influences on the K-capture probability in ^{64}Cu, when solutions of $CuSO_4$ (aq) and $Cu(NH_3)_4^{2+}$ (aq) were measured relative to metallic copper. The observed variations are: $+(1.70 \pm 0.64)\%$ in the first case and $-(0.87 \pm 0.30)\%$ in the second.

Huber *et al.*[74] and St Gagneux *et al.*[75] compared the decay constants of ^{89}Zr diffused into two samples of $BaTiO_3$ maintained at 20 °C and 200 °C, and found that the decay probability was larger in the para- than in ferro-electric phase:

$$\frac{\lambda_{20\,°C} - \lambda_{200\,°C}}{\lambda_{^{89}Zr}} = -(4.4 \pm 0.4) \times 10^{-4}$$

Independently, Odru and Vargas[76], at Grenoble, adopted the same principle of measurement (as the Basle group), which consists in mounting the samples to be compared at the extremities of an arm which exposes automatically and successively each source to a NaI crystal. Each counting step was arranged to proceed for time intervals short compared to the half-life of the isotope. Under these circumstances, the measurements approached the Rutherford differential technique usually employed in studies of half-life variations. The sources to be measured were prepared by thermal neutron irradiation of α-copper phthalocyanine crystals. The decay rates of two aliquots, one of which had been thermally annealed before the measurements, were compared. As a careful radiochemical study has demonstrated 95% of the radioactivity to be separable by Szilard–Chalmers effect, the chemical states of the ^{65}Cu-containing species were certainly very distinct[77]. The results of eight independent measurements, each extending to more than three half-lives, showed the decay rate of the ^{65}Cu recoil atoms to be smaller than that of the normal labelled copper phthalocyanine crystals:

$$\frac{\Delta\lambda}{\lambda} = \frac{\lambda(\mathrm{Cu}) - \lambda(\mathrm{recoil})}{\lambda(\mathrm{Cu})_{\mathrm{metal}}} = (+8.8 \pm 1.3) \times 10^{-4}$$

These last results are of the same order of magnitude as those observed in ^{7}Be, even when the decaying atoms were, in both cases, in rather anomalous lattice sites.

Theoretical aspects of the environmental effects on the electron capture decay rates were discussed by Alder *et al.*[78]. The ^{89}Zr results of St Gagneux *et al.*[75] were specifically treated and attention was called to the interest in measuring K-capture decay rates in nuclei permitting Mössbauer investigations: as the decay rate constants in different chemical environments give the electron densities at the nucleus, the purely nuclear radius factors contained in the Mössbauer expression for the chemical shifts are eliminated.

2.6.2 Isomeric transition

Isomeric transition is another nuclear transformation process in which the nucleus may go from one excited state to another with the emission of atomic electrons. The probability of conversion increases with decreasing energy and increasing difference in angular momentum between the two states. For forbidden transitions (large ΔI and small E), the conversion process may become highly probable and therefore, the total transition rate may be appreciably affected by the state of the electronic surroundings. These ideas were confirmed in a study by Bainbridge *et al.*[79] on ^{99}Tc^{m}. Similar effects were found in chemically different sources of ^{90}Nb^{m} and ^{235}U^{m} (Mazaki and Shimizu[80]). These results are discussed in De Benedetti's article[3]. With the exception of these results and of the variations measured in ^{90}Nb^{m} (3.6%), and in the short-lived 1.6 keV and 14.3 keV levels of ^{193}Pt for which 4 ± 2% differences were found by Marelius[81], the observed influences are of the order of 10^{-3}–10^{-4}. Larger variations than those observed by Mazaki and Shimizu were measured by Nève de Mevergnies[82–84] with ^{235}U^{m} recoils collected under various conditions on metallic backings. A striking correlation was found between the electronegativities of the V, Cu, Ni, Pt and Au supports and the half-life of ^{235}U^{m}. For the pair Pt and V, the half-life is diminished by $(5.7 \pm 1.2)^{-}$, relative to Pt.

$^{90}Nb^{m}$ is the daughter of 5.7 h ^{90}Mo, formed by a (p,4n) reaction on niobium foils, or directly by the $^{90}Zr(d,2n)^{90}Mo^{m}$ reaction.

Using the first method, Cooper *et al.*[85,86] dissolved directly the irradiated foils in a mixture of HF and HNO_3. The dissolution changed the chemical state of the niobium, and whenever an effect actually existed, a new equilibrium was established between ^{90}Mo and $^{90}Nb^{m}$, which was detected through the 122.5 keV γ-ray.

Weirauch *et al.*[87] have chosen to measure the half-life of $^{90}Nb^{m}$ produced by the second reaction. They compared the decay rate of the activity in the zirconium foils with the half-life in fluoride solutions containing niobium carrier. Within an experimental error of 1%, they were unable to find any effect. New measurements were undertaken by Olin[88], in which particular care was taken with source geometry, changes due to diffusion through the sample volume, and the possibility of volatile niobium compounds evolving after the dissolution reactions, etc. The new results confirm those of Cooper *et al.*[89]: on the fluoride complex, the rate variation (relative to metallic niobium) was found to be $(-3.9 \pm 0.8)\%$. Measurements with Nb_2O_5 lead also to an effect of $(-1.9 \pm 0.5)\%$.

The reasons for this high sensitivity to chemical environment is unclear. Free atom calculations, cited by the author, show indeed that conversion is hindered in the d sub-shells, so that the 4d and 5s electrons contribute together less than 0.1% to the internal conversion coefficient. Perhaps effects of Auger cascades associated with the previous ^{90}Mo decay (β^+, E.C.) could survive in solution for times sufficiently long so as to hinder the conversion probability in the 0.2 keV level. Charge neutralisation in the metal sources is expected to be instantaneous.

The other example of an investigation on the influence of chemical binding on I.T. rates is a study by Malliaris and Bainbridge[90] on the decay rate of the highly converted 109 keV transition in 50 d $^{125}Te^{m}$. The decay constants were compared by the double ionisation chamber technique, designed for maximum sensitivity for the K-X rays which followed the conversion. Sources of elemental Te, TeO_2 and Ag_2Te gave:

$$\lambda(\mathrm{Te}) - \lambda(\mathrm{Ag_2Te}) = (2.59 \pm 0.18) \times 10^{-4}$$
$$\lambda(\mathrm{Te}) - \lambda(\mathrm{TeO_2}) = (0.36 \pm 0.17) \times 10^{-4}$$
$$\lambda(\mathrm{TeO_2}) - \lambda(\mathrm{Ag_2Te}) = (2.23 \pm 0.18) \times 10^{-4}$$

all relative to λ(Te).

Since the sources were prepared by neutron irradiation of antimony (^{124}Sb), special care was taken to avoid extraneous isotopic and non-isotopic contamination.

Direct measurements on the change of the conversion ratio O/N of the 23.87 keV transition in $^{119}Sn^{m}$ between cases where the isotope was inplanted in white tin and SnO_2 accompanied by Mössbauer measurements in the same samples have lead Bocquet *et al.*[91] to evaluate the difference in nuclear radius between the excited and ground states. A similar study on the dependence of the internal conversion spectrum of the 8.4 keV level of ^{169}Tm and the environment of the ^{169}Er implanted in tungsten metal,

WO_3 and Tm_2O_3, was carried out by Carlson *et al.*[92]. The P_I conversion in WO_3 was 46% lower than in the metal.

2.6.3 β-decay

As pointed out by De Benedetti[3], β-decay rates could, in principle, be slightly influenced by the chemical environment, via alterations in the Fermi factor for β-emission. After a first note on the influence of the state of aggregation of NaI on the half-life of ^{131}I was published by Bergamini *et al.*[93], surprisingly large effects were reported on the decay rates of ^{131}I and ^{24}Na by Kemeny[73, 94, 95].

The latter investigator has reported that the half-life of ^{131}I decreases in the solid-state with respect to the values found in aqueous solutions, by as much as 0.5%. The possible role $^{131}Xe^m$ (daughter activity of ^{131}I) might play on the earlier observations was examined by the same investigator in a second paper; the rather large variations first reported[73] (4.3±2.1)%, were then reduced[92, 93] to a statistically more satisfactory value (2.64±0.23)%. Parallel measurements on the influence of solidification on the conversion of the 164 keV level of $^{131}Xe^m$ also showed a striking effect:

$$\frac{\lambda \text{ solid} - \lambda \text{ gas}}{\lambda \text{ gas}} = (3.2 \pm 1.1)\%^{94}$$

The observations on ^{131}I were extended to another β-emitter, ^{24}Na. For this isotope, the decay rate in solid NaCl sample was greater than in aqueous solution by (0.52±0.20)%, relative to the solution. The various data on ^{131}I and ^{24}Na were qualitatively rationalised by Kemeny[94], by stating that cations and anions would, as expected, exert opposite influences on the Fermi function.

Zoller *et al.*[101] have determined the half-life of ^{131}I samples to be 7.969±0.014 day under conditions in which the radiations of 12.0 day $^{131}Xe^m$ were not detected and to be 8.117±0.012 day under conditions in which the radiations of $^{131}Xe^m$ were detected. No variations of half-life, due to the physical state of the source, were found within 0.3%. The half-life variations reported by Bergamini *et al.*[93] and Kemeny[94, 95] are thought to be accounted for by the escape of daughter $^{131}Xe^m$ and/or absorption of the radiations of $^{131}Xe^m$.

The importance of the observed effects on β-decay, and even the fact that they were observed at all, constitutes a most important piece of information, and therefore, it is of the utmost importance that such results be verified by other investigators.

A word of caution would probably be in order: reproducible results could, in the author's laboratory, only be obtained when the most stringent care was taken for sample handling. Although half-life measurements may, in the future, probably supply a larger body of interesting information on chemical problems, the only experiments conducted up to now on sufficiently varied chemical environments remain those carried out with ^7Be. Other nuclei, particularly $^{125}Te^m$, $^{235}U^m$, ^{89}Zr and ^{64}Cu, seem promising for future work, mostly the last two, on account of their half-lives and chemical properties.

Other processes which may be influenced by the chemical environment are the strictly forbidden $O^+ \leftrightarrow O^+ \ \gamma$ monopole transitions that can decay almost only by electron conversion – as in the case of the 6.06 MeV transition of ^{16}O studied by Devons *et al.*[96] and which include levels in ^{12}C, ^{40}Ca, ^{42}Ca, ^{70}Ge, ^{72}Ge, ^{90}Zr and ^{240}Pu, presenting competing modes of decay[97].

Finally the environment may also alter the penetrability of the Coulomb barrier to α-emission by changing the screening potential of the atomic electrons. Alder *et al.*[98] calculate that variations of the decay rate of ^{222}Rn, ^{212}Po, ^{147}Sm and ^{241}Am may be as large as 0.96×10^{-3}, for a typical chemically induced screening potential change of 27.21 eV.

2.7 CONCLUSIONS

It is anticipated that the PAC technique will supply valuable information bearing on chemical problems, particularly if selected monocrystalline samples, rather than powders, and much better defined solutions are used in future investigations. The possibility of performing DPAC measurements with ^{57}Fe [99] and $^{119}Sn^{m}$ [100] has now been demonstrated and experiments on selected samples will permit an interesting comparison of the relative merits of Mössbauer and DPAC measurements. The half-life measurements will find a larger place amongst other well-established techniques utilising the hyperfine interactions, when, as demonstrated in the ^{7}Be case, more systematic investigations are carried out with a wider class of compounds and isotopes.

ACKNOWLEDGMENTS

The author has benefited from many discussions with his associates at the Centre d'Etudes Nucléaires de Grenoble. The warm reception extended to him by Professors L. Néel and A. Moussa, and by Dr D. Dautreppe, is gratefully acknowledged. This paper would not have come to light without friendly prodding from A. G. Maddock and the collaboration of Mme A. Tissier, P. Boyer and G. Duplâtre.

References

1. Steffen, R. M. and Frauenfelder, H. (1964). *Perturbed Angular Correlations* (Karlsson, E., Matthias, E. and Siegbahn, K., Eds.). (Amsterdam: North-Holland)
2. Steffen, R. M. and Frauenfelder, H. (1965). *Alpha-, Beta and Gamma-Ray Spectroscopy* (Siegbahn, K., Ed.) Chap. XIX. (Amsterdam: North-Holland)
3. De Benedetti, S., de S. Barros, F. and Hoy, G. R. (1966). *Ann. Rev. Nucl. Sci.*, **16**, 31
4. Bäverstam, U., Johansson, A. and Gerholm, T. R. (1968). *Arkiv. Fys.*, **35**, 451
5. Thun, J. E. (1970). *Int. Conf. Angular Correlations in Nuclear Disintegration.* Delft, Holland, August 17–21
6. Gerholm, T. R. and Holmberg, L. (1970). *Int. Conf. Angular Correlations in Nuclear Disintegrations.* Delft, Holland, August 17–21

7. Matthias, E. and Shirley, D. (1968). *Hyperfine Structure and Nuclear Radiations* (Matthias, E. and Shirley, D. A., Eds.). (Amsterdam: North-Holland)
8. Grodzins, L. (1968). *Ann Rev. Nucl. Sci.*, **18,** 291
9. Grodzins, L. (1969). *Proc. Roy. Soc.*, **A311,** 19
10. Deutch, B. I. (1969). *Proc. Roy. Soc.*, **A311,** 151
11. Murnick, D. E., MacDonald, J. R., Barchers, R. R., Heestand, G. and Herskind, B. (1969). *Proc. Roy. Soc.*, **A311,** 111
12. Cohen, S. G. (1967). *Hyperfine Interactions,* 553 (Freeman, A. J. and Frankel, R. B., Eds). (New York: Academic Press)
13. Steffen, R. M. (1955). *Advan. Phys.*, **4,** 293
14. Heer, E. and Novey, T. B. (1959). *Solid State Physics* (Seitz, F. and Turnbull, D., Eds.), **9**. (New York: Academic Press)
15. Freeman, A. J. and Frankel, R. B. (1967). *Hyperfine Interactions.* (Freeman, A. J. and Frankell, R. B., Eds). (New York: Academic Press)
16. Alder, K., Albers-Schönberg, H., Heer, E. and Novey, T. B. (1953). *Helv. Phys. Acta,* **26,** 761
17. Abragam, A. and Pound, R. V. (1953). *Phys. Rev.*, **92,** 943
18. Béraud, R., Berkes, I., Marest, G. and Rougny, R. (1969,a). *Nucl. Instrum. Methods,* **69,** 41
19. Matthias, E., Schneider, W. and Steffen, R. M. (1963). *Phys. Rev.*, **129,** 1199
20. Andrade, P. da R., Maciel, A. and Rogers, J. D. (1967). *Phys. Rev.*, **159,** 196
21. Lieder, R. M., Buttler, N., Killig, K., Beck, K. and Bodenstedt, E. (1970). *International Conference on Hyperfine Interactions detected by Nuclear Radiations,* Rehovot–Jerusalem, Israël, September 6–11
22. Gerdau, E., Wolf, J., Winkler, H. and Braunsfurth, J. (1969). *Proc. Roy. Soc.*, **A311,** 197
23. Berthier, J., Boyer, P. and Vargas, J. I. (1970). *International Conference on Hyperfine Interactions detected by Nuclear Radiations.* Rehovot–Jerusalem, Israël, September 6–11
24. Alder, K., Matthias, E., Olsen, B., Schneider, W. and Steffen, R. M. (1964). *Perturbed Angular Correlations* (Karlsson, E., Matthias, E. and Siegbahn, K., Eds.). (Amsterdam: North-Holland)
25. Moser, C. M. (1967). *Hyperfine Interactions* (Freeman, A. J. and Frankel, R. B., Eds.). (New York: Academic Press)
26. Bodenstedt, E. (1966). *Lectures on Hyperfine Interactions,* Institut für Strahlen- und Kernphysik der Universität, Bonn
27. Leisi, H. J. (1968). *Hyperfine Structure and Nuclear Radiations,* 379 (Matthias, E. and Shirley, D. A., Eds.). (Amsterdam: North-Holland)
28. Steffen, R. M. (1956). *Phys. Rev.*, **103,** 116
29. Andrade, P. da R., Rogers, J. D. and Vasquez, A. (1969). *Phys. Rev.*, **188,** 571
30. Gabriel, H. (1969). *Phys. Rev.*, **181,** 506
31. Spanjaard, D. and Hartmann-Boutron, F. (1969). *J. Physique.*, **30,** 975
32. Blume, M. (1968). *Hyperfine Structure and Nuclear Radiations* (Matthias, E. and Shirley, D. A., Eds.). (Amsterdam: North-Holland)
33. Blume, M. (1971). *Nucl. Phys.*, **A167,** 81
34. Cameron, J. A., Gardner, P. R., Keszthelyi, L. and Prestwich, W. V. (1969). *Proc. Conf. Appl. Mössbauer Effect,* Tihany, 613
35. Avida, R., Ben-Zvi, I., Gilad, P., Goldberg, M., Goldring, G., Speidel, K. H., Spinzak, A. and Vager, Z. (1970). *International Conference on Hyperfine Interactions detected by Nuclear Radiations,* Rehovot—Jerusalem. Israël, September 6–11
36. Wagner, H. F. and Forker, M. (1969).*Nucl. Instrum. Methods,* **69,** 197
37. Wagner, H. F. and Forker, M. (1970). *International Conference on Hyperfine Interactions detected by Nuclear Radiations,* Rehovot—Jerusalem, Israël, September 6–11
    Wagner, H. F., Forker, M. and Weigand, U. (1970). Ibid.
38. Shirley, D. A. and Matthias, E. (1966). *Nucl. Instrum. Methods.*, **45,** 309
39. De Grazia, A. R. (1968). *Thesis.* Washington State University
40. Gygax, F. N., Egger, J. and Leisi, H. J. (1968). *Hyperfine Structure and Nuclear Radiations* (Matthias, E. and Shirely, D. A., Eds.). (Amsterdam: North-Holland)
41. Leisi, H. J. (1970 a). *Phys. Rev.*, **A1,** 6, 1654
    Leisi, H. J. (1970 c). *Int. Conf. Angular Correlations in Nuclear Disintegration.* Delft, Holland, August 17–21

42. Gygax, F. N. and Leisi, H. J. (1970). *International Conference on Hyperfine Interactions detected by Nuclear Radiation,* Rehovot-Jerusalem, Israël, September 6–11
43. Popp, M. and Wagner, H. F. (1970). *International Conference on Hyperfine Interactions detected by Nuclear Radiations,* Rehovot—Jerusalem. Israël, September 6–11
44. Leipert, T. K., Baldeschwieler, J. D. and Shirley, D. A. (1968). *Nature (London),* **220,** 907
45. Shirley, D. A. (1970). *J. Chem. Phys.,* **53,** 465
46. Leisi, H. J. (1970 b). *Int. Conf. Angular Correlations in Nuclear Disintegration,* Delft, Holland, August 17–21
47. Gardner, P. R. and Prestwich, W. V. (1970). *Can. J. Phys.,* **48,** 1430
48. Bozhko, V. P., Klyucharev, V. A. and Golovchenko, Yu. V. (1968). *Izv. Akad. Nauk. SSSR, Ser. Fiz.,* **32,** 155
49. Oliveira, J. (1970). *Thèse,* Université de Grenoble
50. Falk, F., Linnfors, A. and Thun, J. E. (1968). *Nucl. Phys.,* **A152,** 305
51. Andrade, P. da R. (1968). *Thesis,* University of Pôrto Alegre, Brazil
52. Berthier, J., Jeandey, C., Mathieu, J. P., Oliveira, J., Sette-Camara, A. O. R. and Vargas, J. I. (1971). *Paper submitted to the International Symposium on Hot Atom Chemistry,* Brookhaven National Laboratory (September, 1971)
53. Gerdau, E., Scharnberg, B. and Winkler, H. (1970). *International Conference on Hyperfine Interactions detected by Nuclear Radiations,* Rehovot–Jerusalem, Israël, September 6–11
54. Tissier, Annie (1970). *Thèse,* Université de Grenoble
55. Dezsi, I., Keszthelyi, L., Molnàr, B. and Pòcs, L. (1965). *Phys. Lett.,* **18,** 28
56. Nozik, A.-J. and Kaplan, M. (1967). *J. Chem. Phys.,* **47,** 2960
57. Barret, J. S., Cameron, J. A., Gardner, P. R., Keszthelyi, L., Prestwich, W. V. and Kaplan, M. (1970). *J. Chem. Phys.,* **53,** 759
58. Sato, J., Yokoyama, Y. and Yamazaki, T. (1969). *Radiochem. Acta,* **5,** 115
59. Bădică, T., Dima, S., Gelberg, A., Ianovici, E., Ion-Mihai, R. and Zaitseva, N. G. (1970). *International Conference on Hyperfine Interactions detected by Nuclear Radiations,* Rehovot—Jerusalem, Israël, September 6–11
60. Béraud, R., Berkes, I., Danière, J., Lévy, Michèle, Marest, G., Rougny, R. and Vargas, J. I. (1969). *Proc. Roy. Soc.,* **A311,** 185
61. Vargas, J. I., Berthier, J., Hocquenghem, J. C., Ribot, J. J. and Boyer, P. (1969). *Proc. Roy. Soc.,* **A311,** 191
62. Vulliet, P. (1970). *Thèse,* Université de Grenoble
63. Boyer, P., Tissier, Annie, Vulliet, P. and Vargas, J. I. (1971). *Paper submitted to the International Symposium on Hot Atom Chemistry,* Brookhaven National Laboratory (September, 1971)
64. Bouchez, R. and Depommier, P. (1960). *Rep. Prog. Phys.,* **23,** 395
65. Perlman, M. L. and Emery, G. T. (1969). *BNL,* **13,** 921
66. Segré, E. (1947). *Phys. Rev.,* **71,** 274
67. Daudel, R. (1947). *Rev. Sci.,* **87,** 162
68. Bouchez, R., Daudel, R., Daudel, P. and Muxart, R. (1947). *J. Phys. Radium,* **8,** 336
69. Bouchez, R., Tobailem, J., Robert, J., Muxart, R., Mellet, R., Daudel, R. and Daudel, P. (1956). *J. Phys. Radium,* **17,** 363
70. Kraushaar, J. J., Wilson, E. D. and Bainbridge, K. T. (1953). *Phys. Rev.,* **90,** 610
71. Johlige, H. W., Aumen, D. C. and Born, H. J. (1970). *Phys. Rev.,* **C2,** 1616
72. Siegbahn, K., Nordling, C., Fahlman, A., Nordberg, R., Hamrin, K., Hedman, J., Johansson, G., Bergmark, T., Karlsson, S. E., Lindgren, I. and Lindberg, B. (1967). ESCA-Atomic, Molecular and Solid State Structure studied by means of Electron Spectroscopy. *Nova Acta Regiae Soc. Sci. Upsaliensis Ser. IV,* Vol. 20. Second revised edition in preparation. (Amsterdam–London: North-Holland)
73. Kemeny, P. (1968). *Rev. Roum. Phys.,* **13,** 485
74. Huber, B., St. Gagneux and Leuenberger, H. (1968). *Phys. Lett.,* **27B,** 66
75. St. Gagneux, Huber, P., Leuenberger, H. and Nyikos, P. (1970). *Helv. Phys. Acta,* **43,** 39
76. Odru, P. and Vargas, J. I. (1969). *Rapport d'activité. Laboratoire de Chimie Nucléaire—* C.E.N.–Grenoble
77. Odru, P. and Vargas, J. I. (1971). *Inorg. Nucl. Chem. Lett.,* **7,** 379
78. Alder, K., Hadermann, J. and Raff, U. (1969). *Phys. Lett.,* **30A,** 487
79. Bainbridge, K. T., Goldhaber, M. and Wilson, E. (1953). *Phys. Rev.,* **90,** 430
80. Mazaki, Hiromasa and Shimizu, Sakae (1966). *Phys. Rev.,* **148,** 1161
81. Marelius, A. (1968). *Ark. Phys.,* **37,** 427

82. Nève de Mevergnies, M. (1968). *Phys. Lett.*, **26B,** 615
83. Nève de Mevergnies, M. (1969). *Phys. Rev. Lett.*, **23,** 422
84. Nève de Mevergnies, M. (1970). *Phys. Lett.*, **32B,** 482
85. Cooper, J. A., Hollander, J. M. and Rasmussen, J. O. (1965). *Phys. Rev. Lett.*, **15,** 683
86. Cooper, J. A. (1966). *Thesis,* Berkeley, Calif., Univ. of California
87. Weirauch, W., Schmitt-Ott, W. D., Smend, F. and Flammersfeld, A. (1968). *Z. Physik.*, **209,** 289
88. Olin, A. (1970). *Phys. Rev.*, **1,** 1114
89. Cooper, J. A., Hollander, J. M. and Rasmussen, J. O. (1968). *Nucl. Phys.*, **A109,** 603
90. Malliaris, A. C. and Bainbridge, K. T. (1966). *Phys. Rev.*, **149,** 958
91. Bocquet, J. P., Chu, Y. Y., Kistner, O. C. and Perlman, M. L. (1966). *Phys. Rev. Lett.*, **17,** 809
92. Carlson, T. A., Erman, P. and Fransson, K. (1968). *Nucl. Phys.*, **A111,** 371
93. Bergamini, P. G., Palmas, G., Piantelli, F. and Regato, M. (1967). *Phys. Rev. Lett.*, **18,** 468
94. Kemeny, P. (1969). *Radiochem. Radioanal. Lett.*, **2,** 41
95. Kemeny, P. (1969). *Radiochem. Radioanal. Lett.*, **2,** 119
96. Devons, S., Goldring, G. and Lindsey, G. R. (1954). *Proc. Phys. Soc.*, **67,** 1954
97. Gerholm, T. R. and Pettersson, B. G. (1968). *Alpha, Beta and Gamma Ray Spectroscopy,* 987 (Siegbahn, K., Ed.). (Amsterdam: North-Holland)
98. Alder, K., Baur, G. and Raff, U. (1971). *Phys. Lett.*, **34A,** 163
99. Hohenemmser, C., Reno, R., Benski, C. H. and Lehr, J. (1969). *Phys. Lett.*, **29,** 553
100. Berger, H. L. and Medicus, H. A. (1970). *Int. Conf. Angular Correlations in Nuclear Disintegration,* Delft, Holland, August 17–21
101. Zoller, W. H., Hopke, P. K., Fasching, J. L., Macias, E. and Waters, W. B. (1971). *Phys. Rev.*, **3C,** 1699

3
Trans-curium Elements

R. J. SILVA
Oak Ridge National Laboratory

3.1 INTRODUCTION

The last 5 years have been an extremely active period in the study of trans-curium elements. Although the number of new elements was only increased by two, elements of atomic numbers 104 and 105, the number of known isotopes was nearly doubled from about 50 to over 80. This was due in large part to the availability of increasingly abundant and nearly isotopically pure quantities of some of the heavier actinides (e.g. ^{248}Cm, ^{249}Bk, ^{249}Cf and ^{253}Es) for target materials and to improved isolation, detection and identification techniques.

'Few-atom-at-a-time' chemical studies have been carried out on elements 101 through 104 using both fast gas chromatographic and aqueous chemical separation techniques. These studies have established the chemical relationship of these elements to one another and their positions in the chemical periodic system. Also, the results have increased our confidence in the ability of the chemical periodic system to predict the chemical properties of elements beyond the actinides.

Some of the controversies over certain of the new elements and isotopes have been resolved only to be replaced by new ones. This is not surprising in view of the near Herculean task of attempting to study nuclides produced at rates of only a few atoms per day. Figure 3.1 represents an attempt to bring the reader up to date on the assigned isotopes of the trans-curium elements. It was compiled from the latest published, and in some cases, unpublished results but hopefully will require only minor corrections in the future.

Probably the most exciting development has been the prediction of the existence of an 'island' of relatively stable nuclei near atomic number 114 and mass 300, appropriately referred to as the 'superheavy' elements. On consulting Figure 3.1, one can readily see that α- and spontaneous-fission decay half-lives are becoming ever shorter with increasing atomic number as a result of the disruptive coulomb forces increasing more rapidly than the cohesive nuclear forces. It seems strange then to expect heavier nuclei of even higher atomic number with longer half-lives. This expectation of extra stability against nuclear decay comes from predictions of the increased nuclear binding energy associated with closed (filled) proton and neutron shells.

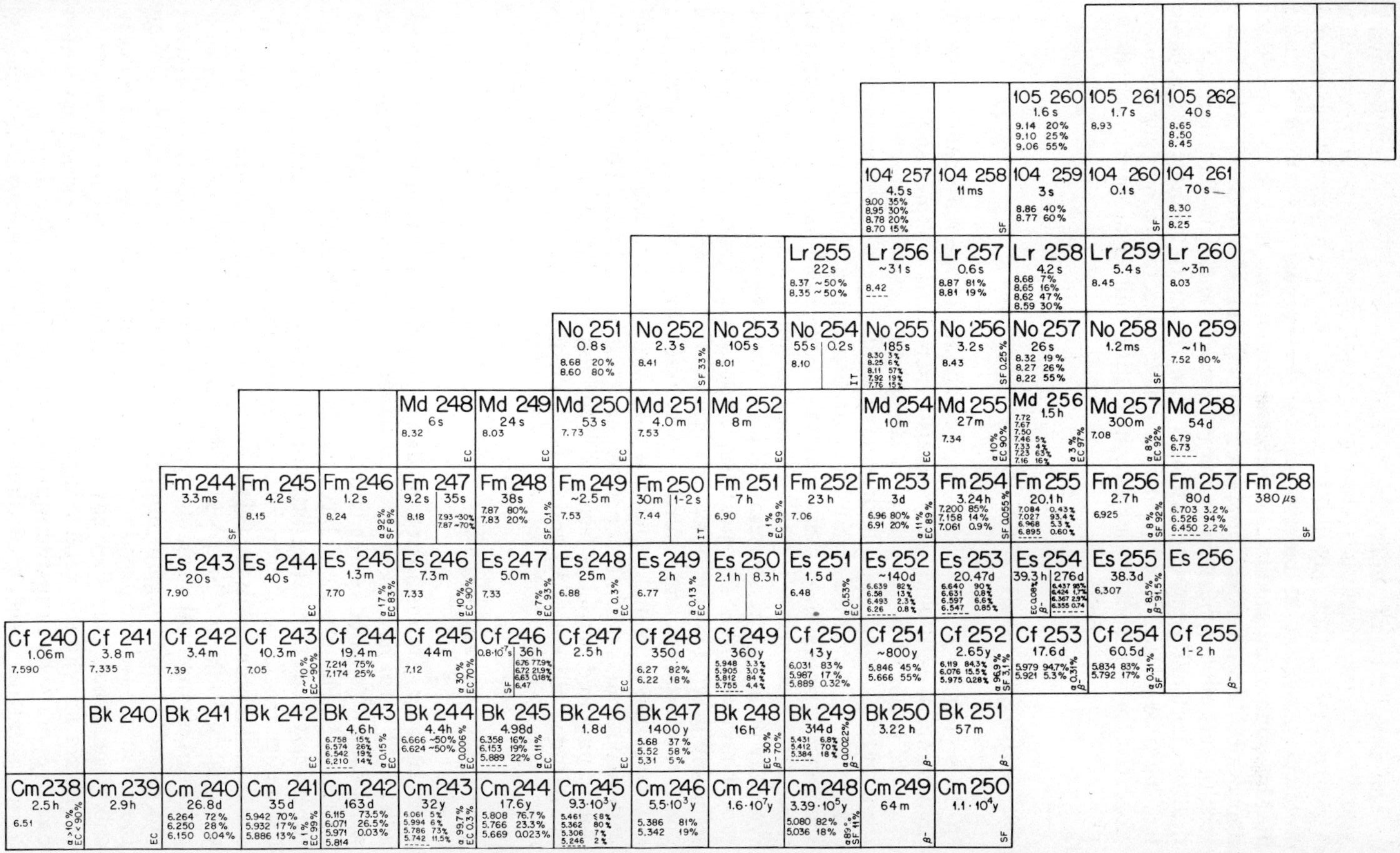

													105 260 1.6 s 9.14 20% 9.10 25% 9.06 55%	105 261 1.7 s 8.93	105 262 40 s 8.65 8.50 8.45		
											104 257 4.5 s 9.00 35% 8.95 30% 8.78 20% 8.70 15%	104 258 11 ms SF	104 259 3 s 8.86 40% 8.77 60%	104 260 0.1 s SF	104 261 70 s 8.30 ---- 8.25		
										Lr 255 22 s 8.37 ~50% 8.35 ~50%	Lr 256 ~31 s 8.42 ----	Lr 257 0.6 s 8.87 81% 8.81 19%	Lr 258 4.2 s 8.68 7% 8.65 16% 8.62 47% 8.59 30%	Lr 259 5.4 s 8.45	Lr 260 ~3 m 8.03		
							No 251 0.8 s 8.68 20% 8.60 80%	No 252 2.3 s 8.41 SF 33%	No 253 105 s 8.01	No 254 55 s | 0.2 s 8.10 | IT	No 255 185 s 8.30 3% 8.25 6% 8.11 57% 7.92 19% 7.76 15%	No 256 3.2 s 8.43 SF 0.25%	No 257 26 s 8.32 19% 8.27 26% 8.22 55%	No 258 1.2 ms SF	No 259 ~1 h 7.52 80%		
					Md 248 6 s 8.32 EC	Md 249 24 s 8.03 EC	Md 250 53 s 7.73 EC	Md 251 4.0 m 7.53	Md 252 8 m EC		Md 254 10 m EC	Md 255 27 m 7.34 α 10% EC 90%	Md 256 1.5 h 7.72 7.67 7.50 7.46 5% 7.33 4% 7.23 63% 7.16 16% α 3% EC 97%	Md 257 300 m 7.08 α 8% EC 92%	Md 258 54 d 6.79 6.73 -----		
		Fm 244 3.3 ms SF	Fm 245 4.2 s 8.15	Fm 246 1.2 s 8.24 α 92% SF 8%	Fm 247 9.2 s | 35 s 8.18 | 7.93 ~30% 7.87 ~70%	Fm 248 38 s 7.87 80% 7.83 20% SF 0.1%	Fm 249 ~2.5 m 7.53	Fm 250 30 m | 1-2 s 7.44 | IT	Fm 251 7 h 6.90 α 1% EC 99%	Fm 252 23 h 7.06	Fm 253 3 d 6.96 80% 6.91 20% α 11% EC 89%	Fm 254 3.24 h 7.200 85% 7.158 14% 7.061 0.9% SF 0.055%	Fm 255 20.1 h 7.084 0.43% 7.027 93.4% 6.968 5.3% 6.895 0.60% -----	Fm 256 2.7 h 6.925 α 8% SF 92%	Fm 257 80 d 6.703 3.2% 6.526 94% 6.450 2.2% -----	Fm 258 380 μs SF	
		Es 243 20 s 7.90	Es 244 40 s EC	Es 245 1.3 m 7.70 α 17% EC 83%	Es 246 7.3 m 7.33 α 10% EC 90%	Es 247 5.0 m 7.33 α 7% EC 93%	Es 248 25 m 6.88 α 0.3% EC	Es 249 2 h 6.77 α 0.13% EC	Es 250 2.1 h | 8.3 h EC	Es 251 1.5 d 6.48 α 0.53% EC	Es 252 ~140 d 6.639 82% 6.58 13% 6.493 2.3% 6.26 0.8% ------	Es 253 20.47 d 6.640 90% 6.631 0.8% 6.597 6.6% 6.547 0.85% -----	Es 254 39.3 h | 276 d EC 0.08% β- | 6.437 93% 6.424 1.7% 6.367 2.9% 6.355 0.74 ------	Es 255 38.3 d 6.307 α 8.5% β- 91.5%	Es 256		
Cf 240 1.06 m 7.590	Cf 241 3.8 m 7.335	Cf 242 3.4 m 7.39	Cf 243 10.3 m 7.05 α ~10% EC ~90%	Cf 244 19.4 m 7.214 75% 7.174 25%	Cf 245 44 m 7.12 α 30% EC 70%	Cf 246 0.8·10⁻⁷ s | 36 h SF | 6.76 77.9% 6.72 21.9% 6.63 0.18% 6.47	Cf 247 2.5 h EC	Cf 248 350 d 6.27 82% 6.22 18%	Cf 249 360 y 5.948 3.3% 5.905 3.0% 5.812 84% 5.755 4.4% -----	Cf 250 13 y 6.031 83% 5.987 17% 5.889 0.32%	Cf 251 ~800 y 5.846 45% 5.666 55%	Cf 252 2.65 y 6.119 84.3% 6.076 15.5% 5.975 0.28% α 96.9% SF 3.1%	Cf 253 17.6 d 5.979 94.7% 5.921 5.3% α 0.31% β-	Cf 254 60.5 d 5.834 83% 5.792 17% α 0.31% SF	Cf 255 1-2 h β-		
	Bk 240	Bk 241	Bk 242 EC	Bk 243 4.6 h 6.758 15% 6.574 26% 6.542 19% 6.210 14% ----- α 0.15% EC	Bk 244 4.4 h 6.666 ~50% 6.624 ~50% α 0.006% EC	Bk 245 4.98 d 6.358 16% 6.153 19% 5.889 22% ----- α 0.11% EC	Bk 246 1.8 d EC	Bk 247 1400 y 5.68 37% 5.52 58% 5.31 5%	Bk 248 16 h EC 30% β- 70%	Bk 249 314 d 5.431 6.8% 5.412 70% 5.384 18% ----- α 0.0022% β-	Bk 250 3.22 h β-	Bk 251 57 m β-					
Cm 238 2.5 h 6.51 α >10% EC <90%	Cm 239 2.9 h EC	Cm 240 26.8 d 6.264 72% 6.250 28% 6.150 0.04%	Cm 241 35 d 5.942 70% 5.932 17% 5.886 13% α 1% EC 99%	Cm 242 163 d 6.115 73.5% 6.071 26.5% 5.971 0.03% 5.814	Cm 243 32 y 6.061 5% 5.994 6% 5.786 73% 5.742 11.5% ----- α 99.7% EC 0.3%	Cm 244 17.6 y 5.808 76.7% 5.766 23.3% 5.669 0.023%	Cm 245 9.3·10³ y 5.461 ≤8% 5.362 80% 5.306 7% 5.246 2% -----	Cm 246 5.5·10³ y 5.386 81% 5.342 19%	Cm 247 1.6·10⁷ y	Cm 248 3.39·10⁵ y 5.080 82% 5.036 18% α 89% SF 11%	Cm 249 64 m β-	Cm 250 1.1·10⁴ y SF					

Figure 3.1 Chart of the trans-curium nuclides. The chemical symbol (or atomic number) and isotopic mass number are shown at the top of each square. Beneath the chemical symbol, the half-life is given in microseconds (μs), milliseconds (ms), seconds (s), minutes (m), hours (h), days (d) or years (y). The α-particle energies and abundances are shown in the left-hand portion of the square (dashed lines indicate other weak α-groups). If decay proceeds by other than α-emission, the decay mode and percentage branching is indicated in the right-hand portion of the square

The possibility of superheavy nuclei has been considered for many years[1, 2], but only began to receive widespread attention about 5 years ago. This renewed interest was due primarily to the predictions by Myers and Swiatecki[3] that fission barriers of superheavy elements would be some 10 MeV high (compared to 5 MeV for uranium), and the results of Meldner and Röper[4] that the next closed proton shell after 82 is probably 114. These results were soon confirmed by many others[5, 6], and since that time a good deal of effort has gone into attempts to make more accurate and reliable half-life predictions. These predictions have triggered many experimenters to search for superheavy elements in Nature and to propose methods to produce them artificially.

3.2 NUCLEAR SYSTEMATICS

In the search for new elements, it is essential to have reasonably accurate predictions of α-, β- and spontaneous-fission decay energies and half-lives. It would be virtually impossible to design productive experiments without them and they usually play a key role in the discovery of new elements.

Figures 3.2, 3.3 and 3.4 show plots of experimentally determined α- and spontaneous-fission half-lives, as well as β-stable isotopes for a number of trans-curium elements. The α- and spontaneous-fission data are most conveniently presented as a series of two-dimensional plots, each for a constant proton number, of the half-lives as a function of neutron number. The most smoothly varying correlations are obtained from plots of nuclei with even numbers of protons and neutrons (even-even). In all cases, varying degrees of retardation in α- and spontaneous-fission decay are exhibited by nuclei with an odd number of either protons or neutrons (odd A), or both (odd-odd), due to unpaired particle effects. Data for some of the odd neutron species are included in Figures 3.2 and 3.3 for comparison.

3.2.1 Decay systematics

In the last 5 years, most efforst to predict α-, β- and spontaneous-fission energies and decay rates have been directed toward the development of improved mass equations based on the liquid-drop model with shell and deformation corrections. The primary object has been to predict decay properties of superheavy nuclei, therefore little, or no improvement over earlier attempts to predict the nuclear properties of the heavy actinides and light trans-actinides has been made. One of the most extensive and successful approaches was that of Viola and Seaborg[7]. They used closed decay cycles and extrapolations of empirical masses to predict properties of undiscovered elements and isotopes in the region immediately adjacent in Z to curium. Though their results have been reasonably successful in predicting α- and β-decay energies and half-lives, they have failed, as have several other attempts[8–10], to predict to within several orders of magnitude the spontaneous-fission half-lives of the recently discovered isotopes ^{258}Fm, ^{258}No and ^{258}104.

In general, the earlier predictions of decay by spontaneous fission suggest a decreasing trend in half-life with increasing Z^2/A of nuclei following the sub-shell at $N = 152$. Beyond $N = 160$, they predict that spontaneous-fission half-lives should increase with increasing neutron number for a given element as the closed shell at $N = 184$ is approached. However, they tend to over-emphasise the stabilising influence of $N = 184$ and fail to predict the observed precipitous drop in half-lives near $N = 158$. The empirical

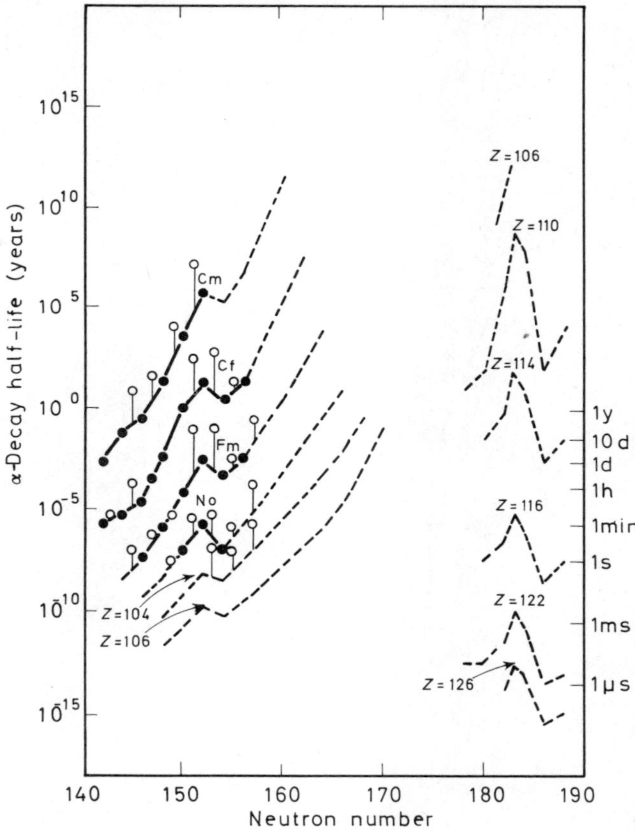

Figure 3.2 Plot of partial α-decay half-lives for even-Z elements versus neutron number. Solid points show experimental values for even-even isotopes; open points show experimental values for even (Z)-odd (N) isotopes. Dashed lines result from the predictions of references 7, 17 and 22

extrapolations of Ghiorso[11] have proven to be more useful. In Figure 3.3, solid lines are drawn through the points obtained from experimental data. The dotted curves shown in the left-hand portion of Figure 3.3 represent an attempted compromise between the predictions of Viola and Seaborg and the straight empirical extrapolations.

Two interesting features are apparent in Figure 3.3. One is the rather large retardation in spontaneous-fission half-lives for nuclei with $N = 157$, i.e. ^{257}Fm, ^{259}No and ^{261}104. Most odd A and odd-odd nuclei have retardation

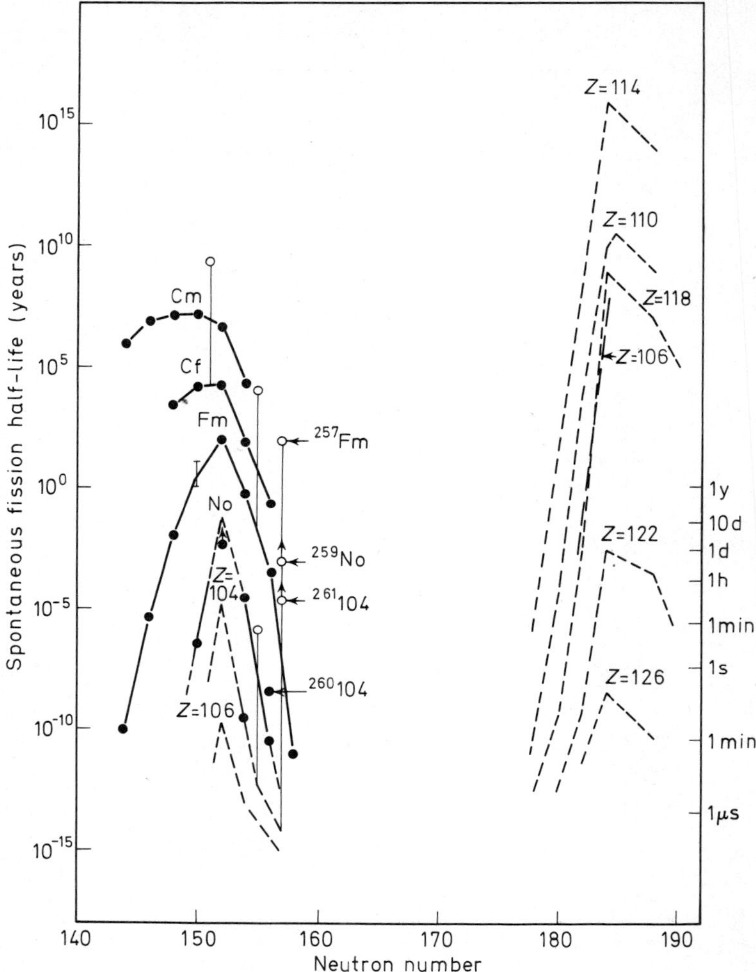

Figure 3.3 Plot of partial spontaneous-fission half-lives for even-*Z* elements versus neutron number. Solid points show experimental values for even-even isotopes; open points show experimental values for even (*Z*)-odd (*N*) isotopes. Dashed lines result from the predictions of references 7, 17 and 22

factors of 10^5 to 10^7 over the trends given by the even-even nuclei. However, ^{257}Fm exhibits a retardation factor of nearly 10^{10} and those of ^{259}No and ^{261}104 may be even larger. This may be due to a large specialisation energy associated with the high spin state, 9/2+[615↓], of the 157th neutron. In such cases, fission may not proceed through the lowest energy fission channel,

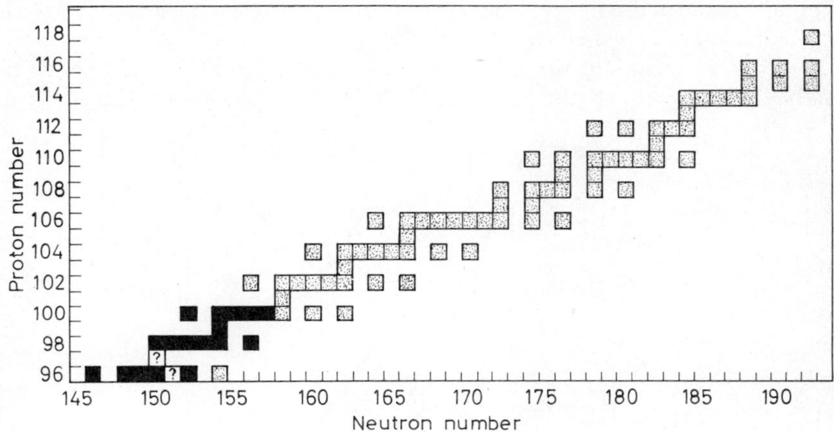

Figure 3.4 *β*-Stable isotopes of the trans-curium elements. Solid squares show known isotopes; dotted squares result from the predictions of references 7, 17 and 22

but may proceed through a higher barrier channel resulting in a longer half-life[12].

The second feature is the anomalously long half-life assigned[13] to ^{260}104 compared to that predicted by the even-even nuclide trends. This half-life appears to be four orders of magnitude too long to be associated with this isotope and must be considered suspect.

3.2.2 Recent theoretical trends

In order to predict decay energies and half-lives for α-, β- and spontaneous-fission decay, reasonably reliable nuclear masses and potential energies must be calculated as a function of nuclear deformation. Potential energy calculations are broadly classified into those treating the nucleus as a liquid drop, and those based on the opposite idea of independent particle (nucleon) motion. The liquid-drop model is able to reproduce the gross properties of nuclei, e.g. total binding energies and other smooth trends. On the other hand, the independent-particle model is able to reproduce local fluctuations due to the influence of shells, but it fails to reproduce the gross properties accurately. Efforts have been made to combine the two approaches and give the total nuclear potential energy as the sum of a smoothly varying liquid-drop part and a fluctuating single-particle correction term. The latter may be as much as 10 MeV for a doubly closed-shell nucleus but it is usually only a few megaelectron volts.

Recent versions of the liquid-drop calculations are based on the well known Bethe–Weizsaecker equation[14]. It describes the ground-state nuclear energy as a sum of volume, surface, symmetry and coulomb energies, plus a pairing term to produce the regular differences in masses due to nuclear type. In more recent versions, surface-symmetry, coulomb exchange and surface diffusion terms are added as adjustable parameters[15]. The 'droplet' model includes surface tension corrections which result from the large curvature of the nuclear surface.

In order to calculate the shell and deformation corrections, one needs reliable predictions of the single-particle energy levels as a function of

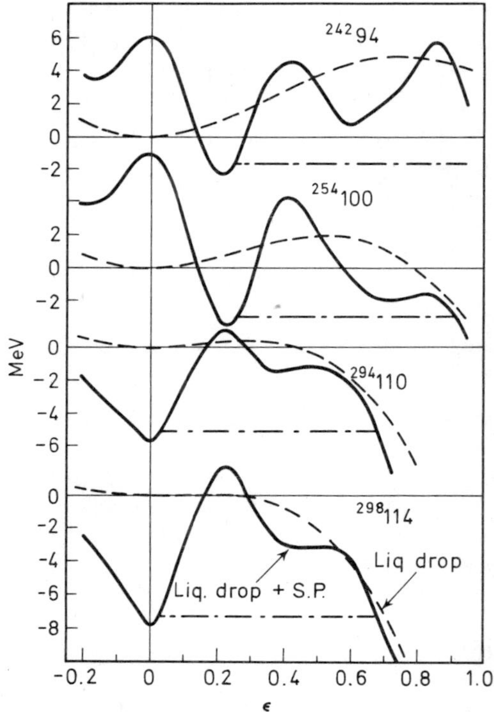

Figure 3.5 Plot of potential energy as a function of the symmetric nuclear deformation, ε, for several heavy and superheavy elements. Dashed curves are the liquid-drop barrier; solid curves are obtained after inclusion of the shell corrections. (From Nilsson *et al.*[16], by courtesy of the North-Holland Publishing Company)

nuclear deformation. Nilsson and co-workers have concentrated on further modifications and generalisations of the deformed harmonic oscillator potential[16, 17] which have been so successful for the calculation of levels in deformed nuclei. However, generalised Woods–Saxon and folded Yukawa potentials have recently received the attention of a number of groups[18], and these more realistic potentials are beginning to be used for larger deformations. The shell correction is assumed to be due to the bunching of a uniform energy distribution of single-particle levels which describes a liquid-drop into one that consists of groups of more closely spaced levels which corresponds to an actual level scheme. The resultant large energy spacings between groups of levels give rise to the closed-shell effects. The currently accepted calculational prescription for this method is due to Strutinsky[19–21], and it has been widely applied to the Nilsson model of energy levels. Figure 3.5 shows plots by Nilsson and Tsang[17] of the liquid-drop, and liquid-drop plus shell-correction, potential energies as a function of nuclear deformation for several heavy nuclei.

As seen in Figure 3.5, the conventional liquid-drop fission barrier has a one-hump shape, but because of the shell effects at moderate deformations, additional structure is found in the corrected spontaneous fission barrier. These shell effects are thought to be due to the same level bunching phenomenon that occurs for spherical shapes. Strutinsky was first to emphasise that these effects will cause a double-humped fission barrier. These peculiar barriers are believed to be responsible for the spontaneously fissioning 'shape-isomers' observed in the lower Z actinides. The curves in Figure 3.5 show that the minimum potential energy for element 114 occurs at zero deformation which corresponds to spherical shape; the minimum in the potential energy of the lower Z actinides corresponds to a permanently deformed shape. Note that there is no liquid-drop barrier for element 114; the entire barrier is a shell-structure effect and depends strongly on the assumed level structure.

The ground-state mass and deformation is obtained by minimising the liquid-drop plus shell-correction expression with respect to nuclear deformation. From the resulting masses, α-decay energies (and thence α half-lives) for the superheavy elements can be estimated and the β-stable isotopes determined. From the potential energy values obtained as a function of deformation, the fission barriers can be estimated and half-lives calculated from a one-dimensional barrier penetration calculation[16]. The dotted curves in the right-hand portions of Figures 3.2 and 3.3 were constructed from the values given in Reference 17 and from further calculations of Nilsson and Tsang[22]. These curves should be taken as an expression of trends, since widely differing values (factors of 10^3–10^6) can be obtained from different nuclear potentials and parameters[12–15]. The values of Nilsson *et al.* represent a more-or-less middle ground.

3.3 METHODS OF PRODUCTION

Two processes have been generally used for the production of trans-curium elements; (1) neutron irradiation of high atomic number targets where successive neutron capture leads to heavier isotopes which β-decay to even higher atomic number elements, and (2) bombardment of appropriate targets in an accelerator with charged particles ranging from protons, deuterons and α particles to heavier ions like carbon, oxygen, neon, argon, etc. The first process is the more efficient for the production of large quantities of some of the neutron-excess trans-curium nuclides. Tens to hundreds of milligrammes of ^{249}Bk and $^{249-252}$Cf, and hundreds of microgrammes of ^{253}Es can be produced per year in the High Flux Isotope Reactor (HFIR) at Oak Ridge, and still larger quantities can be produced at the Savannah River Reactor. These reactor-produced nuclides make excellent targets for the production of even higher atomic number elements in accelerators.

For trans-curium elements through mendelevium, lighter ions (p, d and α) can be used to produce neutron-deficient isotopes a few neutrons from the β stability line, but in order to produce higher atomic number elements and very neutron-deficient isotopes one must use the heavier bombarding ions.

3.3.1 Heavy-ion cross-sections

The ability to predict the production cross-section as a function of bombarding energy for heavy ion reactions is also crucial to modern day alchemists for the design of experiments.

The most successful scheme for predicting heavy-ion cross-sections for the production of the heaviest elements is from Sikkeland[26]. It assumes the formation of an excited compound nucleus (C.N.) from the target nucleus and heavy ion, followed by de-excitation of the C.N. through statistical decay by neutron emission or fission. For reactions of the type (HI,*x*n), the experimental cross-section data are well described by the following formula derived from partial wave analyses:

$$\sigma_x(E_\mathrm{I}) = \prod_{\mathrm{i}=1}^{x} \left[\frac{\Gamma_\mathrm{n}}{\Gamma_\mathrm{n}+\Gamma_\mathrm{f}}\right] \sum_{\mathrm{l}=0}^{l_c} \sigma_\mathrm{l}(E_\mathrm{I})P_x(E^*,\mathrm{l}),$$

where σ_x is the cross-section for the reaction (HI,*x*n) at a given heavy ion bombarding energy of E_I; σ_1 is the partial cross-section for C.N. formation for the lth partial wave at E_I; P_x is the probability for the emission of exactly x neutrons from the compound nucleus of angular momentum 1 and excitation energy E^*. The summation over partial waves is taken from zero angular momentum to a cut-off value, l_c. This upper limit on the angular momentum corresponds to the maximum distance between colliding nuclei which will allow complete amalgamation to form a C.N. The quantities Γ_n and Γ_f are the level widths for neutron emission and fission, respectively. (Charged particle evaporation is assumed negligibly small because of the large coulomb barrier.) The first product term represents the fraction of nuclei that survive fission through the evaporation of exactly x neutrons. This method predicts the shapes of the excitation functions quite well, and the absolute magnitudes usually agree with experiments to within about a factor of three. As an example, Figure 3.6 gives computer results for the reaction ^{249}Cf(^{18}O,*x*n) to produce isotopes of element 106. Comparisons of experimental data with such calculations can be used for assignments of newly discovered nuclides[26], but unambiguous assignments by this method are sometimes difficult to make.

In the heavy-element region, fission competes strongly with neutron emission in the decay of the C. N. The ratio $[\Gamma_\mathrm{n}/\Gamma_\mathrm{n}+\Gamma_\mathrm{f}]_\mathrm{i}$ is usually of the order of 0.1 or less and decreases with increasing Z of the C. N. Therefore, fission annihilates most of the valuable compound nuclei and the neutron-out cross-sections are quite small. The largest yields are usually for $x = 4$. For $x<4$, the yields are lower because the coulomb barrier between the target atom and heavy ion usually exceeds the threshold energies for the reactions; for $x>4$, smaller yields result from the additional opportunity for fission competition. The cross-section for production of a given nuclide is generally largest when the heaviest target nucleus and lightest bombarding ion is used, e.g. for a given number of neutrons out, the cross-section decreases approximately by a factor of 100 as one goes from C to O to Ne. On the other

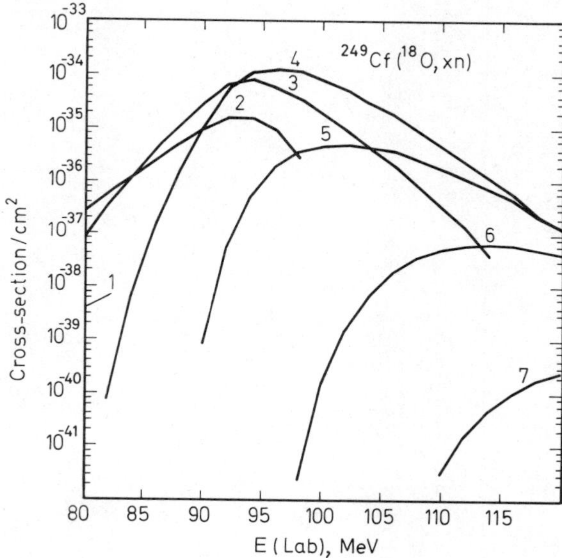

Figure 3.6 Plots of excitation functions for the reactions $^{249}Cf(^{18}O,xn)^{267-x}106$ obtained from computer calculations. The numbered curves correspond to the products which result from the emission of x neutrons from the compound nucleus

hand, the heaviest trans-uranium elements are the least available in quantity so a compromise usually must be made between the amount of target material available and the cross-section.

More complex heavy ion reactions involving the emission of charged particles (particularly α-particles) are proving to be useful, e.g. $^{248}Cm(^{18}O,\alpha 3n)$ ^{259}No. Frequently, these more complex reactions have larger cross-sections than the (HI,xn) reactions, presumably because the charged-particle emission step may involve a different mechanism which is not as strongly affected by fission competition as is a neutron emission step. Preliminary studies[27] indicate that (1) the cross-sections involving charged particle emission are determined primarily by the number of particles transferred from the bombarding ion to the target nucleus rather than the number of particles emitted from the C.N., and (2) that cross-sections decrease rather sharply with increasing numbers of transferred particles.

3.3.2 Production in thermonuclear explosions

Several experiments to produce very heavy elements through the exposure of targets to the intense neutron flux of a thermonuclear explosion have been carried out over the last several years[28]. The synthesis of very heavy nuclides from uranium requires the rapid, successive capture of up to 20 neutrons followed by β-decay. However, there appears to be a limit to the process and the heaviest nuclide identified to date is ^{257}Fm.

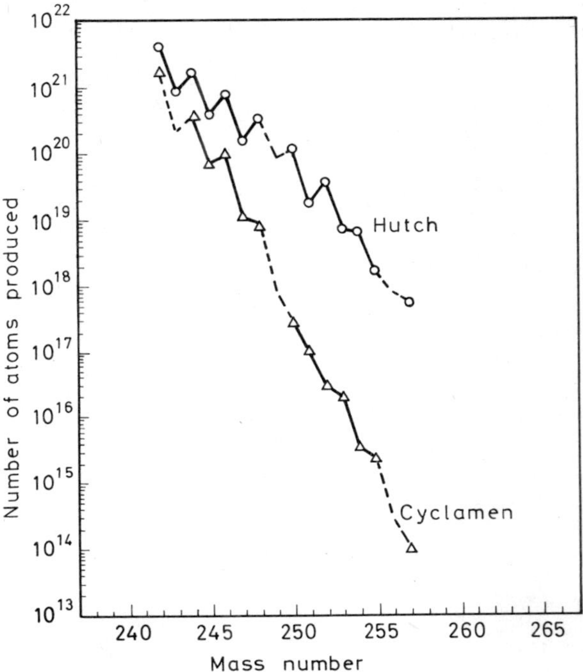

Figure 3.7 Heavy element yields from the thermonuclear experiments, Hutch and Cyclamen

In the most recent experiment called 'Hutch'[29], a remarkable integrated neutron exposure of 40 mol cm^{-2} (equivalent 20 keV neutron flux) was achieved. Figure 3.7 shows the improved yield of heavy nuclei produced in the Hutch experiment of 1969 compared to the Cyclamen experiment of 1967. Analysis of data from Hutch indicated that the amount of ^{257}Fm produced was nearly a factor of 10^{10} higher than current yearly reactor production rates of $\sim 10^8$ atoms. However, only a small fraction, one in 10 million, of the total number of atoms of ^{257}Fm embedded in the tons of debris was recovered. For instance, a 10 kg sample yielded $\sim 10^{10}$ atoms of ^{257}Fm. Table 3.1 compares typical isotope compositions of Cf and Es samples from the HFIR with those from Hutch and illustrates the increased yields of the heavier isotopes from the thermonuclear experiment.

One of the important aspects of the programme is the recovery of enough of the heavy actinides to serve as target material for accelerator or reactor irradiations to produce new elements or isotopes. For example, the ^{257}Fm recovered from Hutch was bombarded with deuterons to synthesise and

Table 3.1

Z	A	*Relative abundance*	
		HFIR	*Hutch*
98	249	0.0012	—
	256	0.0083	—
	251	—	0.48
	252	1.000	1.00
	253	0.0069	0.19
	254	0.0007	0.18
99	253	1.00	1.00
	254^m	0.003	0.0
	255	0.0006	0.231

identify for the first time[30] the elusive isotope ^{258}Fm. This result represented the culmination of a search which extended over a period of several years.

3.3.3 Production of superheavy nuclei by heavy ions

The two processes described above have also been discussed for the production of superheavy nuclei[31, 32]. Although there are opposing opinions[33, 34], evidence seems to indicate that the neutron-capture process cannot be successfully used to produce superheavy elements. The exceedingly short spontaneous-fission half-lives of the nuclei in the β-decay chains are expected to terminate the creation process before the relatively stable region of superheavy nuclei can be reached. This seems to leave heavy ions as the primary hope for producing superheavy nuclei.

There are, however, important difficulties that must be overcome in the production of superheavy nuclei. As seen in Figure 3.5, the fission barrier for a nucleus like plutonium is relatively broad, while the fission barrier predicted for element 114 is much thinner. Thus, superheavy nuclei might be expected to fission at much smaller deformations than plutonium[35]. Also, at high excitation energies of the compound system, where many particles are in excited states, the closed shell structure is disrupted and the fission barrier height is lowered[18]. Indeed, it may be that few superheavy nuclei will survive the traumatic birth by the fairly violent heavy-ion reactions.

Heavy ion reactions are usually classified in three groups:

(1) compound nucleus formation followed by statistical decay, e.g. $^{180}\text{Hf} + ^{132}\text{Xe} \rightarrow ^{309}126 + 3\text{n}$.

(2) grazing or multi-nucleon transfer, e.g. $^{244}\text{Pu} + ^{96}\text{Zr} \rightarrow ^{300}114 + ^{40}\text{Ca}$.

(3) fusion–fission, e.g. $^{238}\text{U} + ^{238}\text{U} \rightarrow ^{476}184^* \rightarrow ^{298}114 + ^{178}\text{Yb}$.

3.3.3.1 *Compound nucleus reactions*

The mechanisms of reaction (1) have been discussed in Section 3.3.1. The formation of the compound nucleus takes place readily only when the bombarding energy exceeds the height of the coulomb barrier which exists between the ion and the target nucleus. However, the less the bombarding energy exceeds the energy necessary to form the compound nucleus, the larger usually will be the yield of product nuclei. This is a result of the lower excitation energy of the C.N. which minimises the number of neutron evaporation (and fission) steps necessary to de-excite the C.N. Generally, this situation is best achieved if the C.N. is formed from both a medium-weight target and projectile rather than a heavy target and light projectile. This apparent advantage may be overshadowed, however, by the disadvantage that larger distortions are probably involved which may lead to fission[18]. In any case, the degree of success in producing superheavy nuclei by this method depends, in large part, on how severe the fission competition is in the neutron cascade, i.e., on the strength of the closed shells.

Unfortunately, we are not certain if the calculational methods described in Section 3.3.1 can be successfully applied to the very heavy ion reactions needed for production of superheavy nuclei. However, recent experiments[36] to produce polonium isotopes, through the bombardment of ^{164}Dy with ^{40}Ar, show no difference in the mechanisms of the (HI,xn) reactions with respect to lighter bombarding ions. Barring unexpected nuclear effects, extrapolations to heavier ions appear to be safe.

Sikkeland[37] and Wong[38] have estimated cross-sections for the production of nuclei in the region $Z = 126$ and $N = 184$ for several different combinations of targets and bombarding ions and they have reported values in the range of tens to hundreds of millibarns. These surprisingly large production cross-sections, compared to elements in the region of curium, are the consequence of the rather large theoretical estimates for the ratio of the neutron to fission widths, i.e. $\Gamma_n/\Gamma_f \sim 1$.

3.3.3.2 *'Grazing' reactions*

In grazing reactions, the target and projectile do not fuse to form a C.N., but a light particle is released during the original interaction while many particles are transferred from the projectile to the target nucleus. The light particle could carry away considerable energy which would otherwise remain as excitation energy. However, experiments indicate that the cross-section decreases quite rapidly with the number of nucleons transferred[39], and estimates of less than microbarns have been made for these reactions to form superheavy nuclei[40].

3.3.3.3 *Fusion-fission reactions*

In the fusion-fission reaction, a heavy target and projectile would fuse to form a massive excited nucleus which would undergo fission. From this cataclysmic event, one of the fission fragments would hopefully be a superheavy nucleus. Experiments on the fission of heavy compound nuclei produced with a variety of ions (ranging from carbon to argon) show the mass yield distributions to have Gaussian shapes centred about symmetric division[82]. The width of the Gaussian was found to increase with both the energy and the Z of the bombarding ion. From extrapolation of this trend

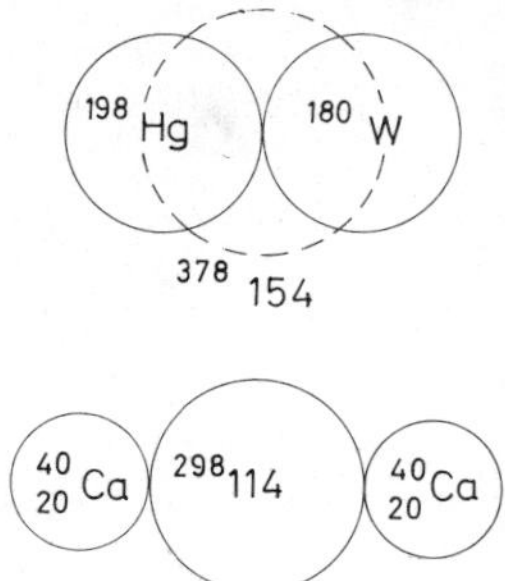

Figure 3.8 The hypothetical 'pinch-off' reaction between ^{198}Hg and ^{180}W to form a superheavy nucleus and two smaller fragments

to even heavier bombarding ions, Oganesyan and associates have estimated cross-sections to be tens of microbarns for the production of superheavy nuclei via asymmetric fission[128]. Unfortunately, Oganesyan's cross-section estimates may be optimistic, since the fission fragments will be very elongated at formation and many of the superheavy nuclei may undergo fission[18, 35].

A slightly different type of reaction, suggested by the observations of actual liquid drops by Thompson's group, is the so-called 'pinch-off' reaction proposed by Swiatecki[35]. When a heavy target and projectile are brought together, the two inner portions of each nuclei could possibly fuse and form a central superheavy nucleus in a nearly spherical shape, as shown in Figure 3.8.

3.3.4 Production of superheavy nuclei by high-energy proton reactions

Recently the suggestion has been made to produce superheavy nuclei by secondary heavy ion reactions generated in very high-energy proton bombardments[41, 42]. For example, when a 24 GeV proton is elastically scattered between the angles of 45 and 180 degrees (relative to the incident beam direction) from a tungsten nucleus, the recoiling tungsten nucleus could attain energies between 1.0 and 5.6 GeV, respectively. Marinov and coworkers also suggest that the (p,n), (p,pn) and (p,d) reactions, or reactions which involve the emission of even higher mass particles, might also produce high-energy, heavy, recoiling nuclei[41]. Such high recoil energies exceed the coulomb barrier height of $\sim$1 GeV which exists between nuclei as heavy as tungsten, and a superheavy nucleus might be produced as a result of the interaction of the recoiling nucleus with a heavy target atom.

Although cross-sections for these processes have not been measured directly and are difficult to estimate, Marinov *et. al.* have suggested[41] a cross-section of 1 μbarn for the production of a beam of high-energy, heavy ions from the irradiation of W with 24 GeV protons. In an attempt to verify this suggestion, Katcoff and Perlman have bombarded W, Au and U foils with 28 GeV protons at the Brookhaven Alternating Gradient Synchroton. They searched for very heavy, recoil nuclei ($Z \geqslant 35$) with sufficient energies (~ 5 MeV/nucleon) to exceed the coulomb barrier of the target nuclei. From their results[43], Katcoff and Perlman have set an upper limit of only 6 nbarns for the production cross-sections of such secondary ions.

Somewhat surprising, but exceedingly interesting results have recently been reported by Maly[44]. He has observed the production of high-energy, heavy nuclei with plastic track detectors in the irradiations of W, Pb and U with 1.3 GeV electrons. Cross-sections of microbarns were reported for the production of secondary nuclei with masses as large as $A = 100$ and energies up to 500 MeV. These nuclei were proposed to come from the photodisintegration or fission of the target atoms by bremsstrahlung and could also conceivably produce superheavy nuclei by secondary reactions.

The use of high-energy protons to produce beams of neutron-excess heavy ions of low Z, i.e. ^{16}C, ^{22}O, ^{26}Ne, etc., has generated considerable interest. On the basis of their measurements of the production cross-sections and energy spectra of light secondary ions, Poskanzer and Hyde have suggested the possibility of producing neutron-rich trans-uranium elements by the irradiation of heavy targets with high-energy protons[42]. Indeed such heavy species as ^{248}Cm and ^{252}Cf have been isolated from ^{238}U targets bombarded with 12 GeV protons at the Argonne National Laboratory[45]. However, there is not yet sufficient experimental data to fully evaluate the potential of this method.

3.4 METHODS OF ISOLATION AND IDENTIFICATION

3.4.1 Recoil and 'gas-jet' methods of isolation

As experimenters attempt to produce elements of higher and higher atomic number, the ever decreasing half-lives and yields of product atoms demand the development of increasingly rapid separation and identification techniques. The discovery experiments on mendelevium were the first in which the recoil momentum imparted to a product atom by the bombarding ion was used to carry out an instantaneous physical separation of the atom from the target material[46]. The recoil atoms were collected on a foil placed behind the target in the evacuated reaction chamber. This greatly shortened the separation time and made it possible to use the same valuable target in repeated bombardments to collect literally one atom at a time.

In later developments, it was found that the recoil atoms could be slowed down and stopped in a gaseous atmosphere, frequently helium. The gas could be pumped out of the reaction chamber through a small orifice to form a 'gas-jet'[47, 48]. If this jet was impinged onto the surface of a foil, some fraction (this can be nearly 100% but is usually less) of the non-volatile product atoms carried along with the gas were deposited on the surface in nearly a monolayer. The foil could be moved in front of a detector periodically, where energy and half-life measurements could be made on the recoil atoms. By this method, it is possible to transport and collect individual product atoms in a fraction of a second.

A more efficient way to use the gas-jet system is to impinge the jet onto the rim of a wheel. After a given deposition time, the wheel is rotated and the deposited atoms positioned in front of a detector, e.g. a Si(Au) surface barrier α-detector. While energy and half-life measurements are being made on that group of atoms, another group of atoms is being collected. By using a number of detectors around the periphery of the wheel, each group of atoms can be observed in sequence to provide further decay data. A typical 'helium gas-jet' system is shown in Figure 3.9.

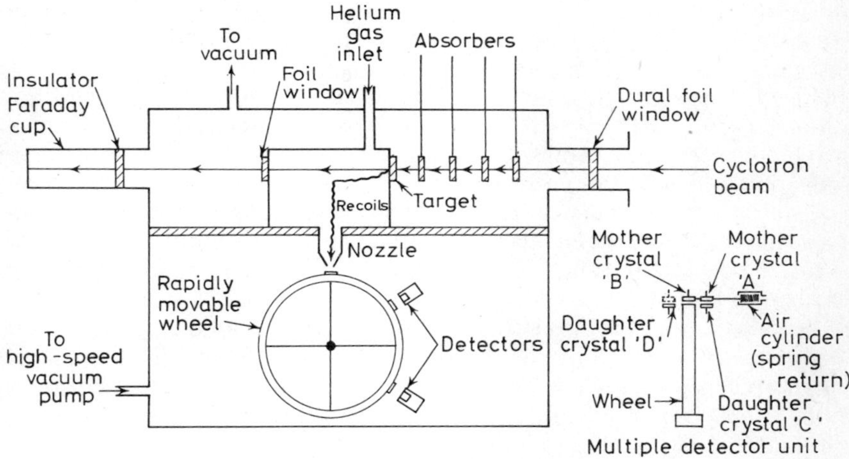

Figure 3.9 A typical 'gas-jet' system for the recoil collection and counting of product atoms. A multiple detector unit for measuring the correlations between parent and daughter atoms is also shown.

An alternative method for collecting recoil atoms is to pump the stopping gas from the reaction chamber through a filtering medium, e.g. filter paper[49]. The atoms, adsorbed onto the surface of the paper, can be analysed as described above. This method is said to yield more consistent collection efficiencies than the gas-jet system[50].

3.4.2 Physical methods of identification

3.4.2.1 Measurement of parent-daughter correlations

The establishment of a genetic linkage between an unknown parent and a known daughter has been used for many years for new element and isotope assignments. It has reached a high degree of sophistication in the so-called 'mother-daughter' experiments used recently for the discoveries[51–54] of isotopes of elements 104 and 105. It is probably best illustrated by a general example involving α-decay.

The technique makes use of the physical separation of a recoiling daughter atom from the mother (and other interfering activities) which occurs when the mother atom undergoes α-decay. The mother atoms, collected on a surface have an equal probability of emitting an α-particle outward from the surface or into the surface. When the α-particle is emitted from the surface, it may be detected with a Si(Au) crystal; when the α-particle is emitted into the wheel, the daughter atom is ejected from the surface and it may be deposited on the face of the detector. If this 'mother' detector is moved over into a position away from these activities (off-wheel), the decay of only the daughter atoms which were deposited on the detector can be recorded. To increase the geometry for detecting the daughter events in the off-wheel position, another 'daughter' detector can be placed facing the 'mother' detector. Obviously there should be a direct relationship between the number of mother events which occur on the wheel and the number of daughter events which subsequently occur in the off-wheel position.

The efficiency for counting the mother activity can be increased with a a double-detector system; one detector can examine the daughter atoms in the off-wheel position while the other detector is looking at the mother atoms on the wheel, and vice versa. This second 'mother' detector requires a second 'daughter' detector for its off-wheel position, thus making a unit of four detectors for each detection station (see Figure 3.9). Other detector *units* can be placed around the periphery of the wheel, as are the single detectors in Figure 3.9.

The genetic relationship is established through the following argument: (1) the energy and half-life of the parent α emitter is obtained from the counts recorded by each set of mother detectors in the on-wheel position; (2) the total yield of daughter atoms transferred to each mother detector unit (detected in the off-wheel position) should decrease in each successive unit, and is directly related to the half-life of the parent; (3) the daughter should exhibit its own half-life in each off-wheel detector unit.

3.4.2.2 Measurement of characteristic x-rays

Identification of the atomic number of an element by its characteristic K-series x-rays was first applied by Moseley to establish the true order of the elements in the chemical periodic system[55]. The direct application of this method to the identification of the trans-fermium elements is beyond present experimental capabilities, since sufficient quantities to carry out direct x-ray fluorescence excitation of these elements are not available. However, with a simple modification, the x-ray method can be extended to atomic-number assignments of many trans-fermium elements even though they are produced only a few atoms at a time.

The technique relies on the coincident observation of characteristic K-series x-rays from the daughter element with α-particles from the decay of the parent element. This is possible if the α-decay of the parent proceeds to excited nuclear states in the daughter nucleus whose energies are above

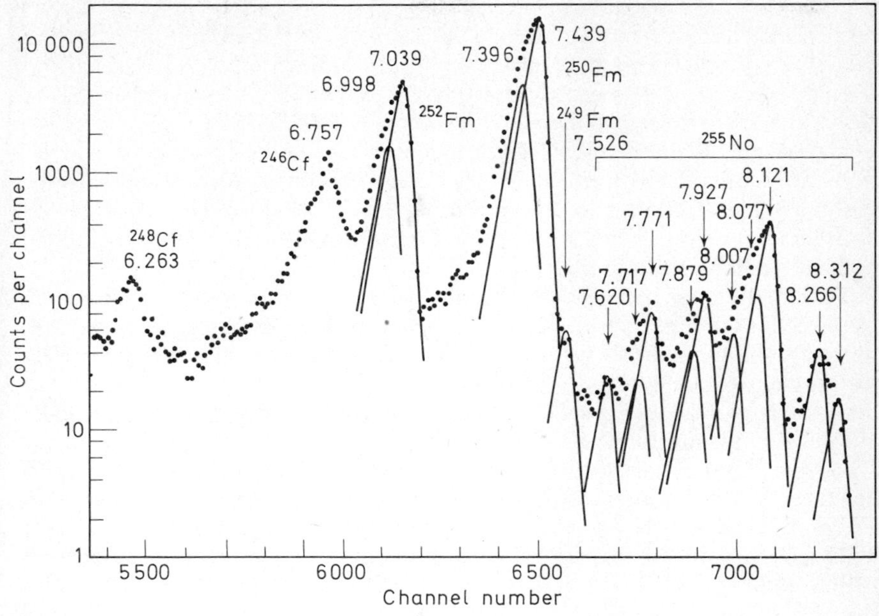

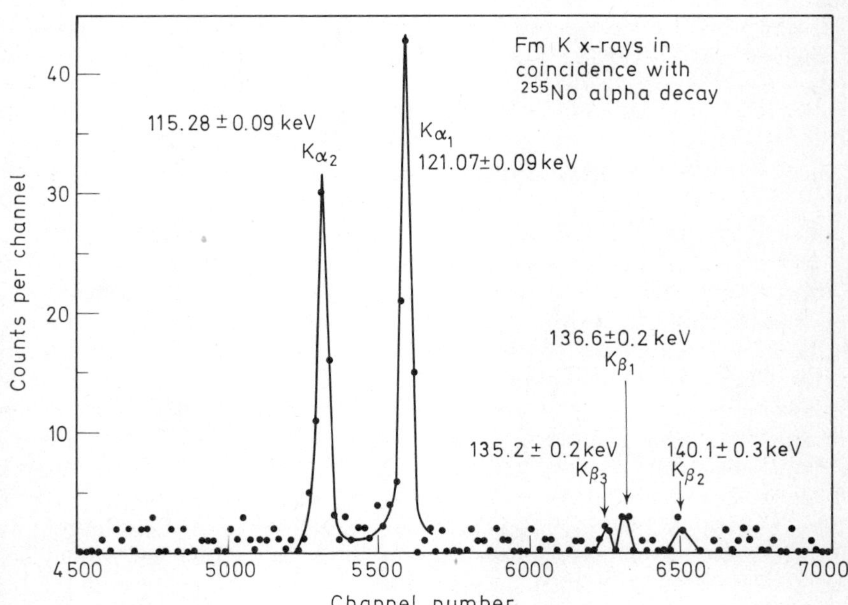

Figure 3.10 Upper α-particle energy spectrum for activities produced in the bombardment of ^{249}Cf with 72 MeV ^{12}C ions (the energies of α groups are indicated in units of MeV). Lower – a portion of the photon energy spectrum observed in coincidence with the indicated α-particles of ^{255}No

the K-binding energy. When de-excitation occurs by the internal conversion process, the emission of characteristic K-series x-rays of the daughter element result from atomic rearrangements following K-shell internal conversion. Since the observed α-particle has $Z = 2$, the identification of K-series x-rays from the daughter element (with atomic number $Z-2$) provides proof of the existence of the parent element with atomic number Z. In cases where the x-ray energies of the daughters have not been measured, theoretical calculations are available with more than sufficient accuracy to establish atomic numbers. This technique is best suited for the study of nuclei with an odd number of protons or neutrons, since α-decay usually takes place to excited nuclear states in the daughter.

This method has recently been applied to the atomic number assignment[56] of element 102. The samples of ^{255}No used in these experiments were prepared by the reaction ^{249}Cf(^{12}C,α2n). The energy spectrum accumulated for the single α-particles with a 200 mm^2 × 300 μm deep Si(Au) surface barrier detector is shown in Figure 3.10. The α groups attributed to the decay of ^{255}No are indicated. Figure 3.10 also shows a portion of the photon energy spectrum obtained with a 16 mm diameter × 5.3 mm deep planar Ge(Li) detector for the photons observed in coincidence with the indicated ^{255}No α groups. The energies of the two largest lines are in excellent agreement with those predicted for Fm K-series x-rays by Carlson *et al.*[57] (predicted values are $K_{\alpha 1} = 121.090$ and $K_{\alpha 2} = 115.320$). The relative line intensities ($K_{\alpha 1}/K_{\alpha 2} \approx 0.65$) are in good agreement with those expected from the calculation of Carlson[58] (predicted value 0.653). Thus an unequivocal identification of the atomic number of the parent, ^{255}No, was obtained. Identification can be made with fewer events than obtained in the nobelium experiments so, hopefully, the method can be applied to isotopes of trans-nobelium elements.

3.4.2.3 Neutron multiplicity counters

As can be deduced from Figures 3.2 – 3.4, superheavy nuclei are expected to decay by spontaneous fission, or by α- and β-decay to daughters which undergo spontaneous fission. For binary fission, the energy release is expected[18] to increase from about 200 MeV for uranium to over 300 MeV for element 114. Some of this additional energy should produce larger excitation energies of the fission fragments, and as it is dissipated primarily through neutron emission, is expected to produce appreciably more neutrons per fission than uranium. The average number of neutrons per fission, $\bar{\nu}$, has been predicted to be about 10 for superheavy nuclei[59]. The emission of such a large number of neutrons for a single fission event would be a distinctive property of these elements, since a $\bar{\nu}$ of 10 is nearly three times greater than that of any known spontaneously fissioning nuclide. This expected increase is the basis for a number of experiments which use neutron multiplicity counters to search for superheavy elements.

The large liquid scintillator[60, 61] provides a sensitive method for detecting events involving multiple neutron emissions. The efficiency can be quite high when the scintillator contains an element, such as gadolinium, which has a large neutron capture cross-section. Neutrons, produced by a source placed in the centre of the chamber, enter the scintillating liquid contained in the surrounding tank and are thermalised by collisions with the hydrogen of the organic solution. They eventually are either captured by the gadolinium or leak out of the tank. The n,γ reaction on gadolinium produces $\sim$9 MeV of energy which is shared by the several γ-rays that are emitted. Electrons, created by the reaction of these γ-rays with the liquid, produce scintillations which are detected by a number of photo multiplier tubes. The wide variation and separation in neutron thermalisation time is used to detect individual neutrons. Such a system, recently constructed and used in a search for superheavy elements in Nature at Berkeley[62], gave an efficiency for detecting four or more neutrons of $\sim$60% (assuming $\bar{\nu} = 10$).

Another type of detector, composed of a cylindrical array of ^{3}He filled proportional counting tubes about a central counting volume, has also been used for detecting multiple neutron emissions[63]. The tubes, of the order of one inch in diameter and two feet long, are imbedded in paraffin to thermalise the fast fission neutrons. The thermal neutrons, on entering the tubes, produce recoil protons from the ^{3}He(n,p) reaction and are recorded. Such a system, operating at Oak Ridge, has an energy resolution of $\sim$5% for the 0.7 MeV recoil protons and a probability of $\sim$35% for the detection of four or more neutrons (assuming $\bar{\nu} = 10$). Unlike the liquid scintillator, this detector is relatively insensitive to γ-rays and so can be used to count moderately radioactive samples from accelerator irradiations.

In order to reduce the cosmic-ray background count rate, both types of detector are usually placed underground or otherwise heavily shielded and they are frequently equipped with anticoincidence shields. With such precautions, background count rates as low as a few counts per day for neutron multiplicities of four or more can be obtained with 100 kg samples of high Z material.

3.5 CURRENT STATUS AND RECENT RESULTS

3.5.1 Fermium

The measurements of kinetic energies and mass distributions, together with the theoretical interpretations, have been an area of interest in the field of nuclear fission for many years. Though the liquid-drop model predicts symmetric fission as the most probable mode, experimentally determined mass-yield distributions for low energy fission have always been asymmetric. The availability of 'relatively' large amounts of ^{257}Fm in recent years has allowed relatively detailed fission studies to be conducted on this heavy nuclide for the first time. The results appear to shed new light on the fission process.

3.5.1.1 Spontaneous and induced fission of ^{257}Fm

In 1970, groups at Los Alamos[64] and Livermore [65] obtained kinetic energy and mass distributions from the spontaneous fission of ^{257}Fm. The data indicate two significant features; (1) the mass-yield distribution, while still primarily asymmetric, exhibits a marked enhancement in the number of symmetric events over that of slightly lighter nuclei, e.g. ^{252}Cf and ^{254}Cf, and (2) the total kinetic energy release increases monotonically with the approach to symmetric fission. However the mass-yield distribution and total kinetic energy suggest a continuation of the trend of a single low-energy mode shown by other actinides.

Kinetic energy measurements[65] of the neutron-induced fission of ^{257}Fm show an even larger amount of symmetric division than in the spontaneous fission. In fact, the main peak in the mass-yield distribution is at symmetry. The increase in the ratio of symmetric to asymmetric fission with increasing excitation energy has been observed for other nuclei; however, even in the induced-fission of a nucleus as heavy as ^{256}Fm, asymmetric division is dominant[66]. Thus for low energy fission, symmetric mass division as a dominant feature appears rather suddenly at ^{257}Fm.

3.5.1.2 *Theoretical interpretations of* ^{257}Fm *fission data*

There are reasons to expect symmetric fission to appear in the region of ^{257}Fm. The probability for symmetric fission might be expected to increase as the mass and charge of the fissioning nuclei approach values such that the symmetric fragments can be doubly closed-shell nuclei. Such a shell structure effect has been used to explain anomalously high yields observed for certain isotopes in the spontaneous fission of lighter nuclides[67].

A possible explanation of asymmetric fission has been proposed by Möller and Nilsson [68], and is supported by the work of Pauli *et al.*[69]. They find a lower energy fission path around the second peak of the fission barrier when asymmetric distortions are included in their potential energy calculations. As seen in Figure 3.5, the second hump becomes progressively lower in energy as the atomic number increases, and recent calculations[70] indicate that the transition from asymmetric to symmetric fission for even-even nuclei should occur at $N \simeq 160$. The spontaneous fission of ^{257}Fm may involve the system proceeding mainly around the second maximum via asymmetric distortions which lead to asymmetric fission, with only occasional tunnelling through the hump to symmetric fission. However, the neutron induced fission of ^{257}Fm involves additional excitation energy which could put the system above the second hump in energy and lead directly to symmetric fission.

3.5.2 Nobelium

Isotopes of element 102 from mass 251 to 258 are now known, and where the same isotopes have been studied, the Dubna and Berkeley groups are substantially in agreement[11]. Both efforts exclude the likelihood of any isotope of element 102 having a half-life of 10 min with the emission of 8.5 MeV α-particles as was first reported in 1957[72]. Furthermore, chemical studies[73] show, that because of its divalency, nobelium could not have exhibited the cation-exchange behaviour attributed to the 10 min activity. The name nobelium was first proposed in the 1957 report, and because of the wide usage over many years, the suggestion[11] has been made that it be retained as the name for element 102.

The recent discovery of a surprisingly long-lived isotope of nobelium, ^{259}No, produced in the bombardment of ^{248}Cm with ^{18}O ions[74, 75], should permit more elaborate chemical studies on element 102 than have been possible with the shorter-lived isotopes. The half-life was determined to be about one hour and the main α-group occurs at 7.52 MeV. Preliminary experiments[73] show the cation-exchange elution behaviour to be the same as has been observed for ^{255}No, and the excitation function to be consistent with that expected for the ^{248}Cm(^{18}O,α3n) reaction.

3.5.3 Lawrencium

In 1961, an 8.6 MeV α-activity with an 8 s half-life was discovered at Berkeley[76] and assigned to element 103. The activity was produced through the bombardment of a Cf target with ^{10}B and ^{11}B ions, but because the target consisted of a mixture of californium isotopes ($^{249-252}$Cf), an unambiguous isotopic assignment from excitation function data was not possible. Though isotopes with masses 255 to 259 could have been produced, the highest yield was expected for mass 257. They suggested the name lawrencium, Lw, for this new element. The name was accepted by the International Union of Pure and Applied Chemistry (IUPAC) but the symbol was changed to Lr.

The results of subsequent studies at Dubna[77, 78], on the production of ^{256}Lr and ^{257}Lr in the irradiation of ^{243}Am with ^{18}O, seemed to conflict with the Berkeley mass assignment. They suggested a half-life and decay energy for ^{257}Lr very nearly the same as ^{256}Lr ($T_{1/2} \simeq 35$ s). More recently, Druin *et al.* have reported preliminary results[79] on the decay properties of ^{255}Lr ($T_{1/2} = 20$ s, $E_\alpha = 8.38$ MeV) which eliminate mass 255 as a possible assignment for the Berkeley activity. This apparent conflict has been discussed by Druin[80].

Very recently six α-emitting isotopes of lawrencium, masses 255–260, have been observed at Berkeley in bombardments of ^{248}Cm and ^{249}Cf targets with ^{14}N and ^{15}N, and ^{10}B and ^{11}B ions, respectively, and the suspected discrepancy appears to be resolved[81] (see Figure 3.1). The mass assignments were based primarily on cross-bombardment techniques and analysis of excitation functions, since genetic linkage studies were hampered due to the large and unknown electron-capture branches in the light mendelevium daughters. The results substantially confirm the Dubna findings for ^{255}Lr and ^{256}Lr while providing more detailed decay data. However, ^{257}Lr was found to have a very short half-life of 0.6 s, contrary to both earlier reports, but ^{258}Lr *was* found to decay by the emission of α-particles in the region of 8.6 MeV with a half-life of 4.2 s. Thus, Ghiorso and associates consider that the mass number assignment of 1961 should have been 258. The difference in this new half-life value, compared to the 1961 value, is attributed to relatively poor statistics in the number of α events obtained in the early work.

In addition, ^{259}Lr and the relatively long-lived ^{260}Lr were discovered. The latter nuclide is the longest lived isotope of element 103 found to date and it should be extremely useful for further chemical studies on lawrencium.

3.5.4 Element 104

In 1964, Flerov and co-workers reported evidence[13, 83] for the production of an isotope of element 104. The bombardment of ^{242}Pu with ^{22}Ne ions produced about one atom per 5 h of a spontaneous-fission activity with a half-life of about 0.3 s. A mass of 260 was assigned on the basis of the measured excitation function. The name kurchatovium, Ku, was suggested for the new element.

During the period 1966–1968, Zvara *et al.* carried out gas-phase chemical studies on the activity and reported a chloride volatility similar to the Group IV elements, Hf and Zr, but different from the actinides[84, 85]; a behaviour expected for the 104th element from trends in the chemical periodic system. The half-life observed for the spontaneous-fission activity was said to be consistent with the earlier physical measurements.

Since that time, from more exact experiments at Dubna, the half-life of the spontaneous-fission activity was lowered to 0.1 s, and a re-evaluation[86] of the original excitation function data led to a relaxation in the mass assignment to 260 ± 1. This new half-life was believed to be too short for any appreciable number of atoms of the nuclide to have survived the period of transport through the chemical steps; therefore a longer lived isotope would have had to have been involved in the original chemical studies[87]. Very recently, 1971, Oganesyan has reported[88] a 20% spontaneous-fission branch in the decay of $^{259}104$. Zvara now feels that the chemical results were probably obtained with this activity[88].

Ghiorso and associates in 1969, after bombardments of ^{246}Cm and ^{248}Cm with ^{18}O and ^{16}O to produce the same compound nucleus ($^{264}104$) as the Dubna experiments, reported that they were unable to confirm the Dubna spontaneous fission results[89]. Also in 1969, the Berkeley group published the results of their bombardments of ^{249}Cf with ^{12}C and ^{13}C ions[90]. Two new α emitting isotopes of element 104 were produced; one with a half-life of 4.5 s and a complex α-particle spectrum centred about 9.0 MeV, and the other with a half-life of 3 s and α-groups at 8.77 and 8.86 MeV. Though the measured excitation functions agreed with those calculated for the production of $^{257}104$ and $^{259}104$, the final proof of mass and atomic numbers came from the genetic linkages to ^{253}No and ^{255}No, respectively, established by parent-daughter correlations. A third, short-lived isotope was also tentatively identified in the same experiments, $^{258}104$, but because it decayed only by spontaneous fission, no genetic relationship could be established to definitely assign the mass and atomic number.

A relatively long-lived isotope was subsequently found by the Berkeley group in irradiations of ^{248}Cm with ^{18}O ions[91]. It was assigned mass 261 from correlations with the ^{257}No daughter. This half-life was sufficiently long to carry out aqueous, cation-exchange column separations which showed behaviour entirely different from the trivalent and divalent actinide elements, but similar to that of Hf and Zr[71]. On the basis of their work, and the failure to confirm the Dubna results, Ghiorso and co-workers suggested the name rutherfordium, Rf, for element 104. At the time of writing, the disagreement between the two groups has not been resolved and no official name has been accepted by IUPAC for element 104.

3.5.5 Element 105

In 1968, Flerov and co-workers reported the synthesis of two short-lived α-emitters of energies 9.4 and 9.7 MeV which were produced in the bombardment of ^{243}Am with ^{22}Ne ions[92]. In delayed coincidence measurements, they reported a significant correlation (10 events) between these α-groups and the 8.35–8.6 MeV α-particles of ^{256}Lr and ^{257}Lr (^{257}Lr was believed at that time to have nearly the same decay properties as ^{256}Lr). A tentative assignment of the 9.4 and 9.7 MeV α-activities to ^{261}105 and ^{260}105, respectively, was made. There were, however, interfering α-activities produced by the interaction of the Ne ions with lead impurity in the Am target, and attention was shifted to spontaneous-fission studies.

Early in 1970, the Dubna group published several reports[93, 94] on the production of a 2 s spontaneously-fissioning nuclide in the irradiation of ^{243}Am with ^{22}Ne ions to produce the compound nucleus ^{265}105. From comparisons of the measured excitation function with those expected for the ^{243}Am (^{22}Ne,4n) reaction, the activity was assigned to ^{261}105; however, the assignment to ^{260}105, produced by the 5n reaction, was not excluded. Angular distribution studies of the recoiling atoms indicated that the activity was not due to a short-lived spontaneously fissioning isomer of a nuclide in the region of americium. Zvara and associates performed chemical experiments on the 2 s fission activity using gas-phase techniques and reported[95] that its chloride was less volatile than niobium chloride but more volatile than hafnium chloride; a property predicted for eka-tantalum, i.e. element 105.

A new ^{243}Am target was prepared, which contained 100 times less lead impurity, and further α-particle studies were carried out in 1970. A time correlation was reported[96] between observed α-particles of 8.9 and 9.1 and the 8.25–8.6 MeV α-particles of the 35-s daughter, ^{256}Lr and/or ^{257}Lr. Thus, the α-activity was assigned to ^{260}105 or ^{261}105. The observed half-life of 1.4 s agreed with the half-life of 2 s measured for the spontaneous-fission activity.

Also in early 1970, Ghiorso and colleagues published data from experiments to produce isotopes of element 105 via the ^{249}Cf(^{15}N,*x*n) reactions[97]. They were unable to confirm the earlier Dubna assignment of 9.4 and 9.7 MeV α-particles to element 105, but reported the discovery of an α-emitter of 9.06, 9.10 and 9.14-MeV particles with a half-life of 1.6 s. The activity was assigned to ^{260}105 on the basis of parent-daughter recoil atom correlations with ^{256}Lr. Later, the Berkeley group synthesised two additional isotopes of element 105 and assigned[98] them to masses 261 and 262. The former was produced by the reactions ^{249}Bk(^{16}O,4n) and ^{250}Cf(^{15}N,4n), and it was found to have a half-life of 1.7 s and decay with emission of 8.93 MeV α-particles. However, the daughter nuclide, ^{257}Lr, was found to exhibit such a short half-life (0.6 s) that it was not possible to obtain the parent-daughter recoil atom correlation. The mass assignment was made on the basis of the time correlation between parent and daughter α-particles emitted from the same source. The $^{262}105$ was produced by the $^{249}Bk(^{18}O,5n)$ reaction and was found to decay with a 40 s half-life and with the emission of 8.45–8.50 MeV α-particles. Genetic linkage to ^{258}Lr was established from both the time correlation of α's and the recoil-atom correlations. On the basis of their work on $^{260}105$, the Berkeley group proposed to IUPAC that element 105 be named hahnium, Ha.

An exchange of letters on the priority of discovery of element 105 has recently appeared in *Science*[99]. As yet no official name has been accepted for element 105.

3.5.6 Chemical properties of the heavy actinides and light trans-actinides

3.5.6.1 Mendelevium

Hulet and co-workers first noticed an anomalous chemical behaviour for Md in certain chemical systems involving reducing conditions. With 10^5–10^6 atoms per experiment of ^{256}Md, co-precipitation and extraction chromatography experiments were carried out in the presence of various reducing agents, and they showed that Md^{3+} could be readily reduced to a fairly stable Md^{2+} in aqueous solution[100]. Further, an estimate was obtained for the oxidation potential of the half-reaction, Md^{II}–$Md^{III}+e^-$, of $E^\circ \simeq +0.2$ V (relative to the normal hydrogen electrode). Similar results were obtained by Maly and Cunningham[101].

3.5.6.2 Nobelium

In over 600 experiments, Maly *et al.*[73] subjected ~50 000 atoms of ^{255}No to cation-exchange and co-precipitation studies. These tracer experiments showed that nobelium exhibits a chemical behaviour substantially different from the trivalent actinides but similar to the divalent alkaline earth elements, Sr, Ba and Ra. Thus the divalent ion of nobelium is the most stable species in aqueous solution. This result confirmed a prediction[102] made by Seaborg in 1949 that a relatively stable 2+ state might exist for element 102 due to the special stability of the filled 5f electronic configuration. Additional experiments which used the technique of solvent-extraction chromatography were performed under a variety of oxidising conditions, and allowed an estimate for the oxidation potential of the half-reaction, No^{II}–$No^{III}+e^-$, of $E^0 \simeq -1.45$ V[103] (relative to the normal hydrogen electrode).

3.5.6.3 Lawrencium

With only ten atoms per experiment of ^{256}Lr, tracer chemical studies of element 103 have been conducted to determine the most stable oxidation state in aqueous solutions[104] By comparing the solvent-extraction behaviour of Lr with other actinides, as well as lanthanides and alkaline earth elements, the most stable species was determined to be the trivalent ion. This result is in agreement with the prediction that lawrencium, the last member of the actinide series ($5f^{14}7s^26d^1$ electronic configuration), should return to the 3+ oxidation state as the most stable in aqueous solution.

3.5.6.4 II–III Oxidation potential correlations for the actinides

Nugent, Baybarz and Burnett have made correlations of the II–III oxidation potentials for the actinides and lanthanides through the use of Jorgensen's refined spin-pairing electronic energy theory[105]. In Figure 3.11, the known II–III oxidation potentials of the actinides are compared with the lan-

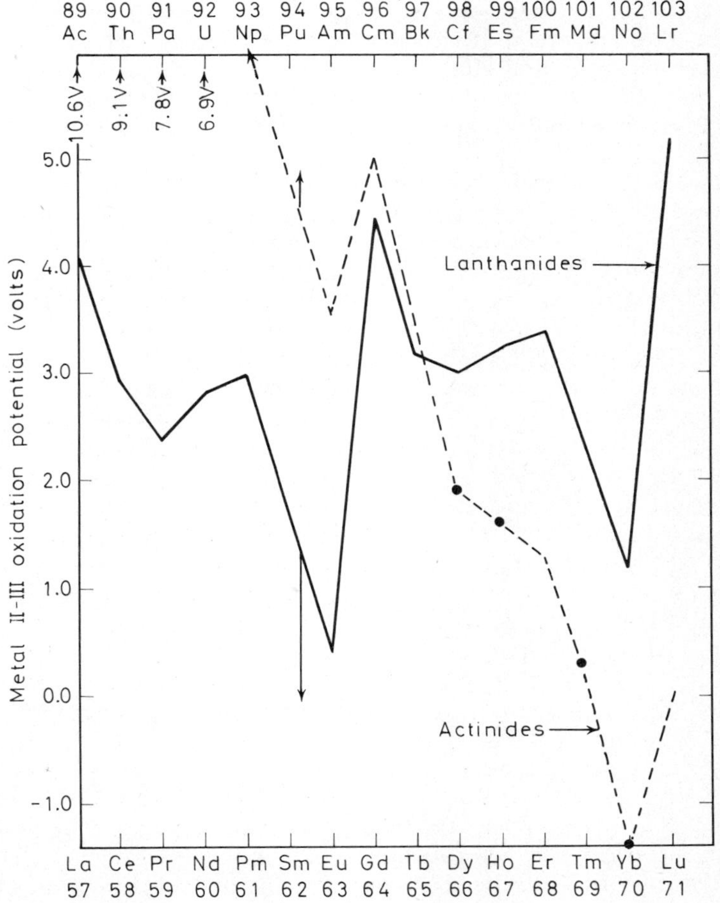

Figure 3.11 Correlations of the II–III oxidation potentials for the actinides and lanthanides. Points show experimental values for the actinides

thanides and the calculated curves. The potentials for Cf and Es were obtained from the measured energies of the electron-transfer bands of the chlorides and bromides in anhydrous ethanol and the linear relationship with the II–III potentials. The excellent correlation of the measured potentials with the curve calculated on the basis of the actinide hypothesis supports the notion of the approaching end of the 5f electron series.

All the current evidence supports the original prediction that element 103 is the last member of the actinide series, and that element 104 would be expected to fall in the next group, i.e. the IVB elements, of the periodic system. This conclusion was also reached by the Dubna group from the results of the gas-phase studies of the volatilities of the chlorides of the heavy actinides and trans-actinides[106].

3.5.6.5 *Element 104*

Element 104 is expected to belong to the Group IVB elements of the chemical Periodic Table. It is predicted[107,108] to have a valence and ionic radius similar to Zr and Hf, and to exhibit similar chemical properties to these elements. Experiments, using quite different chemical techniques, have been carried out at Berkeley and Dubna to test this prediction of chemical similarity.

The Dubna researchers have concentrated their efforts to separate element 104 from the actinides on a rapid gas-phase technique[84,85]. The basis for the separation scheme is the observed correlation between the adsorbability of molecules from a gas phase onto solid surfaces and the macroscopic property of volatility. Atoms of a spontaneously fissioning activity, produced in the bombardment of ^{242}Pu with ^{22}Ne ions (the compound nucleus would be element 104), were collected and mixed with the chlorinating agents and carrier gases thionyl chloride, niobium pentachloride and zirconium tetrachloride. After passing through a 2 m long glass column, the heated gases flowed between mica track detectors for recording fission events. Retention times (adsorbabilities) were measured for a number of elements and correlated with known heats of vaporisation. The readily volatile chlorides of Hf and Zr passed through the system with little delay, while the non-volatile chlorides of the actinide elements were adsorbed on the walls of the column. From comparative measurements, the Dubna experiments reported that the chloride volatility of the spontaneous-fission activity was similar to Hf and Zr but quite different from those of the actinides[109,110].

In a collaborative experiment, Berkeley and Oak Ridge experimenters employed the technique of ion-exchange chromatography for their chemical studies[71] of element 104. The atoms of $^{261}104$, produced in the $^{248}Cm(^{18}O,5n)$ reaction, were passed through a small cation-exchange resin column. The positions of elution from the column were determined for the α-activity assigned to $^{261}104$, as well as a number of other elements. In several hundred repetitive experiments, approximately 100 atoms of element 104 were produced for study. From these comparative measurements, the ion-exchange behaviour of element 104 was found to be entirely different from the trivalent and divalent actinides but was similar to Hf and Zr.

3.5.6.6 *Element 105*

Experiments[95] have been performed on the chemical separation of radioactive products formed in the bombardment of ^{243}Am with ^{22}Ne ions (the compound nucleus is element 105). Gaseous chlorides of the products were allowed to flow through a glass chromatographic column over which a temperature gradient of 300–50 °C was established. The products were adsorbed from flowing carrier gas onto the walls and formed zones in different temperature ranges. Contained in the tube were mica detectors for recording spontaneous-fission events. From the position of the zone of an unknown nuclide relative to the zones of known elements, it is possible to estimate the volatility of its chloride. On the basis of 18 events, the spontaneous fission activity was reported to form a chloride less volatile than niobium pentachloride but more volatile than hafnium tetrachloride. As mentioned previously, such a property is predicted for element 105, eka-tantalum, on the basis of the periodic system[95].

3.5.7 Superheavy elements

3.5.7.1 *Search in Nature*

The estimates presented in Figures 3.2 and 3.3 for even-even superheavy elements indicate half-lives for both α- and spontaneous-fission decay greater than 10^8 years for several nuclides near $Z = 110$. Therefore, if such nuclei had been formed in nucleosynthesis, residual superheavy nuclei could still be present in Nature. Since uncertainties of as much as a factor of 10^6 are possible in these estimates of half-lives, and since odd-even and odd-odd nuclides generally exhibit retardation in decay compared to even-even nuclei, the surviving nuclides could extend over several atomic numbers, e.g. 108–115. In addition, a search[111] for very heavy particles in cosmic rays at high altitudes indicates the presence of uranium and adjacent elements, and even suggests a possible contribution from elements with $Z > 100$. Thus, the possibility exists that superheavy nuclei might have been produced in more recent cosmic events, and shorter half-life nuclides (10^4–10^5 years) could conceivably be deposited in small amounts on the surface of the earth.

Searches, based on the straightforward prediction[31, 32, 112, 113] that elements 108, 109, 110, 111, 112, 113 and 114 will be the chemical homologues of osmium, iridium, platinum, gold, mercury, thallium and lead, have been directed toward investigation of pure metal and ore samples of the elements with atomic numbers 76 to 82 (see Figure 3.12). It is not certain that the superheavy elements will exhibit completely analogous chemical behaviour to their homologues; in particular, the electronic configurational energies of elements 111–112 are calculated to be different from the lighter homologues[114]. However, a prominent chemical characteristic of elements 108–114 appears to be their predicted nobility and one might expect them to be found with the noble metals[115, 129].

The first attempts to detect superheavy elements in terrestrial samples were by Thompson *et al.*[116], and Flerov and Perelygin[117, 118]. Thompson's group searched for element 110 in natural platinum ores but found no evidence for its presence. Using a variety of techniques, e.g. low background neutron, γ-ray and spontaneous-fission counting, they set limits of less than 2×10^8 years or greater than 10^{15} years for the half-life. Using x-ray fluorescence, mass spectrometric and activation analysis techniques, a limit of $<10^{-10}$ g

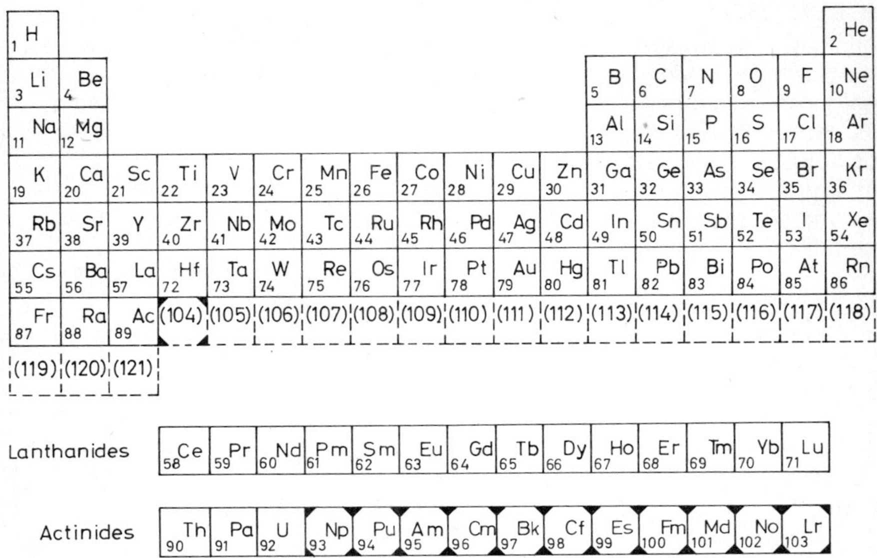

Figure 3.12 A chemical Periodic Table of the elements which shows the predicted placement of elements 106–118

of superheavy element/gramme of platinum was established for a half-life longer than 10^{15} years.

Flerov and Perelygin reported results from spontaneous-fission measurements on lead-bearing samples, in particular lead glass, which they felt could be explained as due to the presence of superheavy nuclei. Their earliest results came from scanning plastic track detectors (~ 1 m^2 of mylar foils) which had been in contact with lead foils for 100 days. Employing etching techniques, further investigations on lead glass samples (including a fragment from an eighteenth century glass vase) showed excess spontaneous-fission events above that expected from the small amounts of U and Th present; however, similar experiments on lead-bearing minerals were inconclusive. The results on lead glass appeared to be confirmed by fission counting the samples in large area (1.4 m^2) proportional counters. Assuming a half-life of 10^9 years for the spontaneously fissioning nuclei, the concentration would be $\sim 10^{-12}$ g/g of lead.

Using a very elegant and more sensitive method, Price and Fleischer[119] feel they have obtained results in conflict with those of Flerov and Perelygin. Price and Fleischer searched for fission tracks accumulated over millions of years in ancient minerals ($>10^8$ years old). Their results for Pb (hardystone) and Au (quartz) bearing samples gave no evidence for the presence of superheavy elements, and concentration limits of $<10^{-}15$ g/g of Pb and $<10^{-17}$ g/g of Au were assigned. However, as the samples were of different origin and composition, it might be possible for superheavy nuclei to be present in the lead samples of Flerov and not in the lead minerals of Price.

Recently Thompson's group, utilising a liquid scintillator for neutron multiplicity counting, examined kilogramme samples of tungsten, lead and lead ore (galena)[62]. They were unable to obtain any evidence for superheavy nuclei in any of the samples and set a concentration limit of $<\sim 10^{-15}$ g/g of sample; a value in agreement with Price and Fleischer.

Grimm, Herrmann and Schüssler have made the most general search for superheavy elements in Nature[120]. Employing a ^{3}He-filled proportional-type neutron counter, they have examined multikilogramme samples of pure and ore grade minerals. The samples were selected to include the eka-osmium to eka-bismuth elements on the basis of both homologous chemical behaviour and geochemical rules. Some 101 samples were counted including: sulphidic ores and concentrates of lead, mercury, molybdenum, zinc and copper; by-products from industrial treatment of such ores, i.e. slags, etc.; raw and pure metals, e.g. gold, platinum, silver, lead and adjacent metals. No events attributable to superheavy nuclei were observed, but a concentration limit of $<10^{-11}$ g/g of sample was estimated.

To date, only one direct claim for the discovery of a superheavy element in Nature has been made. From a vast body of experimental data accumulated over many years, Cherdyntsev has assigned an α-emitter observed in osmium bearing samples to element 108, and suggested the name Sergenium (Sg)[121]. However, after a careful analysis of the data, Kulakov concluded that the claim was unfounded and premature[122], an opinion generally shared by others in the field.

Several other groups are involved in a general search for superheavy elements in Nature[123, 124], but so far their results are negative. At present the prospects do not appear encouraging; however, the question 'Do superheavy elements exist in Nature?' is still unanswered.

3.5.7.2 Search with accelerators

A few attempts have been made to produce superheavy elements through the use of accelerators. Unfortunately, very heavy ions of high energy and high-beam currents have not been available, and the results from less favourable nuclear reactions cannot provide a definite conclusion about the new 'island of stability'.

The first endeavours[125, 126] involved attempts to observe the spontaneous fission of short-lived nuclei produced by the reaction, ^{248}Cm(^{40}Ar,4n)^{284}114. The results of these experiments were negative, but consistent with the half-life of less than 10^{-15} s which is predicted for such a neutron deficient isotope.

More recently, Lefort *et al.*[127] have attempted to produce element 126 by bombarding ^{232}Th with 400-MeV ^{84}Kr. All the nuclei which can be produced by the compound nucleus process are far off the β-stability line, but one hopes to form a more stable product of lower Z by α- or β-decay. Preliminary experiments have failed to reveal any α- or spontaneous-fission activity attributable to superheavy nuclei. Though the integrated beam flux of $10^{11}-10^{12}$ particles is rather low, these early results would seem to be in conflict with the predictions of multimillibarn cross-sections for (HI,*x*n) reactions to produce superheavy nuclei.

The only report of the possible production of superheavy elements in an accelerator has come from a rather surprising direction. Marinov and co-workers reported the possible production of element 112, eka-mercury, by secondary reactions in tungsten targets irradiated with 24 GeV protons[41]. Two tungsten targets were exposed to a total of $\sim 10^{18}$ protons over a period of several months and were chemically processed. A number of samples (Pt, Au, Hg, Tl and Pb fractions) were isolated and a few counts per day of both a spontaneous-fission and an α-activity were observed in certain of the Hg sources. They suggested this activity could be attributed to the presence of one or more isotopes of element 112. Unfortunately, because of the low count rates involved and the thickness of the sources (~ 2 mg/cm^2), it was not possible to measure a property more characteristic of superheavy nuclei, e.g. neutron multiplicity, characteristic x-rays or total energy release in fission. Alternative explanations for the observed activities in terms of existing nuclides are possible, and further experiments seem necessary to establish the discovery.

References

1. Scharff-Goldhaber, G. (1957). *Nucleonics,* **15,** 126
2. Werner, F. G. and Wheeler, J. A. (1958). *Phys. Rev.,* **109,** 126
3. Myers, W. D. and Swiatecki, W. J. (1966). *Nucl. Phys.,* **81,** 1
4. Meldner, H. and Röper, P. (1965). Unpublished results quoted in Ref. 3.
5. Sobiczewski, A., Gareev, F. A. and Kalinkin, B. N. (1966). *Phys. Lett.,* **22,** 500
6. Gustafson, C., Lam, I. L., Nilsson, B. and Nilsson, S. G. (1967). *Ark. Fys.,* **36,** 613
7. Viola, V. E. and Seaborg, G. T. (1966). *J. Inorg. Nucl. Chem.,* **28,** 697
8. Johansson, S. A. E. (1959). *Nucl. Phys.,* **12,** 449
9. Dorn, D. W. (1961). *Phys. Rev.,* **121,** 1740
10. Viola, V. E. and Wilkins, B. D. (1965). *Nucl. Phys.,* **82,** 65
11. Ghiorso, A. and Sikkeland, T. (1967). *Physics Today,* **20,** 25
12. Wheeler, J. A. (1956). *Physica,* **22,** 1103
13. Flerov, G. N., Oganesyan, Yu. Ts., Lobanov, Yu. V., Kuznetsov, V. I., Druin, V. A., Perelygin, V. P., Gavrilov, K. A., Tretiakova, S. P. and Plotko, V. M. (1964). *Phys. Lett.,* **13,** 73
14. Bethe, H. A. and Backer, R. F. (1936). *Rev. Mod. Phys.,* **8,** 82
15. Comay, E., Liran, S., Wagman, J. and Zeldes, N. (1970). *International Conference on the Properties of Nuclei Far from the Region of Beta-Stability,* 165. (Genève: CERN Report 70–30)
16. Nilsson, S. G., Tsang, C. F., Sobiczewski, A., Szymanski, Z., Wycech, S., Gustafson, C., Lamm, I., Möller, P. and Nilsson, B. (1969). *Nucl. Phys.,* **A131,** 1
17. Tsang, C. F. and Nilsson, S. G. (1970). *Nucl. Phys.,* **A140,** 289
18. Nix, J. R. (1970). *International Conference on the Properties of Nuclei Far from the Region of Beta-Stability,* 605. (Genève: CERN Report 70–30)
19. Strutinsky, V. M. (1967). *Ark. Fys.,* **36,** 629
20. Strutinsky, V. M. (1967). *Nucl. Phys.,* **A95,** 420
21. Strutinsky, V. M. (1968). *Nucl. Phys.,* **A122,** 1
22. Tsang, C. F. (1971). (Berkeley: private communication)
23. Musychka, Yu. A. (1968). *Phys. Lett.,* **28B,** 539
24. Wong, C. Y. (1967). *Phys. Lett.,* **19,** 328
25. Rost, E. (1968). *Phys. Lett.,* **26B,** 184

26. Sikkeland, T., Ghiorso, A. and Nurmia, M. (1968). *Phys. Rev.*, **172,** 1232
27. Sikkeland, T. and Trautmann, N. (1971). (Berkeley: private communication)
28. Eccles, S. F. and Hulet, E. K. (1969). *UCRL-50767.* (Livermore: University of California Lawrence Radiation Laboratory Report)
29. Hoff, R. W. and Hulet, E. K. (1969). *UCRL-72165.* (Livermore: University of California Lawrence Radiation Laboratory Report)
30. Hulet, E. K., Wild, J. F., Lougheed, R. W., Evans, J. E., Qualheim, B. J., Nurmia, M. and Ghiorso, A. (1971). *Phys. Rev. Lett.*, **26,** 523
31. Seaborg, G. T. (1968). *Ann. Rev. Nucl. Sci.*, **18,** 53
32. Seaborg, G. T. (1969). *The Robert A. Welch Foundation Conferences on Chemical Research, XIII. The Transuranium Elements,* 5. (Houston, Texas)
33. Berlovich, E. E. and Novikov, Yu. N. (1969). *JETP Letters,* **9,** 445
34. Viola, V. E. (1969). *Nucl. Phys.*, **A139,** 188
35. Swiatecki, W. J. (1970). *International Conference on Nuclear Reactions Induced by Heavy-Ions,* 729. (Amsterdam: North-Holland Publishing Co., Inc.)
36. Sikkeland, T., Silva, R. J., Ghiorso, A. and Nurmia, M. J. (1970). *Phys. Rev.*, **C1,** 1564
37. Sikkeland, T., Clarkson, J. E., Lebeck, D. F. and Ghiorso, A. (1966). *Inorg. Nucl. Chem. Lett.*, **2,** 141
38. Wong, C. Y. (1967). *Nucl. Phys.*, **A103,** 625
39. Grochulshi, W., Kwiecinska, T., Go-Chan, L., Lozynski, E., Maly, J., Tarasov, L. K. and Volkov, V. V. (1963). *3rd Conference on Reaction Between Complex Nuclei,* 120. (Berkeley and Los Angeles: University of California Press)
40. Sikkeland, T. (1966). *Ark. Fys.*, **36,** 539
41. Marinov, A., Batty, C. J., Kilvington, A. I., Newton, G. W. A., Robinson, V. J. and Hemingway, J. D. (1971). *Nature (London),* **229,** 464
42. Poskanzer, A. M., Butler, G. W. and Hyde, E. K. (1971). *Phys. Rev.*, **C3,** 882
43. Katcoff, S. and Perlman, M. L. (1971). *BNL-15829.* (Upton, N. Y.: Brookhaven National Laboratory Report)
44. Maly, J. (1971). *Phys. Lett.*, **35B,** 148
45. Fields, P. (1971). (Argonne National Laboratory: private communication)
46. Ghiorso, A., Harvey, B. G., Choppin, G. R., Thompson, S. G. and Seaborg, G. T. (1955). *Phys. Rev.*, **98,** 1518
47. Ghiorso, A. (1959). *UCRL-1814,* 18. (Berkeley: University of California Lawrence Radiation Laboratory Report)
48. Macfarlane, R. D. and Griffioen, R. D. (1963). *Nucl. Inst. Methods,* **24,** 461
49. Friedman, A. M. and Mohr, W. C. (1962). *Nucl. Inst. Methods,* **17,** 78
50. Druin, V. A. (1970). *International Conference on Nuclear Reactions Induced by Heavy Ions,* 657. (Amsterdam: North-Holland Publishing Co., Inc.)
51. Ghiorso, A., Nurmia, M., Harris, J., Eskola, K. and Eskola, P. (1969). *Phys. Rev. Lett.*, **22,** 1317
52. Ghiorso, A., Nurmia, M., Eskola, K. and Eskola, P. (1970). *Phys. Lett.*, **32B,** 95
53. Ghiorso, A., Nurmia, M., Eskola, K., Harris, J. and Eskola, P. (1970). *Phys. Rev. Lett.*, **24,** 1498
54. Ghiorso, A. (1969). *The Robert A. Welch Foundation Conferences on Chemical Research, XIII. The Transuranium Elements,* 107. (Houston, Texas)
55. Moseley, H. G. (1913). *Phil. Mag.*, **26,** 1024
56. Dittner, P. F., Bemis, C. E., Hensley, D. C., Silva, R. J. and Goodman, C. D. (1971). *Phys. Rev. Lett.*, **26,** 1037
57. Carlson, T. A., Nestor, C. W., Malik, F. B. and Tucker, T. C. (1969). *Nucl. Phys.*, **A135,** 57
58. Lu, C. C., Malik, F. B. and Carlson, T. A. (1971). Submitted to *Nucl. Phys.* for publication
59. Nix, J. R. (1969). *Phys. Lett.*, **30B,** 1
60. Ashby, V. J., Catron, H. C., Newkirk, L. L. and Taylor, C. J. (1958). *Phys. Rev.*, **111,** 616
61. Chodil, G., Jopson, R. C., Mark, H., Swift, C. D., Thomas, R. G. and Yates, M. K. (1967). *Nucl. Phys.*, **A93,** 648
62. Cheifetz, E., Giusti, E. R., Bowman, H. R., Jared, R. C., Hunter, J. B. and Thompson, S. G. (1971). *UCRL-19957.* (Berkeley: University of California Lawrence Radiation Laboratory Report)

63. Macklin, R. L., Glass, F. M., Halperin, J., Roseberry, R. T., Stoughton, R. W. and Tobias, M. (1971). *Chemistry Division Annual Progress Report*. (Oak Ridge: Oak Ridge National Laboratory Report)
64. Bologna, J. P., Ford, G. P., Hoffman, D. C. and Knight, J. D. (1971). *Phys. Rev. Lett.*, **26,** 145
65. John, W., Hulet, E. K., Lougheed, R. W. and Wesolowski, J. J. (1971). Submitted to *Phys. Rev. Lett.*
66. Unik, J. P. (1971). Unpublished data quoted in Ref. 65
67. Hyde, E. K. (1964). *The Nuclear Properties of the Heavy Elements, III. Fission Phenomena*, 112. (Englewood Cliffs, New Jersey: Prentice-Hall, Inc.)
68. Möller, P. and Nilsson, S. G. (1970). *Phys. Lett.*, **31B,** 283
69. Pauli, H. C., Ledergerber, T. and Brock, M. (1971). *Phys. Lett.*, **34B,** 264
70. Gustafsson, C., Möller, P. and Nilsson, S. G. (1971). *Phys. Lett.*, **24B,** 349
71. Silva, R. J., Harris, J., Nurmia, M., Eskola, K. and Ghiorso, A. (1970). *Inorg. Nucl. Chem. Lett.*, **6,** 871
72. Fields, P. R., Friedman, A. M., Milstead, J., Atterling, H., Forsling, W., Holm, L. W. and Aström, B. (1957). *Phys. Rev.*, **107,** 1460
73. Maly, J., Sikkeland, T., Silva, R. J. and Ghiorso, A. (1968). *Science*, **160,** 1114
74. Silva, R. J., Mallory, M. L., Dittner, P. F., Hammons, P. J. and Keller, O. L. (1971). (Oak Ridge: unpublished data)
75. Ghiorso, A., Eskola, K., Eskola, P. and Nurmia, M. (1971). (Berkeley: unpublished data)
76. Ghiorso, A., Sikkeland, T., Larsh, A. E. and Latimer, A. E. (1961). *Phys. Rev. Lett.*, **6,** 473
77. Donets, E. D., Shchegolev, V. A. and Ermakov, V. A. (1965). *Atomn. Energiya*, **19,** 109
78. Flerov, G. N., Korotkin, Yu. S., Mikheev, V. L., Miller, M. B., Polikanov, S. M. and Shchegolev, V. A. (1967). *Nucl. Phys.*, **A106,** 476
79. Druin, V. A. (1970). *Yadern. Fiz.*, **12,** 268
80. Druin, V. A.(1971). *Sov. J. Nucl. Phys.*, **12,** 146
81. Eskola, K., Eskola, P., Nurmia, M. and Ghiorso, A. (1971). *UCRL-20442*. (Berkeley: University of California Lawrence Radiation Laboratory Report)
82. Oganesyan, Yu. Ts. (1968). *Nuclear Structure, Dubna Symposium*, 489. (Vienna: IAEA Press)
83. Flerov, G. N., Oganesyan, Yu. Ts., Lobanov, Yu. V., Kuznetsov, V. I., Druin, V. A., Perelygin, V. P., Gavrilov, K. A., Tretiakova, S. P. and Plotko, V. M. (1964). *Atomn. Energiya*, **17,** 310
84. Zvara, I., Chuburkov, Yu. T., Caletka, R., Zvarova, T. S., Shalaevsky, M. R. and Shilov, B. V. (1966). *Atomn. Energiya*, **21,** 83
85. Zvara, I., Chuburkov, Yu. T., Caletka, R. and Shalaevsky, M. R. (1969). *Radiokhimiya*, **11,** 163
86. Oganesyan, Yu. Ts., Lobanov, Yu. V., Tretyokova, S. P., Lasarev, Yu. A., Kolesov, I. V., Gavrilov, K. A., Plotko, V. M. and Poluboyarinov, Yu. V. (1969). *JINR P7-4797*. (Dubna: Joint Institute for Nuclear Research Report)
87. Zvara, I. (1969). *The Robert A. Welch Foundation Conferences on Chemical Research, XIII. The Transuranium Elements*, 165. (Houston, Texas)
88. Oganesyan, Yu. Ts. (1971). *International Conference on Heavy Ion Physics*. (Dubna: to be published)
89. Ghiorso, A., Nurmia, M. and Harris, J. (1969). *UCRL-18714*. (Berkeley: University of California Lawrence Radiation Laboratory Report)
90. Ghiorso, A., Nurmia, M., Harris, J., Eskola, K. and Eskola, P. (1969). *Phys. Rev. Lett.*, **22,** 1317
91. Ghiorso, A., Nurmia, M., Eskola, K. and Eskola, P. (1970). *Phys. Lett.*, **32B,** 95
92. Flerov, G. N., Druin, V. A., Demin, A. G., Lobanov, Yu. V., Skobelev, N. K., Akapiev, G. N., Fefilov, B. V., Kolesov, I. V., Davrilov, K. A., Kharitonov, Yu. P. and Chelnokov, L. P. (1968). *JINR P7-3808*. (Dubna: Joint Institute for Nuclear Research Report)
93. Flerov, G. N., Oganesyan, Yu. Ts., Lobanov, Yu. V., Lazarev, Yu. A. and Tretiakova, S. P. (1970). *JINR P7-4932*. (Dubna: Joint Institute for Nuclear Research Report)
94. Flerov, G. N., Oganesyan, Yu. Ts., Lobanov, Yu. V., Lazarev, Yu. A. and Tretiakova, S. P. (1970). *JINR P7-5108*. (Dubna: Joint Institute for Nuclear Research Report)
95. Zvara, I., Belov, V. Z., Korotin, Yu. S., Shalaevsky, M. R., Shchegolev, V. A., Hussonnois, M. and Zager, B. A. (1970). *JINR P12-5120*. (Dubna: Joint Institute for Nuclear Research Report)

96. Druin, V. A., Demin, A. G., Kharitonov, Yu. P., Akapiev, G. N., Rud, V. I., Sun-Tzin-Yan, G. Ya., Cholnokov, L. P. and Gavrilov, K. A. (1970). *JINR P7-5161*. (Dubna: Joint Institute for Nuclear Research Report)
97. Ghiorso, A., Nurmia, M., Eskola, K., Harris, J. and Eskola, P. (1970). *Phys. Rev. Lett.*, **24**, 1498
98. Ghiorso, A., Nurmia, M., Eskola, K. and Eskola, P. (1971). *UCRL-20487*. (Berkeley: University of California Lawrence Radiation Laboratory Report)
99. Flerov, G. N. (1970). *Science*, **170**, 15; Ghiorso, A. (1971). *Science*, **171**, 127
100. Hulet, E. K., Lougheed, R. W., Brady, J. D., Stone, R. E. and Coops, M. S. (1967). *Science*, **158**, 486
101. Maly, J. and Cunningham, B. B. (1967). *Inorg. Nucl. Chem. Lett.*, **3**, 445
102. Seaborg, G. T., Katz, J. J. and Manning, W. M. (1949). *The Transuranium Elements, IV 14B*, 1492. (New York: McGraw-Hill, Inc.)
103. Silva, R. J., Sikkeland, T., Nurmia, M., Ghiorso, A. and Hulet, E. K. (1969). *J. Inorg. Nucl. Chem.*, **31**, 3405
104. Silva, R. J., Sikkeland, T., Nurmia, M. and Ghiorso, A. (1970). *Inorg. Nucl. Chem. Lett.*, **6**, 733
105. Nugent, L. J., Baybarz, R. D. and Burnett, J. L. (1969). *J. Phys. Chem.*, **73**, 1177
106. Flerov, G. N. and Zvara, I. (1969). *The 10th Jubilee Mendeleev Assembly Commemorating the 100th Anniversary of the D. I. Mendeleev Periodic Table*, 115. (Leningrad)
107. Seaborg, G. T. (1969). *American Chemical Society, Mendeleev Centennial Symposium*. (Minneapolis, Minn.)
108. Cunningham, B. B. (1969). *The Robert A. Welch Foundation Conferences on Chemical Research, XIII. The Transuranium Elements*, 307. (Houston, Texas)
109. Zvara, I. (1968). *The 155th American Chemical Society National Meeting*. (San Francisco)
110. Zvara, I., Chuburkov, Yu. T., Belov, V. Z., Buklanov, G. V., Zakhvataev, B. B., Zvarova, T. S. Maslov, O. D., Caletka, R. and Shalaevsky, M. R. (1970). *J. Inorg. Nucl. Chem.*; **32**, 1885
111. Price, P. B., Fowler, P. H., Kidd, J. M., Kobetech, E. J., Fleischer, R. L. and Nichols, G. E. (1971). *Phys. Rev.*, **D3**, 815
112. Waber, J. T. (1969). *The Robert A. Welch Foundation Conferences on Chemical Research, XIII. The Transuranium Elements*, 353. (Houston, Texas)
113. Cowan, R. D. and Mann, J. B. (1970). *2nd International Conference on Atomic Physics*. (Oxford: in press)
114. Carlson, T. A., Lu, C. C., Tucker, T. C., Nestor, C. W. and Malik, F. B. (1970). *ORNL-4614*. (Oak Ridge: Oak Ridge National Laboratory Report)
115. Keller, O. L., Burnett, J. L., Carlson, T. A. and Nestor, C. W. (1970). *J. Phys. Chem.*, **74**, 1127
116. Nilsson, S. G., Thompson, S. G. and Tsang, C. F. (1969). *Phys. Lett.*, **28B**, 458
117. Flerov, G. N. and Perelygin, V. P. (1969). *Atomn. Energiya*, **26**, 520
118. Flerov, G. N., Skobelev, N. K., Ter-Akopyan, G. M., Subbotin, V. G., Gvozdev, B. A. and Ivanov, M. P. (1969). *JINR D6-4554*. (Dubna: Joint Institute for Nuclear Research Report)
119. Price, P. B., Fleischer, R. L. and Woods, R. T. (1970). *Phys. Rev.*, **1**, 1819
120. Grimm, W., Herrmann, G. and Schüssler, H. D. (1971). *Phys. Rev. Lett.*, **26**, 1040
121. Nikitin, A. (1970). *Nauka i Zhizn (Science and Life)*, **5**, 102
122. Kulakov, V. M. (1970). *Atomn. Energiya*, **29**, 401
123. Cartwright, D. R., Martin, H. R. and Miller, H. W. (1971). *U.S.A.E.C. Division of Research, Transplutonium Program Committee Special Session on Superheavy Elements*. (Germantown, Md.)
124. Halperin, J., Cheifetz, E., Drury, J. S., Gentry, R. V., Giusti, E. R., Jared, R. C., Milton, R. M., Macklin, R. L., Stoughton, R. W. and Thompson, S. G. (1971). *Chemistry Division Annual Progress Report*. (Oak Ridge: Oak Ridge National Laboratory Report)
125. Ghiorso, A., Sikkeland, T. and Silva, R. J. (1967). Unpublished data
126. Thompson, S. G., Swiatecki, W. J., Bowman, H. R., Gatti, R. C., Moretto, L. G. and Jared, R. C. (1968). *UCRL-17989*. (Berkeley: University of California Lawrence Radiation Laboratory Report)
127. Lefort, M. (1971). (Orsay: private communication to C. Y. Wong)
128. Karamian, S. A. and Oganesyan, Yu. Ts. (1969). *JINR P7-4339*. (Dubna: Joint Institute for Nuclear Research Report)
129. Keller, O. L. (1969). *The Robert A. Welch Foundation Conferences on Chemical Research, XIII. The Transuranium Elements*, 245. (Houston, Texas)

4
Medium- and Lower-Energy Nuclear Chemistry

M. LEFORT
Université Paris-Sud

4.1 INTRODUCTION

Nuclear chemistry has developed from the discoveries of natural and artificial radioactivity; just as chemistry is concerned with the structure of molecules and atoms, with reactions between molecules and with methods of synthesis, so nuclear chemistry is essentially concerned with reactions between nuclei, with the transformation of one nuclear species into another and with the structure of stable and unstable nuclei and with nuclear synthesis.

Beams of various projectiles have become available for the study of nuclear reactions; protons, deuterons, helions and heavier nuclei at low or high energies. From the bombarded target a large amount of information may be collected, either on the new nuclei that have been produced, or on the particles emitted during the interaction or very shortly after it. The development of new techniques has greatly increased the possibilities for obtaining more and more elaborate data.

The main purpose of nuclear chemistry is the synthesis of new species: new isomers, new isotopes of very short half-life and new elements. A knowledge of the mechanisms of nuclear reactions is most useful in such syntheses for the problem of the identification of new species is difficult when the half-life is short or when the characteristics of the product are not known. Important developments in the methods of identification have occurred during recent years, particularly in the use of new detectors, magnets, velocity measurements on recoiling products, on-line mass spectrometry, etc.

Nuclear chemistry has been undertaken with many kinds of bombarding particles ranging from thermal neutrons to protons of 25 GeV. At very high energies, a bombarding particle may produce a very large number of species in a given target and the number of channels open for the interaction of a GeV projectile and a complex nucleus is very large. For this reason, a limit has been fixed at around 60 MeV dividing low- and medium-energy nuclear reactions. A less arbitrary distinction might be made between high-energy and low-energy nuclear chemistry. Thus, at high energies the associated wavelength of the bombarding particle is short enough to permit interaction with a single nucleon or with a small number of nucleons of the target nucleus. This direct process is subsequently followed by a more collective interaction for which many random possibilities exist. At low energies, on the other hand, interaction takes place in the first stage of the collision process in a collective manner involving most of the nucleons of the projectile and of the target, although even at low energies the possibility of direct interactions, such as occur in transfer reactions, cannot be ignored.

In this latter energy region (below 60 MeV for protons and below 10 MeV per nucleon for complex particles), the principal reaction mechanism involves the formation and decay of a compound nucleus. Although this mechanism has been known for light projectiles for some 40 years, the advent of heavy ion accelerators has widened the field and many new developments have occurred in recent years.

4.2 FUNDAMENTALS OF NUCLEAR REACTIONS

4.2.1 The concept of cross-section

When a beam of accelerated particles traverses a target consisting of a certain number of identical nuclei, only a certain proportion of these particles will react with the target nuclei. First, it must be recognised that since the dimensions of the nucleus are small in relation to those of the atom, there is a strong possibility that the particle will be slowed down by interaction with orbital electrons rather than by nuclear collision, and although each nucleus must be considered as a centre of interaction for scattering or absorption of incident particles, the probability of such interaction is quite small. The probability of a particular reaction occurring may be defined in terms of the *cross-section.*

If α is the nuclear reaction channel concerned, it is possible to define a ratio σ_α where

$$\sigma_\alpha = \frac{\text{Number of events of type } \alpha \text{ per nucleus per unit of time}}{\text{Number of incident particles per unit area per unit of time}}$$

For all particles crossing one square centimetre of the target surface, σ_α per nucleus will lead to the event α.

The differential cross-section in a solid angle $d\Omega$ at an angle θ from the beam direction is defined as $d\sigma/d\Omega$, and the angular distribution may be obtained by plotting the different values of $d\sigma/d\Omega$ obtained at different angles θ. The total cross-section is obtained by integrating the differential cross-section over all angles from 0 to π, and may be expressed as

$$\sigma_t = 2\pi \int_0^\pi \frac{d\sigma}{d\Omega} \sin\theta d\theta$$

When the bombarded target is sufficiently thin, the number of events occurring in a particular reaction channel during an irradiation time t is $n = \sigma I \mathcal{N} t$, where I is the beam intensity expressed as the number of particles per second, and $\mathcal{N}$ is the number of nuclei per cm^2 in the target. Using the flux, ϕ, the number of particles per cm^2/second, the relation becomes $n = \sigma\phi Nt$ where N is now the number of nuclei in the target. When the target thickness is such that there is a noticeable weakening of the beam intensity because of the relatively large number of nuclear reactions occurring, then for a thin layer de located at a depth e, $dI = n = I\sigma N de$ and integration yields the well-known 'thick target' relation:

$$n = I_0 - I = I[1\text{-exp}(-\mathcal{N}\sigma)]$$

$\mathcal{N}\sigma$ is particularly large in the case of thermal or intermediate energy neutrons because of the very big cross-sections involved.

The neutron flux on a target placed in a neutron bath can be written $\nu_0 V_0$, where ν_0 is the neutron density and V_0 the velocity. Under these circumstances the number of events becomes $\nu_0 V_0[1\text{-exp}(-\mathcal{N}\sigma)]$. The introduction of the target modifies the neutron flux from $\nu_0 V_0$ to νV and $\sigma = \frac{1}{\mathcal{N}} \ln \left| \frac{\nu_0 V_0}{\nu V} \right|$.

For a fixed velocity, the cross-section may be determined by measurement of the ratio ν_0/ν of the neutron densities with and without the target.

4.2.2 Calculation of the reaction cross-section. Partial waves

A beam of particles at a large distance from the target nucleus may be described by the Schrödinger equation, i.e.,

$$\nabla^2 \psi_{\text{inc}} + k^2 \psi_{\text{inc}} = 0 \tag{4.1}$$

where $k = 1/\lambda\!\!\!^-$ is the wave number and $\lambda\!\!\!^-$ the de Broglie wavelength,

$$k = \frac{mv}{\hbar} = \frac{\sqrt{2\mu\varepsilon}}{\hbar}$$

ε being the kinetic energy and μ the reduced mass of the particle. When collision between the particle beam and a suitable target occurs, such collision has to be considered in the centre of mass system where the kinetic energy is entirely usable for excitation energy. Thus, in this system the momentum is $p = \mu v$, where $\mu = m_1 m_2\ (m_1 + m_2)$ is the reduced mass and the kinetic energy $\varepsilon = \varepsilon_L m_2/(m_1 + m_2)$, if ε_L is the kinetic energy in the laboratory system.

One solution of the Schrödinger equation is the plane wave $\psi_{\text{inc}} = e^{ikz}$, where z is the direction of motion.

Nuclear reactions will be considered in terms of an interaction of the wave ψ_{inc} with a nuclear potential. In order to describe this interaction, it is useful to express the plane wave in spherical polar coordinates, since the nuclear potential has spherical symmetry. The classical expression of e^{ikz} then becomes

$$e^{i\vec{kr}} = \sum_{l=0}^{\infty} (2l+1)\, i^l j_l(kr)\, P_l(\cos\theta) \tag{4.2}$$

an expansion in spherical waves of different angular momentum $l\hbar$, where $j_l(kr)$ are spherical Bessel functions and $P_l(\cos\theta)$ are Legendre polynomials.

Physically, each term of the summation corresponds to particles with an angular momentum $l\hbar$. In this so-called decomposition into 'partial waves', particles from the beam which strike the nucleus with an impact parameter

b, belong to a partial wave of order l, and their orbital angular momentum, $L = bmv$ is equal to $\sqrt{l(l+1)}\,\hbar$. These particles strike the nucleus within a ring between $(l+1)\dfrac{\hbar}{mv} = (l+1)\,\lambda\!\!\!^{-}$ and $l\lambda\!\!\!^{-}$, the cross-section for this partial wave being $\sigma_l = \pi\lambda\!\!\!^{-2}\,(l+1)^2 - l^2 = \pi\lambda\!\!\!^{-2}(2l+1)$.

4.2.3 The transmission coefficient

In such an approximate description, the nucleus appears like a black obstacle for all the partial waves. However, this is not always the case and only part of the incident wave penetrates the nucleus. This is expressed by a transmission coefficient T_l, the magnitude of which depends on the angular momentum, and on introducing this transmission coefficient the usual expression for the reaction cross-section becomes the well-known relationship[1]

$$\sigma_{\mathrm{R}} = \pi\lambda\!\!\!^{-2} \sum_l (2l+1)\, T_l \tag{4.3}$$

The problem of calculating total reaction cross-sections hinges on the calculation of transmission coefficients. At present this is undertaken in most of the sophisticated calculations by using an optical model to describe the nuclear potential. However, the *continuum theory*[1] is still useful for obtaining a first estimation for T_l and hence σ_{R}. The two main assumptions employed in this theory are:

(a) A particle which has penetrated the nucleus is strongly absorbed through collision with the nucleons, i.e. the probability of re-emission through the entrance channel is very small, and the number of open reaction channels is very large.

(b) The effective potential outside the nucleus, when $r > R$, is given by

$$\mathrm{V}_l(r) = \mathrm{V}_c(r) + \left(\frac{l(l+1)\hbar^2}{2\mu r^2}\right) \tag{4.4}$$

where $\mathrm{V}_c(r)$ is the Coulombic repulsion, $Z_1 Z_2 e^2/r$, and the second term is the 'centrifugal barrier' which arises in the treatment of the Schrödinger equation when separating the wave function into radial and angular functions, and which express the fact that the easiest way for a particle to penetrate a nucleus is to travel along a radius towards the centre.

At the nuclear boundary, the potential suddenly becomes negative as the incident particle comes within range of the nuclear forces and under these circumstances we may write that

$$\mathrm{V}_l(r) = -\mathrm{V}_0 \text{ for } r \leqslant R \tag{4.5}$$

The wave number of the incident particle that has penetrated the nucleus may then be written as

$$K = \frac{\sqrt{2\mu(\varepsilon + \mathrm{V}_0)}}{\hbar} \tag{4.6}$$

Two extreme cases which lead to simple approximations of equation (4.3) are well known.

In the sharp cut-off approximation, which may be applied to energetic neutrons, T_l is strongly dependent on the product $kR = R/\lambdabar$, i.e. on the orbital angular momentum for a grazing trajectory. By putting $T_l = 1$ up to a maximum value of l, given by $l_{max} = R/\lambdabar = kR$, and $T_l = 0$ for $l > kR$, summation of equation (4.3) gives:

$$\sigma_R = \pi\lambdabar^2 \sum_{l=0}^{R/\lambda} (2l+1) = \pi(R+\lambdabar)^2 \tag{4.7}$$

For determining cross-sections for charged particles, the deviation of the incident beam by the Coulomb potential must be taken into account. The maximum impact parameter for incident particles reaching the nuclear surface is smaller than R, and under these circumstances the maximum angular momentum is $R\sqrt{2\mu(\varepsilon - V_c(R))}$ instead of $R\sqrt{2\mu\varepsilon}$, and

$$\sigma_r = \pi(R+\lambdabar)^2\left(1 - \frac{V_c(R)}{\varepsilon}\right) \tag{4.8}$$

The second simple case is that of an incident s-wave neutron ($l = 0$). In this case it is unnecessary to convert the incident plane wave into spherical waves, and the wave may be expressed as $\psi_{inc} = Ae^{ikz}$.

Since there is partial reflection on the surface of the nucleus, the total wave function outside the nucleus ($z > R$) is:

$$\psi = Ae^{ikz} + Be^{-ikz} \tag{4.9}$$

The wave number of the transmitted wave is, according to equation (4.6), equal to K, and $\psi_{trans} = Ce^{+ikz}$ ($z < R$).

The amplitudes of the various wave functions are A, B and C, and they are related to each other by boundary conditions, in particular by the requirements of continuity at $z = R$ for both the wave function and its derivative. Therefore, the condition $\psi = \psi_{tr}$ for $z = R$ implies that $A + B = C$, and the condition $\frac{\partial\psi}{\partial z} = \frac{\partial\psi_{tr}}{\partial z}$ for $z = R$ implies $ikA - ikB = iKC$. Combining these two conditions gives $C = A\frac{2k}{k+K}$

Since the transmission coefficient $T_{l=0}$ is the ratio of the transmitted to the incident neutron flux, and since a flux is the product of the velocity and the probability of finding the particle ψ, ψ^*, then,

$$T_{l=0} = \frac{\psi_{tr}\psi_{tr}^* K}{\psi_{inc}\psi_{inc}^* k}$$

K and k being proportional to the velocities inside and outside the nucleus.

Since $e^{ikz}.e^{-ikz} = 1$, this equation becomes

$$T_{l=0} = \frac{C^2}{A^2}\frac{K}{k} = \frac{4kK}{(k+K)^2} \tag{4.10}$$

At high energies, $K \simeq k$ and $T_{l=0}$ approaches 1. For very low energies, $k \ll K$ and $T_{l=0} \simeq 4k/K$. Since the wavelength is very great, the partial wave $l = 0$ is the only one which interacts with the target nucleus and:

$$\sigma_R = \pi \lambda\!\!\!^{-2} T_{l=0} = 4\pi \lambda\!\!\!^{-2} \frac{k}{K}$$

or

$$\sigma_R = \frac{2\pi\hbar^2}{m} (\varepsilon V_0)^{-\frac{1}{2}}$$

where V_0 is the nuclear potential and m the neutron mass. When σ_R is expressed as a function of the neutron velocity v, this equation acquires the well-known form

$$\sigma_R = \frac{4\pi\hbar^2}{\sqrt{2mV_0}} \cdot \frac{1}{mv} \tag{4.11}$$

4.2.4 Reaction cross-section and elastic scattering cross-section

The fact that a nucleus is placed in the path of an incident wave corresponding to a particle beam, leads to a scattering effect, even if there is no reaction. On the other hand, even when the reaction cross-section is a maximum, i.e. for $T_l = 1$, over all the partial waves, so that the nucleus is a black obstacle, there is still an elastic scattering cross-section. The total cross-section for the interaction occurring between the incoming wave and the nucleus must therefore be expressed as the sum of two contributions,

$$\sigma_{tot} = \sigma_{el.\,sc.} + \sigma_r$$

only outgoing waves which are coherent with the incoming beam interfering with it and contributing to true elastic processes.

As is well known, the maximum reaction cross-section is obtained when $\sigma_r = \sigma_{el.\,sc.} = \pi \lambda\!\!\!^{-2} \Sigma_l (2l+1)$, corresponding to $T_l = 1$ for all values of l. Under these circumstances the total cross-section becomes twice the geometric area of the black obstacle when

$$\pi \lambda\!\!\!^{-2} \sum_{l=0}^{R/\lambda\!\!\!^{-}} (2l+1) = \pi (R + \lambda\!\!\!^{-})^2$$

Elastic scattering, which is unexpected from the simple picture of the interaction, may be explained by analogy to an optical situation. Thus, if the black sphere of radius R may be assumed to cast a shadow, this shadow may be accounted for by saying that just enough 'light' is scattered in the forward direction to cancel the incident beam. For particles of a few MeV, $\lambda\!\!\!^{-}/R$ is of the order of 1/3 and the scattering shadow will be confined to an angle $\lambda\!\!\!^{-}/R \simeq 1/3$ of a radian from the forward direction, so that the region of complete shadow extends only a short distance behind the nucleus, and not as far back as the measuring instruments.

The scattering cross-section has, however, been measured experimentally, since it is possible to derive general expressions for the total and elastic scattering cross-sections, the treatment having been given in detail by Blatt and Weisskopf[1]. Starting from equation (4.2), when the incident beam is far from the nucleus $kr \gg l$, so that the asymptotic values of the spherical Bessel functions may be expressed as

$$j_l(kr) = \frac{i\left\{\exp\left[-i\left(kr-\frac{l\pi}{2}\right)\right]-\exp\left[+i\left(kr-\frac{l\pi}{2}\right)\right]\right\}}{2kr}$$

Furthermore, in equation (4.2), ψ_{inc} is composed of two exponential terms, the first referring to that part of the wave moving toward the nucleus, while the second refers to the outgoing wave. If a nuclear reaction occurs, the outgoing wave should be affected since the reaction reduces its amplitude relative to that of the incoming wave. This argument introduces a scattering amplitude η_l, which is a complex number satisfying the relationship $|\eta_l|^2 \leqslant 1$, and if this is included in the above equations the scattered wave $\psi - \psi_{\text{inc}}$, as well as the scattered flux can be calculated so that the expression

$$\sigma_{\text{el.s.}} = \pi\lambda\!\!\!^{-2} \sum_l (2l+1)\,|\,1-\eta_l\,|^2 \tag{4.12}$$

may be written for the scattered flux, while the expression

$$\sigma_r = \pi\lambda\!\!\!^{-2} \sum_l (2l+1)(1-|\,\eta_l\,|^2) \tag{4.13}$$

applies to the absorbed flux, i.e. the reaction cross-section.

The transmission coefficient T_l can be identified with the factor $(1-|\,\eta_l\,|^2)$ and the cross-section is a maximum for $\eta_l = 0$, and $\sigma_r = \sigma_s = \pi\lambda\!\!\!^{-2} \sum_l (2l+1)$.

4.2.5 Angular momentum and partial cross-section

The inclusion of spin does not change the magnitude of the total reaction cross-section given in equations (4.3) or (4.13), but it is obvious that the value of the partial cross-section σ_l will be affected. Since most of the recent work on nuclear reaction cross-sections has been concerned with heavy ion reactions in which the orbital angular momentum brought in by the bombarding particle may be high, it seems worth while to outline very briefly the mathematical arguments concerning the coupling of the angular momenta.

Let j and s be the spin of the target and projectile respectively. The vector sum, referred to as channel spin S, is

$$\vec{S} = \vec{j} + \vec{s} \tag{4.14}$$

Therefore, the total angular momentum of the nucleus resulting from the entrance channel for the reaction is

$$\vec{J} = \vec{l} + \vec{S} \tag{4.15}$$

$\vec{l}$ being the orbital angular momentum vector.

The various angular momenta are coupled by a Clebsch–Gordon coefficient, the square of which gives the probability that a target and a projectile with channel spin S and orbital angular momentum l will couple to give a total angular momentum J. The partial reaction cross-section is obtained after summing over all the projections of J on the direction of motion and averaging over the projections of S on the same axis, i.e.,

$$\sigma_r(S,l,J) = \pi \lambda\!\!\!^{-}{}^2 T_l \frac{(2J+1)}{(2S+1)} \tag{4.16}$$

This expression shows that in the case of heavy ions where l, and consequently J, may reach high values, the contribution of partial cross-sections of high l numbers is predominant.

Summing over all values of l and averaging over S yields the cross-section for a target of spin j, a bombarding particle of spin s and a resulting total angular momentum J

$$\sigma_r(j,s,J) = \frac{\pi \lambda\!\!\!^{-}{}^2(2J+1)}{(2s+1)(2j+1)} \sum_{S=|j-s|}^{j+s} \sum_{l=|J-S|}^{J+S} T_l \tag{4.17}$$

If $s = j = S = 0$, then the only contribution to the angular momentum is the orbital momentum l. Making $J = l$ in equation (4.17) gives $\pi \lambda\!\!\!^{-}{}^2(2l+1)T_l$.

The total reaction cross-section obtained by summing equation (4.17) over all values of J is:

$$\sigma_r = \frac{\pi \lambda\!\!\!^{-}{}^2}{(2s+1)(2j+1)} \sum_{S=|j-s|}^{j+s} \sum_{l=0}^{\infty} T_l \sum_{J=|l-S|}^{l+S} (2J+1) \tag{4.18}$$

which reduces to $\pi \lambda\!\!\!^{-}{}^2 \sum_{l=0}^{\infty} (2l+1)T_l$ when the summations are evaluated.

4.3 MODERN EXPERIMENTAL METHODS

When Rutherford bombarded nitrogen with α-particles from the active deposit of radon, the fact that protons were emitted and caused scintillations on a zinc sulphide screen first indicated the existence of a nuclear reaction. The identification of reaction products is, however, much more satisfactory when they are radioactive and can be observed by their decay. The first artificial radioisotope was the positron emitter ^{32}P, discovered in the bombardment of aluminium by α-particles and the first experiment in nuclear chemistry was that undertaken by F. Joliot and I. Curie at that time.

In the main, experiments in nuclear chemistry are still of the same type as those of Rutherford and of Joliot and Curie. Complementary information, necessary for an understanding of nuclear reaction mechanisms, are supplied by measurements of particle spectra and of reaction products. Conversely, the use of known nuclear reactions permits the production of many new species either as residual nuclei or as emitted fragments.

4.3.1 The study of the residual nuclei

Measurements of residual nuclei are still important in low-energy nuclear chemistry since only a few reactions are possible and the number of reaction products is not too numerous. Although most of the measurements are restricted to cases where the products are radioactive and have suitable and known decay schemes, stable isotopes have also been detected and counted using mass spectrometry. Nowadays, nuclei with short half-lives may also be detected with the help of on-line techniques utilising the very important progress made in radiation detectors with good energy resolution. We shall briefly mention three new techniques: on-line mass spectrometry, helium gas removal of the recoiling atoms and characterisation of products by x-ray measurements.

4.3.1.1 On-line mass spectrometry

The development of on-line mass spectrometry and on-line isotope separation has been described elsewhere[2,3], and a schematic representation of a typical installation is shown in Figure 4.1. The precise arrangement depends on the nature of the bombarding particles (protons, neutrons or heavy ions). The particle beam traverses the ion source in which the target has been placed,

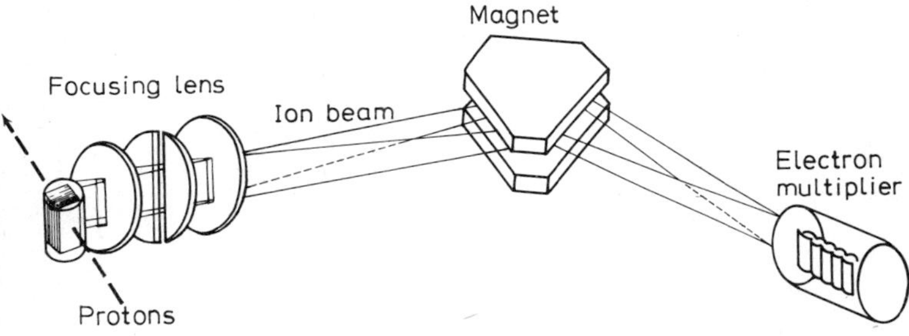

Figure 4.1 Schematic representation of the mass spectrometer (on-line) operating in the beam of an accelerator. (From Bernas [2], by courtesy of Annual Reviews)

and the product nuclei, recoiling after the nuclear reaction, are brought to rest in the source itself. Diffusion is relied upon to extract them from the filament and surface ionisation to lead to an efficient and selective emission. Alkali elements diffuse fast and are the only ones efficiently ionised on a hot metallic surface. At higher temperatures, alkaline earth elements might also be emitted. So long as the half-life is longer than the diffusion time, any product, even if not radioactive, can be detected.

The method has a great advantage which is also very disadvantageous. Since the experimental conditions are specific for one element, or at least of one part of the Periodic Table – essentially the alkaline metal, alkaline earth and perhaps some rare earth elements – there are many reaction products which cannot be analysed in this way. Further development of the method is, however, still possible and interesting new ideas might yet appear. The technique is rather cumbersome, especially for measurement of low yields, where the background radiation is always at a high level even at some distance from the beam.

4.3.1.2 *The helium gas jet technique*

This has been used at Berkeley by MacFarlane *et al.*[4] and by Hyde *et al.*[5], and is, in principle, very simple and permits one to collect the species recoiling from the target very quickly. It is particularly useful for heavy ion induced reactions for which the recoil momentum is large.

The recoiling nuclei are first slowed down in helium gas, and the gas is then pumped through a capillary tube at a rate close to the velocity of sound, carrying the heavy atoms with the same velocity and therefore with a much larger momentum. At the exit of the capillary, which can be as much as several metres away, the heavy atoms strike a collector placed in front of the aperture (Figure 4.2). Their activity can be measured less than 10^{-3} s after

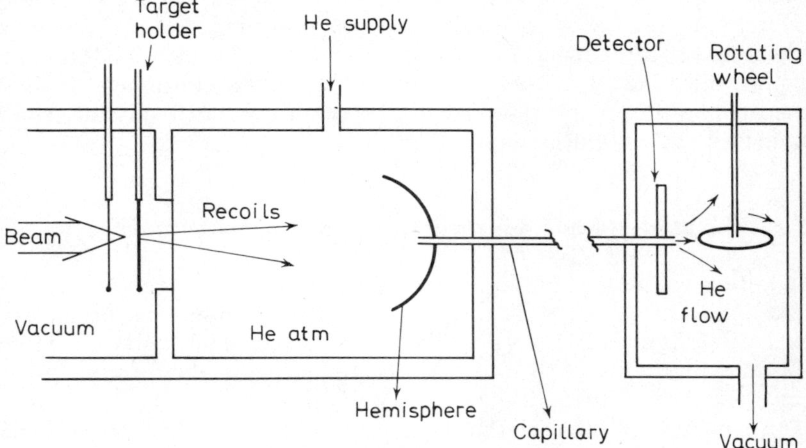

Figure 4.2 Helium jet device

their production in the target. Several ingenious systems have been proposed for following the daughters and measuring their half-lives[6] (rotating wheel, tape, sets of many solid-state detectors etc). Many new neutron-deficient isotopes of francium, actinium and protactinium, and new isotopes of elements 102, 103 and 104 have been identified after collection with such helium gas devices[5,7].

4.3.1.3 *X-ray measurements*

Identification of the atomic number of an element by its characteristic K-series x-rays was first applied by Moseley[8]. The method has been used more recently[9] to confirm the identification of technetium and promethium, but extensive application only became possible with the recent manufacture of lithium-drifted germanium detectors of very good energy resolution. Since it is now possible to distinguish between two x-rays separated by about 0.5 keV, the identification of medium and heavy elements is relatively simple provided that these atoms emit K x-rays during their decay.

This is often the case because soft gamma transitions decay by internal conversion. For this reason, x-ray measurements may be used instead of chemical analysis for characterising an element. The method is suited to on-line data storage and also to coincidence techniques. It is of particular interest for nuclear reaction studies with the trans-uranium elements, where a change in Z by one unit corresponds to a change in energy of about 3 keV for a given K x-ray series, and where calculated values for electron binding energies are available and in good agreement with measured values.

One of the recent illustrations of the efficiency of the technique is the identification of element 102 produced in a reaction $^{249}_{98}$Cf($^{12}_{6}$C,α2n) at the Oak Ridge Cyclotron[78]. A coincidence between alpha rays of 8.121, 8.077 and 8.007 MeV and K_{α_2} and K_{α_1} x-rays of 115.28 and 121.07 keV shows that the alpha decay is followed by K x-rays uniquely associated with the element Fermium 100 and therefore the parent is element 102. A detailed study of the level schema of fermium shows that the isobar is 251 and therefore the element 102 had the mass number 255. It is clear that the new techniques, some of which have been briefly described above, do not prevent the more classical procedures of radiochemical separation and, milking studies yielding a large amount of data.

4.3.2 The study of nuclear reactions by recoil techniques

In addition to the recognition of the product of a reaction, considerable interest has been shown in the recoil energies of the residual nuclei. Measurements of the angular distribution and ranges of the recoils provide information for kinematic studies of the reaction and therefore for the formulation of a mechanism.

The simplest example is the formation of a completely fused nucleus between the projectile and the target. If the compound system has a relatively long life-time, it will have a laboratory velocity equal to the velocity of the

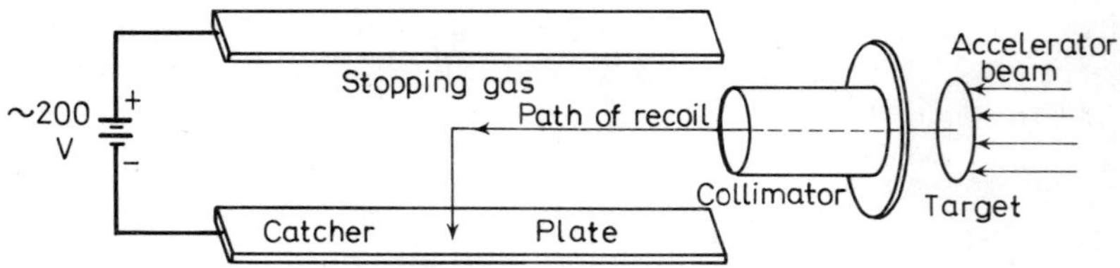

Figure 4.3 Schematic diagram of apparatus for differential range measurements in gases by electrostatic collection. (From Harvey[11], by courtesy of Annual Reviews)

centre of mass. The corresponding range in a material can be calculated, and the measurement of the range of the product nuclei resulting from de-excitation of the compound system should be in agreement with the calculated value. A shorter or a longer range can be interpreted as proof for another mechanism. An extensive survey of such data and their treatment has been made by Alexander[10]. Several ingenious and simple experimental arrangements have been described; only one such arrangement is given here. This relates to range and range distribution measurements, and is shown schematically in Figure 4.3, (after Harvey[11]). It consists of a gaseous 'stopper', the collection of the decelerated nuclei being made on charged plates.

4.3.3 Measurements of emitted particles

The study of the emitted particles has contributed greatly to our knowledge of reaction mechanisms. Now that heavy ions are accelerated and used as complex projectiles, the variety of exit channels is large even for low- and medium-energy nuclear reactions, and the possibility of a wide variety of light products exists. It is necessary, therefore, not only to measure the energy of the emitted particle, but also to check that a particular detected energy does indeed correspond to a particle of mass m and charge Z. Very efficient methods of identification have been developed during the last 10 years. Most of them rely on simultaneous measurement of at least two characteristics of the particle, i.e. kinetic energy and energy loss, or kinetic energy and momentum, or kinetic energy and time of flight.

Because of the progress in fast electronics, it is now possible to measure very short intervals of time, of the order of 10^{-9} s, during which a particle travels only a short distance. Generally, one semi-conductor detector is used for energy measurement and the pulse produced completes the time measurement. The initial timing pulse comes either from an r.f. signal from a pulsed accelerator, or from a thin foil, from which secondary electrons or light are emitted when the particle passes through, and are collected with an electron lens or a photomultiplier. Measurements of the time of flight $t = L/v$ along a path L gives the velocity, and the mass of a particle is obtained from the relationship $m = 2(Et^2/L^2)$. An ambiguity remains, however, regarding the charge Z.

Measurements of both the energy ΔE and the energy loss in an entirely depleted solid-state detector determine a quantity proportional to mZ^2, since an approximation to the Bethe formula shows that the product $E\Delta E$ is proportional to mZ^2. A further ambiguity rests in the relationship $\Delta m/\Delta Z = -2m/Z$, for ${}^{11}_{6}C$ and ${}^{16}_{5}B$, for example, have the same product mZ^2.

Magnetic analysis is now used for measuring the momentum mv of a recoiling particle, the motion of the ionised particle in a magnetic field B being described by the equation $mV = BRz_e$ where z_e is the effective charge of the ion and R the radius of curvature. The combination of a momentum measurement and an energy measurement, made by a detector placed in the focal plane of the magnet, determines the ratio z_e^2/m, which is proportional to $E/(BR)^2$, which still does not resolve the ambiguity since $\Delta m/\Delta Z = +2m/z_e$.

For light particles, protons, deuterons, tritons, helions and even ionised atoms up to O or Ne, the $E,\Delta E$ method is very convenient and detector telescopes are widely used. Improvements in identification may be obtained by combining time-of-flight and $E,\Delta E$ techniques, or by time-of-flight, momentum and E measurements[12].

4.4 COMPOUND NUCLEAR REACTIONS

4.4.1 Definition of the compound nucleus

In the energy region in which we have defined low- and medium-energy nuclear chemistry, the principal reaction mechanism involves the formation and decay of a compound nucleus. Let us recollect what is referred to as a compound state and a compound nucleus. Consider a nucleon entering the nucleus. It can collide with a target nucleon and excite it above the highest stable level of the ground state, i.e. in terms of the Fermi gas model, above the Fermi sea. It is likely that the incident nucleon will also be left above the Fermi level. Both incident and stricken nucleons are left in excited levels as a result of their collision, and thus a hole is created in the ground-state levels. This situation corresponds to a *compound state.* If one or both nucleons happens to leave the nucleus without further interaction, it must have been in an excited state exceeding the separation energy. Such a process is referred to as a *direct reaction.* The time-scale is short, since it must be of the same order as the time required for a nucleon to traverse the nucleus, i.e. of the order of 10^{-22} s for a medium size atom.

Alternatively, the compound state may be followed by further collisions between nucleons prior to particle emission, a process which leads to the excitation energy being distributed more and more widely over the nucleus. As a result of this process, a *compound nucleus* is formed. After the formation of several hole–particle pairs, there is still a chance of particle emission, under pre-equilibrium conditions. However, there is also a tendency towards the achievement of statistical equilibrium. In the latter case an intermediate nucleus is formed, in which the energy of the incident particle is distributed among all possible modes of excitation. This nucleus has a long life-time in comparison to the nuclear transit time. It can pass through many different configurations, and has no memory of its mode of formation. When configurations arise in which a particular nucleon or a correlated group of nucleons (d,α) randomly acquire sufficient energy, the excited nucleus then decays.

The independence of the modes of formation and decay is expressed by the splitting of the expression for the cross-section for a particular reaction into two components

$$\sigma_{a.b} = \sigma_{\text{CN}} \times G_b \tag{4.19}$$

where σ_{CN} is the cross-section for the formation of the compound nucleus and G_b is the probability or branching ratio for decay via a particular channel b.

It therefore follows that calculation of a reaction cross-section for any compound nucleus process is performed in two stages. First, the compound formation cross-section, which in many cases might correspond to the total reaction cross-section, is calculated, followed by the branching ratio for decay via a particular channel, i.e. gamma de-excitation, emission of a given particle at a given energy, etc.

Two cases are of particular interest:

(a) When the excitation energy of the compound nucleus is low – that is to say, less than a few MeV in the case of light elements, and less than a few keV for heavy elements – the levels are well separated, and each level width

Γ is much narrower than the level spacing D, i.e. $\Gamma/D \ll 1$. This is the favoured region for radiative captures, (n,γ) (p,γ) etc., since in most cases, the excited levels are still bound. If the bombarding energy corresponds to a single excited state, whose energy above the ground state of the compound nucleus is the sum of the incident particle binding and kinetic energies, then resonance occurs in the capture cross-section. If the bombarding energy is so low that the compound nucleus has *no states*, then there is no facility for penetration and T_l, the transmission coefficient may be calculated as in equation (4.10) by general definition. Since only *s*-wave ($l = 0$) neutrons can interact at these low energies, the $1/v$ law is predicted.

(b) When the excitation energy increases, the level width at a given energy broadens, because it is the sum of more and more partial widths since the number of channels open increases. In addition, the number of possible excited configurations increases, since with a higher energy new levels of various spins and parities become available. For this reason the level spacing decreases, and when $\Gamma/D > 1$ different levels overlap. The compound nucleus is no longer formed in a single state and the statistical theory may apply to both level densities and particle evaporation.

4.4.2 The reciprocity theorem

The statistical theory predicts the rate of emission of a particle b with channel energy ε from a given compound nucleus CN. The basic theorem for the calculation of the de-excitation process of a compound nucleus is well known as the reciprocity, or 'detailed' balance, principle. It may be used both for deriving the Breit and Wigner formula describing the resonances and for deriving the width of any particular channel in the decay from a distribution of compound nuclear states with a statistical distribution of phases. The probability W_{AB} with which the system A, with energy between E and $E+\mathrm{d}E$ decays to system B, is related to the probability of the time-reversed transition from B to A, $W_{-\mathrm{B}-\mathrm{A}}$, i.e.,

$$\rho_{\mathrm{A}} W_{\mathrm{AB}} = \rho_{\mathrm{B}} W_{-\mathrm{B}-\mathrm{A}} \tag{4.20}$$

where ρ_{A} and ρ_{B} are the densities of the initial and final states of the system. More precisely, A may be regarded as an excited compound nucleus with a density of states $\rho_{\mathrm{CN}}(E^*)$, (i.e. with a phase space factor $g_{\mathrm{CN}} = \rho_{\mathrm{CN}}(E^*)\mathrm{d}E$), W_{AB} being the probability $W(\varepsilon)\mathrm{d}\varepsilon$ of disintegration into states of the residual nucleus Y with energy E^*_{Y} above the ground state by the emission of a particle b of energy between ε and $\varepsilon+\mathrm{d}\varepsilon$. Then $W_{-\mathrm{B}-\mathrm{A}}$ is the probability $P_{\mathrm{inv}}(\mathrm{b,Y})$ that nucleus Y will capture a particle b of energy between ε and $\varepsilon+\mathrm{d}\varepsilon$ and produce a compound state of excitation E^*. $\rho_{\mathrm{B}}\mathrm{d}\varepsilon$ is the phase space factor g_{B} of the system B formed by Y and b. It is related to the number of states available for the outgoing particle in the 'phase space' multiplied by $\rho_{\mathrm{Y}}(E^*_{\mathrm{Y}})$, the density of states of the residual nucleus Y at an excitation energy given by $E^*_{\mathrm{Y}} = E^*_{\mathrm{CN}} - \varepsilon - S$, when S is the binding energy in the unexcited nucleus A of the particle b.

The number of states per unit volume in the phase space may be obtained from the volume of the momentum element $4\pi p^2 dp$, recalling that the elementary cell has a volume h^3. The application of the detailed balance principle gives:

$$W(\varepsilon)\mathrm{d}\varepsilon.\rho_{\mathrm{CN}}(E^*) = P_{\mathrm{inv}}(\mathrm{b,Y}).\rho_{\mathrm{Y}}(E^*_{\mathrm{Y}})(2s+1)\frac{4\pi}{h^3}p^2\mathrm{d}p \tag{4.21}$$

since there are $(2s+1)$ spin states.

Now, $P_{\mathrm{inv}}(\mathrm{b,Y})$, the capture probability for the inverse process, may be expressed in terms of the capture cross-section σ_{inv} and of the velocity v of particle b by the expression

$$P_{\mathrm{inv}}(\mathrm{b,Y}) = \sigma_{\mathrm{inv}}\, v = \sigma_{\mathrm{inv}}.\frac{p}{\mu}$$

Since $p = (2\mu\varepsilon)^{\frac{1}{2}}$ and $\mathrm{d}p = \mu(2\mu\varepsilon)^{-\frac{1}{2}}$, the following well-known equation may be obtained

$$\mathrm{W}\,(\varepsilon)\mathrm{d}\varepsilon = \frac{8\pi\mu\varepsilon\sigma_{\mathrm{inv}}}{h^3}.\frac{(2s+1)\rho(E^*_{\mathrm{Y}})}{\rho_{\mathrm{CN}}(E^*)}\mathrm{d}\varepsilon \tag{4.22}$$

A more general spin-dependent form of this equation may be derived for the decay probability of a compound nucleus of angular momentum J into a residual nucleus of angular momentum j, by considering that any level with angular momentum J has a degeneracy of $(2J+1)$ states since there are $2J+1$ projections on some arbitrary axis. Thus, the phase space factor must be multiplied by $(2J+1)$ on the compound nucleus side and by $(2j+1)$ on the residual nucleus side.

The probability that a compound nucleus of excitation energy E^*_{CN}, spin J, decays to a residual nucleus E^*_{Y} with spin j, after emission of a particle b of energy ε and spin s, is given by:

$$W(\varepsilon,s,J,E^*_{\mathrm{CN}},j,E^*_{\mathrm{Y}})\mathrm{d}\varepsilon = \frac{8\pi\mu\varepsilon}{h^3}\sigma_{\mathrm{inv}}\,(j,s,E^*_{\mathrm{Y}},\varepsilon,J)\frac{(2s+1)(2j+1)}{(2J+1)}\times\frac{\omega(E^*_{\mathrm{Y}},j)}{\omega(E^*_{\mathrm{CN}},J)}\mathrm{d}\varepsilon \tag{4.23}$$

where the ω terms represent the level densities.

If the cross-section for the inverse process is evaluated from the equation (4.23) on the assumption that all reactions lead to compound nucleus formation, the factors $(2J+1)$, $(2s+1)$, $(2j+1)$ cancel and

$$W(\varepsilon,J,E^*_{\mathrm{CN}},j,E^*_{\mathrm{Y}})\mathrm{d}\varepsilon = \frac{1}{h}\frac{\omega(E^*_{\mathrm{Y}},j)}{\omega(E^*_{\mathrm{CN}},J)}\sum_{S=|j-s|}^{j+S}\;\sum_{l=|J-S|}^{J+S} T_l \tag{4.24}$$

However, during recent years, as a result of a large number of studies it has been shown that (a) if the bombarding particle enters with a high orbital angular momentum, the compound reaction cross-section is smaller than the reaction cross-section, and (b) the evaluation of the inverse cross-section may be made with better accuracy using the optical model.

The general derivation of the probability of disintegration of a compound nucleus, taking into account the angular momentum, has been one of the outstanding developments[13, 14] of the statistical theory in the last 2 years.

4.4.3 Compound nuclei at low energies. Resonance. Radiative capture

An interesting application of the detailed balance principle is the development of a relation between the average separation between neighbouring states and the transmission coefficient T_l, for the case of $\Gamma/D \ll 1$.

Setting $l = 0$ for simplicity (other l values give similar results), and considering the disintegration probability into a *definite state* E_Y^* of the residual nucleus, let this particular channel be β. W_{AB} then becomes $\Gamma_\beta/\hbar$, where Γ_β is the partial level width for decay by channel β. Since the compound nucleus has been formed at low energy where levels are separated, $\rho_{CN}(E^*)$ may be expressed as 1/D, where D is the average separation between states (level spacing). $\rho_Y d\varepsilon$ should be put equal to unity since a single state is assumed to be reached, and the reaction cross-section σ_{inv} is given by equation (4.3) putting $l = 0$

$$\sigma_{inv} = \pi\lambda\!\!\!^{-2} T_{l=0} = \pi\frac{\hbar^2}{2\mu\varepsilon} T_{l=0}$$

The reciprocity theorem then gives the simple relation

$$\frac{\Gamma_\beta}{\hbar}\times\frac{1}{D} = \frac{8\pi\mu\varepsilon}{h^3}\times\frac{\hbar^2}{2\mu\varepsilon} T_{l=0}$$

$$\Gamma_\beta = \frac{D}{2\pi} T_{l=0} \tag{4.25}$$

The disintegration probabilities $\Gamma_\beta/\hbar$ are related to the width of the well-known resonances, since the total decay probability may be expressed as $\Gamma/\hbar = (1/\hbar)\Sigma_\beta\Gamma_\beta$. The width Γ of a nuclear level is related to the decay lifetime τ by the relationship $\tau = \hbar/\Gamma$, and numerically

$$\tau = \frac{6.6\times10^{-16}}{\Gamma(\text{eV})}\text{ s}$$

The fact that there is decay, makes it necessary to write the energy $E = E_0 - i\Gamma/2$, in such a way that the time-dependence of the wave function should be of the form $\psi = \exp\left\{-i\left(E-\frac{i\Gamma}{2}\right)\frac{t}{\hbar}\right\}$ so that $\psi\psi^*$ may decay with time. Since the absolute square of the Fourier transform is equal to $1/[(E-E_0)^2+(\Gamma/2)^2]$, the functional form of the resonance over a given level is $f(E) = C/((E-E_0)^2+(\Gamma/2)^2)$.

Since the reaction cross-section ($\sigma_{l=0} = \pi\lambda\!\!\!^{-2} T_{l=0}$) varies smoothly with energy, the resonance behaviour must be superimposed upon it. In order to conserve the expression of $\sigma_{l=0}$ over a large number of resonances with average spacing D, it is necessary to write

$$\sigma_{l=0} = \pi \lambda\!\!\!^{-2} T_{l=0} \times f(E) \tag{4.26}$$

and

$$\frac{1}{D}\int_{-\infty}^{+\infty} f(E)\mathrm{d}E = 1 \tag{4.27}$$

The constant C can be evaluated from equation (4.27):

$$\mathrm{C} = \frac{\Gamma D}{2\pi} \tag{4.28}$$

Combining equations (4.25), (4.26), (4.27) and (4.28) gives

$$\sigma_{l=0} = \pi \lambda\!\!\!^{-2} \frac{\Gamma_\beta \Gamma}{(E-E_0)^2 + \left(\frac{\Gamma}{2}\right)^2} \tag{4.29}$$

Taking $\Gamma_\beta = \Gamma_0$, the partial width for particle emission leaving the residual nucleus in the ground state and identical to the target nucleus, the expression for the resonance reaction cross-section for the *s* wave becomes

$$\sigma_{l=0} = \pi \lambda\!\!\!^{-2} \frac{\Gamma_0 \Gamma}{(E-E_0)^2 + \left(\frac{\Gamma}{2}\right)^2} \tag{4.30}$$

The process whereby the compound nucleus decays to *its own ground state* via γ-emission, called radiative capture, can now be calculated since the branching ratio G_γ is related to the emission width by the relationship $G_\gamma = \Gamma_\gamma/\Gamma$.

With the inclusion of target and projectile spins the Breit–Wigner single-level formula is obtained for the case of radiative neutron capture.

$$\sigma_{(\mathrm{n},\gamma)} = \frac{(2J+1)}{2(2j+1)} \pi \lambda\!\!\!^{-2} \frac{\Gamma_0 \Gamma_\gamma}{(E-E_0)^2 + \left(\frac{\Gamma}{2}\right)^2} \tag{4.31}$$

since $2s+1 = 2$ for the neutron.

4.4.4 Radiative neutron, capture and gamma-ray spectroscopy

Radiative capture occurs with large cross-sections, especially when they are induced by neutrons, since at energies up to a few keV the only channel open to neutron emission is the incident channel Γ_0, while by contrast γ-ray de-excitation can populate a large number of states. In recent years, neutron capture has been studied by means of γ-ray spectroscopy since the high excitation results in complex spectra[15]. The introduction of the lithium-drifted germanium detector has improved the resolution of the gamma spectra measured. In addition, prompt coincidence techniques have been used to resolve individual transitions and cascades.

Compound nucleus theory predicts entirely random amplitudes for γ-rays. This has been observed[16] for a large number of cases although some departures have been reported for nuclei like ^{104}Rh, ^{142}Pr, ^{176}Lu and ^{239}U [17]. Non-statistical populations in resonances suggest the possibility of some direct capture effects which are at present under study.

4.4.5 Recent experimental results relating to the compound nucleus hypothesis

At incident neutron energies of a few keV, overlapping of the levels occurs, i.e. $\Gamma/D > 1$, and the statistical region is reached. The basic feature of compound nuclear reactions is that the life-time of the excited nucleus is sufficiently long prior to decay that all correlation between the entrance and exit channels is lost. The independence hypothesis was first tested by means of the well-known experiments of Ghoshal[18] utilising excitation function measurements for reactions involving the formation of a particular compound nucleus in two different ways. More sensitive verifications of the independence hypothesis have been obtained recently from measurements of angular distributions and the energy spectra of the evaporated particles[19, 20].

The concept of complete independence of the entrance and exit channels is illustrated by considering a compound nucleus CN, at excitation energy E^*, which can be formed by two reaction systems, and which can decay through two or more distinguishable exit channels, i.e.

$$\begin{matrix} a+A & \searrow & & \nearrow & b+B \\ & & CN & & \\ a'+A' & \nearrow & & \searrow & b'+B' \end{matrix}$$

B and B′ being residual nuclei, which in the Ghoshal-type experiments were used to identify the exit channels. As an example, Porile *et al.*[21] have measured the ratio $\sigma(\alpha,n)/\sigma(\alpha,2n)$ and $\sigma(p,n)/\sigma(p,2n)$ in the case of the system

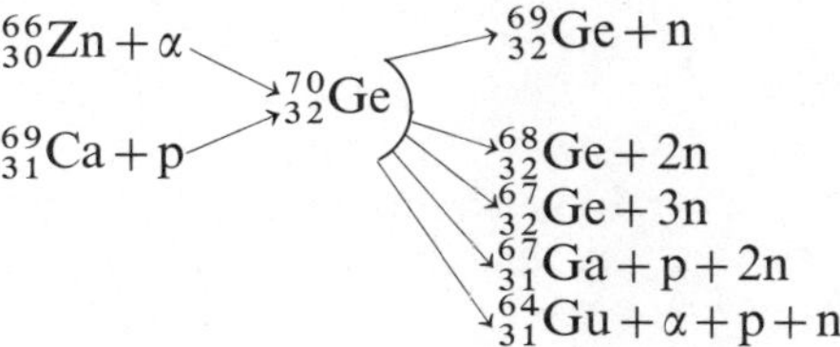

by measuring the cross-section for the residual nuclei. The quantity that is measured is the *integral* of the differential cross-sections for the emitted particles leading to a particular radioactive product, ^{69}Ga, ^{68}Ge, etc. Since the residual nuclei B and B′ are not generally left in their ground states, and the particles b and b′ are emitted with a spectrum of energies, it would appear preferable to verify the independence hypothesis by observing the energy spectra at several angles, i.e. $d^2\sigma/d\Omega d\varepsilon$. Miller *et al.*[19] have compared the energy spectra and angular distributions of emitted α-particles and protons from the reaction system ^{63}Cu + ^{12}C and ^{59}Co + ^{16}O, both leading to the compound nucleus ^{75}Br at the same excitation energy and with the same angular momentum. It was found that the relationship $\sigma(^{12}C,p)/\sigma(^{12}C,\alpha) = \sigma(^{16}O,p)/\sigma(^{16}O,\alpha)$ applies for all parts of the energy spectrum, and also that

$$\frac{\sigma(^{12}C,p)}{\sigma(^{16}O,p)} = \frac{\sigma(^{12}C,\alpha)}{\sigma(^{16}O,\alpha)}$$

It follows, therefore, that if careful attention is given to matching not only the excitation energy but also the angular momentum distributions of the compound nuclei formed by two different entrance channels, then the emission of particles conforms to the independence hypothesis. On the other hand, if the orbital angular momentum differs[20] for two different entrance channels, then the requirement for the conservation of angular momentum leads to differences in the shapes of the spectra of the emitted particles and in the angular distribution. The study of neutron emission from the ^{64}Zn compound nucleus formed in the two ways, $p+{}^{63}Cu$ and $\alpha+{}^{60}Ni$, has shown[22] that the angular distribution is essentially isotropic for the $p+{}^{63}Cu$ system, while an average anisotropy of 1.4 was observed for $\alpha+{}^{60}Ni$. The decay of ^{64}Zn was precisely studied in Goshal's experiment. The observed effect is due to the difference of angular momentum.

At the present time, although many experiments have been devised and many arguments against the independence hypothesis have been presented, the results strengthen the conviction that the only 'memory' that a compound system has is that of the constants of the motion, i.e. momentum, angular momentum, energy, parity and nuclear number.

4.4.6 Particular aspects of compound nucleus formation by heavy ion projectiles

Heavy ions are interesting probes for compound reactions. Bombardment of a target nucleus with heavy ions having from 5 to 10 MeV per nucleon kinetic energy is the best method for producing compound nuclei with excitation energies of 50–200 MeV which has been devised in recent years. The mean free path in a nucleus for such particles is short and complete absorption is expected. Excitation to such energies by lighter projectiles, such as protons or α-particles, is accompanied by an important contribution from reactions proceeding by direct interaction. These produce a large variety of nuclei of unknown energy that subsequently de-excite in various ways. The situation is much simpler, in principle, for heavy ion reactions, since there is only a single excitation energy and only a single class of compound nuclei.

However, some complications arise because of the high values of the angular momentum. The heavy mass of the projectile brings to the compound nucleus substantial orbital angular momentum, the maximum being

$$l(l+1)\,\hbar = \sqrt{2\mu(\bar{\varepsilon}-B)}\,(R_1+R_2)$$

For example, a 100 MeV carbon ion incident on gold is calculated to produce ^{209}At nuclei with up to about 55 units of angular momentum. By calculating the transmission coefficients T_l according to the optical model, the distribution of angular momenta, due to Miller *et al.*[19], is obtained as shown in Figure 4.4 for $^{16}O + ^{59}Co$ at 60 MeV.

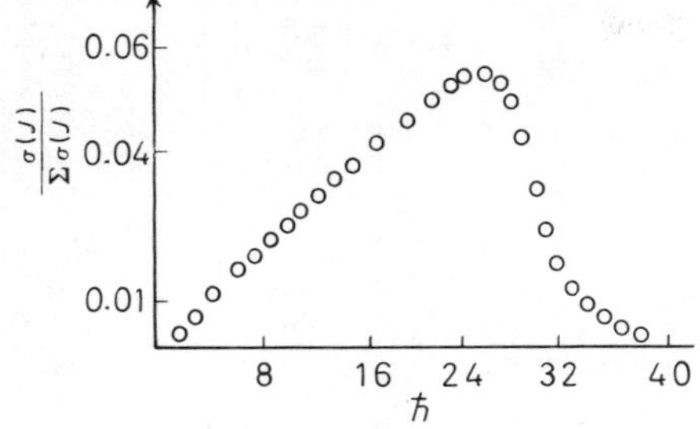

Figure 4.4 Angular momentum distribution for the reaction ^{16}O on ^{59}Co. (From Miller[19], by courtesy of American Institute of Physics)

From equation (4.3), it follows that high values of J are predominant in the contribution to the cross-section. However, these high values introduce some complications. For example, in some cases the lowest level with a particular momentum $J \# l(l+1)\hbar$ is at a higher energy than the excitation energy of the compound nucleus. Thus, the bombardment of Co by oxygen$^+$ ions at 110 MeV produces the compound nucleus ^{75}Br with an excitation energy of 88 MeV, and the highest value of J is about 66$\hbar$. The simplest relationship between the energy and the angular momentum J is given by the rotational formula,

$$E_J = \frac{J^2}{2\zeta} = \frac{l(l+1)\hbar^2}{2\zeta} \tag{4.32}$$

where ζ is a moment of inertia and E_J the energy for a spin J. This is the lowest energy for a particular spin J since it is assumed that for a non-deformable drop all the excitation is rotational. Assuming that the moment of inertia is that of a rigid body, the energy of this so called 'Yrast level' (Grover[23]) can be estimated. In the case of ^{75}Br, 56 units of $\hbar$ correspond to 88 MeV for E_J. For this reason, no compound nuclei of energy greater than 56 $\hbar$ may be formed, and incident heavy ions with angular orbital momenta between 66 and 56 $\hbar$ do not contribute to the compound nucleus cross-section. A variety of models have been proposed, based on a deformable liquid drop, by considering single particle states or other possibilities, in order to give a more accurate estimate of E_J[24, 25]. In Figure 4.5 Yrast energies, calculated by Thomas[13] for 20 and 60 $\hbar$, are illustrated.

Attempts to ascertain the fraction of nuclear reactions that lead to compound nucleus formation have been based on angular distributions of charged particles, measurements of the momentum deposited in the product nuclei and direct measurements of the cross-section for the recoiling products formed after complete fusion. A large amount of work has been done during the last couple of years and the status of such measurements has been discussed by Thomas[13], Natowitz[26] and Lefort[27]. Theoreticians have also built models for calculating the critical value of the angular momentum above which the existence of the compound nucleus should not be possible. Kalinkin and Petkov[28] have shown that a deformed liquid drop if of ellipsoidal shape has limited eccentricity because of the surface energy of the nucleus.

The calculation of the total energy of the system (surface energy, volume binding energy and rotational energy) shows that there is a limit to the rotational energy and therefore the angular momentum. If l_c is the critical

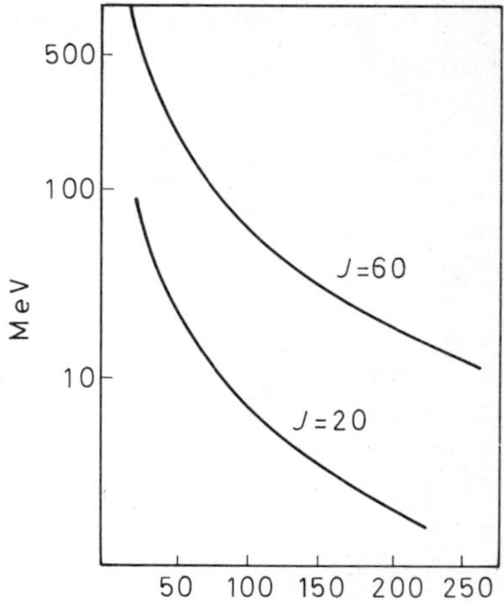

Figure 4.5 Yrast energies calculated for angular momentum 20 and 60 $\hbar$ plotted against mass number. (From Thomas[13]. by courtesy of Annual Reviews)

value, then from the sharp cut-off model assumption, the cross-section for complete fusion is given by

$$\sigma_{CF} = \pi\lambda\!\!\!^{-2} \sum_{l=0}^{l_c} (2l+1) \cong \pi\hbar^2 \frac{l_c^2}{2\mu\varepsilon} \tag{4.33}$$

On the experimental side, measurements of complete fusion cross-sections have been used to derive values of the critical momentum l_c, and a comparison between σ_{CF} and σ_R (reaction cross-section) gives

$$l_c = \left(\frac{\sigma_{CF}}{\sigma_R}\right)^{\frac{1}{2}} l_{max} \tag{4.34}$$

Examples of experiments of this nature are those concerned with the fact that in the oxygen ion bombardment of uranium nearly all interactions lead to fission. Sikkeland[29] and Viola have measured the coincidence rate between two detectors as a function of angle. Since the angles at which detectors should be placed for observing the two fragments depends on the momentum of the fissioning nucleus, it becomes possible to obtain the recoil energy, and therefore the mass of the nucleus, before fission. Sikkeland *et al.* found that a large fraction of the events corresponded to full momentum transfer, but that an appreciable fraction results from a transfer reaction in which, instead of ^{16}O, an α-particle induces fission in the uranium target.

There have also been measurements[30] of the yield of non-compound processes, i.e. transfer reactions on medium A targets. By subtraction from the total reaction such measurements lead to values of l_c.

At present the following statements can be made,

(a) σ_{CF}/σ_R goes through a maximum when the bombarding energy is increased, i.e. when the orbital angular momentum increases,

(b) σ_{CF}/σ_R decreases when the projectile becomes heavier.

Both results clearly indicate that a critical value is reached, as shown in Figure 4.6, which has been constructed from the results of many experiments[26, 29–31] and theoretical calculations[28]. It has been shown that direct interactions take place at the expense of high *l* waves in the incoming beam. Recently, Lefort *et al.*[32] have shown that for high *l* waves, corresponding to grazing trajectories, there is the possibility of the existence of a short-lived rotating system which disintegrates before a complete rotation.

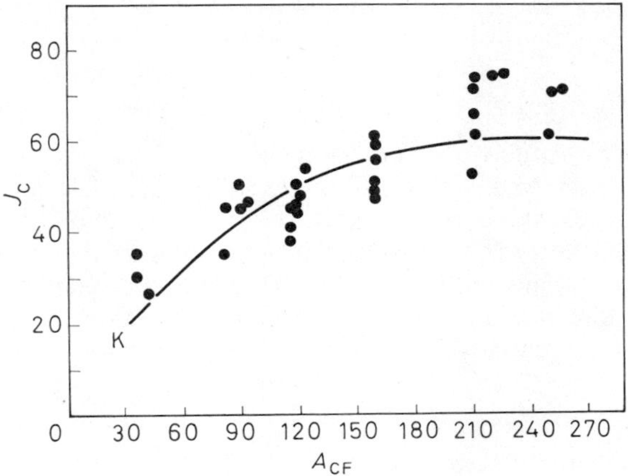

Figure 4.6 Critical angular momentum as a function of the mass of the fusion nucleus. Closed circles are experimental determinations. (The curve calculated by Kalinkin[28])

The fact that the ratio of complete fusion cross-section to reaction cross-section decreases very appreciably when the mass of the projectile increases, is of importance in attempts to synthesise new heavy elements by bombardment with heavy projectiles[27]. As an example, it is quite easy to calculate that the angular momentum brought to a compound nucleus produced by the fusion of a Kr ion and a Th target is already 81 $\hbar$ for an excitation energy of 90 MeV, i.e. just above the Coulomb barrier. It reaches 150 $\hbar$ for ^{186}W bombarded by heavy ions around mass 130 and accelerated to 700 MeV in order to overcome the Coulomb barrier. A compilation of cross-sections measurements for compound nuclei produced in the region of the rare earths by carbon, oxygen, neon, argon and, very recently, krypton beams has shown that there is a drastic drop in the ratio σ_{CN}/σ_R: at 30 MeV above the Coulomb barrier; it is around 0.8 for ^{12}C, 0.2 for ^{40}Ar and less than 0.1 for ^{84}Kr.

4.4.7 De-excitation of a compound nucleus

The rate at which particles are emitted from the compound nucleus depends on the nuclear level density, as is shown in equations (4.22) and (4.23). A comprehensive survey of the variation of level density with excitation energy and angular momentum was given in 1960 by Ericson[33]. Most analyses at high energies have been based on Fermi gas models, and explicit derivations of the level density have been presented in many papers. A statistical thermodynamic derivation of the formula for the state density can be made on the basis of the partition function of a system consisting of a number of quantum states. The relation between the level density $\omega(E)$ and other thermodynamic variables, i.e. the temperature T and entropy S is given by

$$\omega(E) = \frac{\exp(S)}{\left(2\pi T^2 \frac{dE}{dT}\right)^{\frac{1}{2}}}$$

The nuclear temperature τ is defined as $\frac{d}{dE}(\ln \omega(E)) = \frac{1}{\tau} \simeq \frac{dS}{dE}$; τ and T approach the same value with increasing excitation energy.

In the equidistant spacing model, which is the simplest model devised, all the single particle states have a constant energy spacing. Many sophisticated expressions for $\omega(E)$ as a function of the nuclear temperature and of the excitation energy have been derived. A simple approach is possible if it is remembered that the thermal heat capacity, $c = dE/d\tau$, is proportional to the temperature, and therefore the energy is proportional to τ^2,

$$E = a\tau^2 \tag{4.35}$$

a being the level density parameter.

Then $$\frac{d}{dE}(\text{Ln}\,\omega(E)) = \sqrt{\frac{a}{e}}$$

and $$\omega(E) \sim \exp 2\sqrt{aE} \tag{4.36}$$

which gives the general behaviour of the density of states per unit energy internal.

Ericson[33] has obtained the expression

$$\omega(E) = \frac{1}{12}\left(\frac{\pi^2}{a}\right)^{\frac{1}{4}} E^{-\frac{5}{4}} \exp(2\sqrt{aE}) \tag{4.37}$$

in a more exact derivation of equation (4.36).

For free nucleons confined within a nucleus of radius $R = r_0 A^{\frac{1}{3}}$, the parameter a is given by

$$a = 2\left(\frac{\pi}{3}\right)^{\frac{4}{3}} \frac{mr_0^2}{\hbar^2} A \tag{4.38}$$

where m is the mass of a nucleon. For $r_0 = 1.2$ f, a $= A/13.5\ \text{MeV}^{-1}$, and for $r_0 = 1.5$ f, a $= A/8.7\ \text{MeV}^{-1}$.

The level density $\omega(E)$ describes the density of states at a given excitation energy E, summed over all values of angular momentum J of the excited nucleus. If the angular momentum is assumed to be *randomly* coupled, the density of states is expressed by the equation:

$$\omega(E,J) = \frac{(2J+1)}{2\sigma^2\sqrt{\pi}}\exp(-J(J+1)/2\sigma^2)\omega(E) \tag{4.39}$$

The parameter σ^2 is the mean square projection of the nuclear angular momentum on a fixed axis (spin cut-off parameter). For a free Fermi gas model, σ^2 is related to the rigid-body moment of inertia: $\sigma^2 = \zeta_0\tau/\hbar^2$, where $\zeta_0 = (2/5)\mu R^2$. With this identification, $\exp(-J(J+1)/2\sigma^2)$ equals $\exp(-E_{rot}/\tau)$ where E_{rot} is the classical rotational energy. This can be interpreted approximately as representing a replacement of the actual excitation energy E by a smaller value, i.e. by $E-E_{rot}$ which is equivalent to saying that the level density for spin J and energy E is the same as the level density independent of spin at the lower energy $E-E_{rot}$.

A widely used form can be obtained by expanding the logarithm of the level density function in a Taylor's series. The maximum excitation energy left in the residual nucleus is $E^*_{y(max)}$; consider an evaporation process with channel energy ε for emitted particles. Taylor's expansion can be performed on the logarithm:

$$\ln\omega[E^*_{y(max)}-\varepsilon)] = \ln\omega(E^*_{y(max)}-\varepsilon\left(\mathrm{d}\frac{\ln\omega}{\mathrm{d}E}\right)_{E-E_{max}} + \ldots$$

Recalling the definition of τ and making the substitution $E_y = E_{y(max)}-\varepsilon$,

$$\omega(E) = \mathrm{C}eE_y/\tau = \mathrm{C}\exp\left[\frac{E_{y(max)}-\varepsilon}{\tau}\right]$$

$E_{y(max)}$ has a fixed value and therefore $\omega(E)\infty\exp\left(-\frac{\varepsilon}{\tau}\right)$.

Since $\int W(\varepsilon)\mathrm{d}\varepsilon = 1$ for a particle b, it is easy to show that

$$\int_0^\infty \varepsilon e^{-\varepsilon/\tau}\mathrm{d}\varepsilon = 1/\mathrm{C}$$

and therefore $\mathrm{C} = 1/\tau^2$. The equation giving the differential cross-section, i.e.

$$\frac{\mathrm{d}\sigma}{\mathrm{d}\varepsilon} = \sigma_{inv}\frac{\varepsilon}{\tau^2}\exp\left(-\frac{\varepsilon}{\tau}\right)\mathrm{d}\varepsilon \tag{4.40}$$

is one of the most familiar expressions for the energy spectrum of the emitted particles.

There have been a great number of experiments to determine energy distributions of emitted particles. From the results the function $((\mathrm{d}^2\sigma)/\mathrm{d}\varepsilon\mathrm{d}\Omega)/\sigma_{inv}\varepsilon$ is usually plotted as a function of either ε, the kinetic energy of the emitted particle, or E_y^*, the excitation of energy of the residual nucleus ($E_y^* = E_{y(max)}^* - \varepsilon$). Figure 4.7 presents an example[34] in which σ_{inv} was calculated by the optical

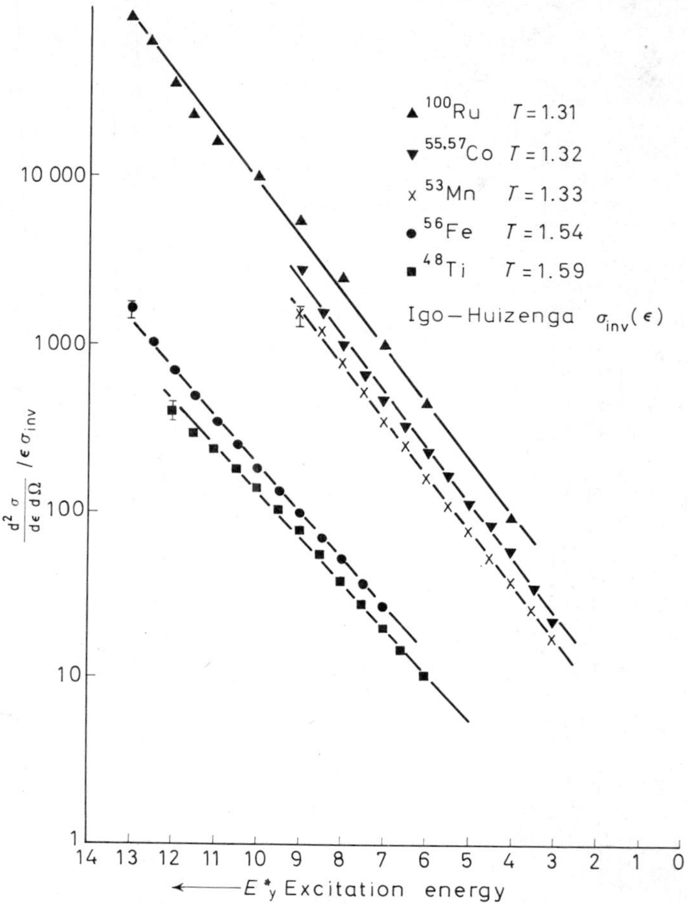

Figure 4.7 Relative level density as a function of excitation energy calculated from (p.α) spectra. The inverse cross-section σ_{inv} (ε) are from an optical model calculation of Igo and Huizenga. (From Sherr and Brady[34], by courtesy of the American Institute of Physics)

model. From the slope of the lines, the nuclear temperature may be obtained and, assuming $E = a\tau^2$, it is possible to obtain the level density parameter a.

In order to estimate the *absolute* level density, Huizenga[35] has proposed that the level density be normalised in a fixed low energy range where the number of levels is known by high resolution spectroscopy. Typical results have been obtained for the α-particle spectrum emitted in the $^{59}Co(p,\alpha)^{56}Fe$ reaction[36].

Knowledge of the level density is most important in the statistical approach to the compound nucleus mechanism. This is why several models have been proposed, e.g. the modified Fermi gas model, the superconductor model, the Rosensweig model, etc. Vonach *et al.*[37] have, for example, applied the superconductor model to the determination of a nuclear state density function. The basic hypothesis is the assumption of a residual interaction between pairs of nucleons in an unfilled shell. This suggestion is made by analogy with a superconducting metal for which a residual interaction exists between pairs of electrons of equal but opposite momentum.

4.4.8 The life-time of the compound nucleus

At low excitation energies, the widths of resonances can be measured and hence the life-time may be deduced. This is of the order of 10^{-14} s. In the statistical region, it is possible to estimate the life-time from the rate of decay by emission of neutrons with energy ε, and then integrating over the neutron energy between 0 and $E^*_{CN} - S_n = \varepsilon_{max}$,

$$\theta = \frac{1}{W(\varepsilon_n)} = \frac{\pi^2 \hbar^3}{(2s+1)\mu} \frac{\omega(E^*_{CN})}{\int_0^{E^*_{CN}-S_n} \varepsilon \sigma_{inv} \omega(E^*_y) d\varepsilon} \tag{4.41}$$

Using equation (4.41), a rough approximation may be made by putting $\sigma_{inv} = \pi R^2$ and then evaluating the integral, having expressed $\omega(E_{CN})$ and $\omega(E^*_y)$ by means of equation (4.36) in which the level density parameter is $a = A/10$. The life-time is then given in seconds by

$$\theta = 2.2 \times 10^{-21} \cdot \frac{A^{\frac{1}{3}}}{E_{CN} - S_n} \exp\left(\frac{S_n}{\tau}\right) \tag{4.42}$$

For $A = 50$ excited to 40 MeV, S_n of the order of 10 MeV and a nuclear temperature of 2 MeV, θ is found to be around 4×10^{-20} s, i.e. more than 500-times the duration of a collision.

There are many very rough approximations in this evaluation and for checking the statistical theory a direct measurement of θ would be most desirable. In particular, it would be very interesting to obtain data on the level density, on the moment of inertia at various excitation energies, on the pairing effect and closed shell effect. Nowadays, such a measurement is possible, in the range of $10^{-16} - 10^{-19}$ s with the help of solid state physics. This is the shortest lifetime measured directly by any technique and it seems worthwhile to give some details. When protons or α particles are emitted by a nucleus placed in a normal lattice position in a single crystal, there is a strong hindrance for the directions parallel to the low index atomic planes or axes. This 'blocking' effect is due to the coulomb repulsion induced by the rows of atoms along the low index axial directions. This has been described in detail by Lindhard's theory[38], assuming an electrostatic potential which scatters the particles. If a detector is placed exactly along an axial direction at a large distance from the crystal, very few particles are observed. Therefore a measurement of the particles emitted over all angles shows shadows corresponding to the main axes of the crystal. The depth of the shadow in an axial direction has been measured by counting α particles or protons elastically scattered from the beam. Its width is large enough to be observed with photographic plates or with position sensitive detectors (of the order of 1 cm if the detector is around 1 meter from the crystal). Then, if the point of origin of the emitted particles is displaced from the atomic rows, the emergent yield distribution is changed and the dip partly disappears. If the displacement arises from recoil of a compound nucleus formed between a lattice atom and a particle incident on the crystal, the recoil energy and the recoil velocity perpendicular to the row are known. Then a relation exists between this velocity, the decay of the compound nucleus (for the emission of a particle) and the minimum yield χ in the dip. If r_c is the effective cut-off of the inter-atomic potential, the contribution of the particles emitted is proportional to $\exp(-r_c/v\theta)$ where v is the perpendicular velocity and θ the decay half-life of the compound nucleus.

Applying this method to the detection of fission fragments emitted from ^{238}U bombarded by 12 MeV protons, Gibson and Nielsen[39] have established a value of $2.5.10^{-16}$ s for the partial lifetime of ^{238}Np compound nuclei at an average excitation energy of 7 MeV. But de-excitation through a fission process is a special case. Very recently the method has been applied[40] for the evaporation of protons from the compound nucleus of ^{71}As produced by bombardment of a germanium crystal by protons. For an excitation energy of 10 MeV, a half-life of 2.10^{-17} s was measured, corresponding to a width $<\Gamma> = 34$ eV. The level density parameter a which might be derived from this value was around 11 MeV^{-1}, i.e. A/6.5. Another experiment was made with a silicon crystal bombarded by ^{12}C heavy ions in order to produce a recoiling ^{40}Ca compound nucleus and to measure the evaporation of protons.

4.4.9 Effects of high angular momenta on the de-excitation of compound nuclei

During the last 2 years, one of the main refinements of the statistical theory has been to study the effect of high spins. As has been noticed already, heavy ions beams are a very well adapted tool for producing compound nuclei, and the theory relates the angular distribution to the angular momentum of the compound nucleus J, the orbital angular momentum of the compound nucleus J, the orbital angular momentum of the emitted particle l, and the spin cut-off parameter σ^2. The prediction is a symmetrical angular distribution peaking at 0 and 180 degrees to the beam, which can range from $P(\theta) = 1$ (isotropy) to $P(\theta) = 1/\sin\theta$. A recent example[41] is given in Figure 4.8 which shows a comparison of the $^{56}Fe(\alpha,p)^{59}Co$ and $^{59}Co(p,p')^{59}Co$ angular distributions. The first shows a much more pronounced anisotropy due to the larger angular momentum brought by alpha particles. This is because the orbital angular momentum of the incident particle is perpendicular to its direction of motion. Since the direction of the emitted particle is perpendicular to its angular momentum, it follows that preferential emission occurs along the beam axis. A detailed calculation has been performed by Ericson and Strutinski[49] and is described in many reviews. For $Jl/\sigma^2 \ll 1$, so-called weak coupling,

$$W(\theta) = 1 + \frac{\bar{J}^2 \cdot \bar{l}^2}{8\sigma^4} \cos^2\theta \tag{4.43}$$

and for $(Jl/\sigma^2) \gg 1$, strong coupling

$$W(\theta) = \frac{1}{2\pi^2 \sin\theta}$$

The weak coupling case is the more commonly encountered, since the product nucleus has usually kept a high angular momentum as compared to l. By evaluating $\bar{l}^2 = 2\tau\mu_e R^2/h^2$, one derives from (4.43):

$$W(\theta) = 1 + \frac{\mu_e E_{rot}}{4\tau} R \cos^2\theta \tag{4.44}$$

where it is interesting to notice that the anisotropy increases with the mass μ_e of the emitted particle.

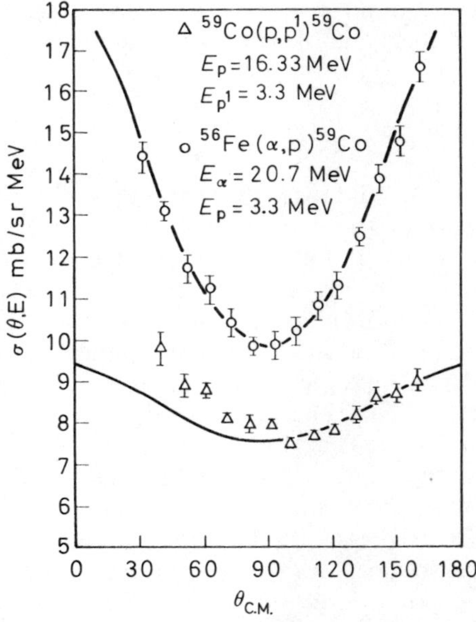

Figure 4.8 Comparison of $^{56}Fe(\alpha,p)$, ^{159}Co and $^{59}(Co(p,p')$ ^{59}Co angular distributions. (From Benveniste *et al.*[41], by courtesy of American Institute of Physics)

Another property of high-spin compound nuclei has been observed in de-excitation sequences where γ-ray emission competes with particle emission in the last step of the process. The effect has been considered in detail by Grover, Grover and Nagle[43] and Gilat and Grover[44]. In these de-excitation sequences, the shapes of excitation functions are affected and the values of level density parameters required to fit the data have to be changed. The explanation lies in the fact that most of the excitation energy removed by neutrons or protons is in the form of binding energy and that the angular momentum carried away by them is small. This implies that the angular momentum distribution of de-excited nuclei would not differ very much from that of the compound nucleus. It has, however, been shown that high spins cannot be obtained for low energies (Yrast levels). Hence, the decay should proceed via neutrons with large l values, and consequently low probabilities, since T_l is small for large $l\hbar$. On the other hand, a cascade of dipole or quadrupole photons may remove a large amount of angular momentum without carrying away too much energy. There are well-known experiments which show the effect of this competition between γ-rays and neutrons. Direct measurements of the γ-rays have been made by Mollenauer[45] in the case of the system $^{165}Ho + ^{12}C$.

An indirect proof that all the excitation energy is not lost by particle emission has been found in the study of excitation functions for (HI,xn) reactions. An extensive body of results by Alexander and Simonoff[46] is often quoted. It shows that the excitation functions for the production of isotopes 149 to 151 of dysprosium by reactions like (^{12}C, xn), ($^{16}0$,xn), (^{20}Ne,xn), x ranging between 5 and 9, peak at much higher excitation energies than would be expected if neutron emission were the only mode of de-excitation. The authors found that the available energy per neutron,

$$\left(E^* - \sum_{i=1}^{x} S_i\right)/x$$

is around 6 MeV, i.e. around twice the average kinetic energy. The conclusion is that a large part of the excitation energy appears as electromagnetic radiation. Results showing essentially the same features have been obtained[47] very recently with argon ions on dysprosium, and the same value of 6 MeV per neutron was derived from excitation functions of polonium isotopes.

Another fundamental effect of angular momentum is the enhancement of the evaporation of heavy particles, mainly α-particles, as compared to neutrons and protons. This is because the release of high angular momentum is easier when the kinetic energy is removed by a heavy mass. The average change in angular momentum Δ has been calculated[48] as a function of the mass of the emitted particle and also the width ratio Γ_α/Γ_n. It should also be noted that when the excitation energy approaches the Yrast level, the emission of α-particles becomes very important. Hence the behaviour of the ratio Γ_α/Γ_n as a function of the spin of the compound nucleus shows a very strong enhancement for a particular range of $\hbar l$ values, depending on the excitation energy, as shown in Figure 4.9 at 60 $\hbar$ for the system Cd + ^{14}N excited to 50 MeV.

Many details of the theoretical aspects of this question may be found in a series of paper by Thomas *et al.*[13, 48]. The point which is emphasised here is one of practical interest. When the energy of separation of an α-particle becomes low or positive (in the region of lead), nuclear reactions induced by heavy ions (and therefore with high angular momenta), and proceeding through a compound nucleus mechanism, produce rather large yields of

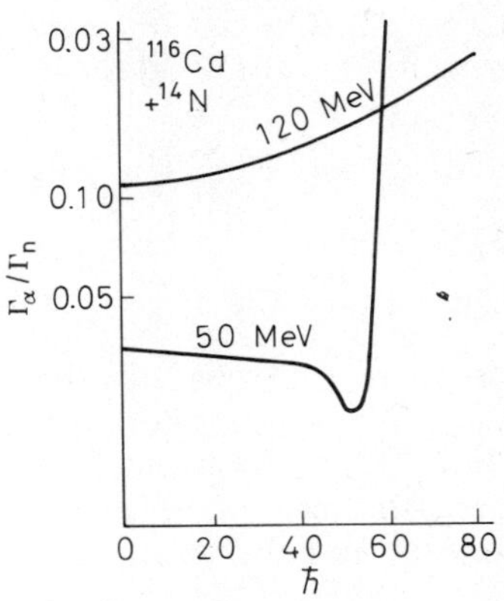

Figure 4.9 Alpha/neutron emission ratios for (^{116}Cd+^{14}N) reaction as a function of angular momentum. (From Williams and Thomas[48], by courtesy of North-Holland Publ. Coy.)

α-particle emission as compared to neutron emission. In these circumstances the (HI,αxn) reactions compete with (HI,xn) reactions. For example, the bombardment of $^{209}_{83}$Bi by ^{12}C produces nearly as many francium isotopes decaying from the compound nucleus $^{221}_{89}$Ac by prompt α-particle emission as light actinium (^{218}Ac and lighter ones) which also lead to francium daughter nuclei by decay. The same effect is observed for the bombardment of heavier targets like ^{242}Pu or ^{244}Cm where the yield for complete fusion followed by neutron emission is decreased both by α-particle competition and by fission. The same effect is expected to occur in the synthesis of superheavy elements, and some advantage might be taken of this since the complete fusion nucleus seems to be neutron deficient. If, for example, the compound nucleus $Z = 126$, $A = 316$ is synthesised by the reaction $^{84}_{36}$Kr + $^{232}_{90}$Th, prompt α-decay competing with neutron evaporation will increase the ratio N/Z up to a value closer to β stability and to the 'magic' shell at $Z = 114$. Residual nuclei with $A = 304$ or 308 and $Z = 120$ or 122 might be produced by such an 'overshoot' method[27].

4.4.10 Isomer ratios

A fairly large number of determinations of the cross-sections for the production of a low-lying metastable state and the ground state of a particular nuclide have been based on isomer ratio measurements. The *isomer ratio* σ_H/σ_L is defined as the ratio of the cross-section in the high-spin state to that of the low-spin state. For example, such a ratio has been measured for the ^{137}Ce ground ($J = 3/2$) and metastable states ($J = 11/2$), and for ^{214}Fr, ^{196}Au, and ^{197}Hg etc. The relative probability of forming a residual nucleus of spin J_Y, starting with a compound nucleus of spin J_{CN} is given by a simplified form of equation (4.24) when J_{CN} is large in comparison to S:

$$P(J_Y) = C.\omega(J_Y,E_Y) \sum_{J_{CN}-J_Y}^{J_{CN}+J_Y} T_l \tag{4.45}$$

where $$\omega(J_Y,E_Y) = 2(J_Y+1)\omega(E_Y)\exp\left[-\frac{J_Y(J_Y+1)}{2\sigma^2}\right]$$

The ratio $P(J_{YH})/P(J_{YL})$ for the two isomeric states is then governed by

$$\frac{P(J_{YH})}{P(J_{YL})} \propto \frac{J_{YH}+1}{J_{YL}+1}\exp\left[-\frac{J_{YH}^2-J_{YL}^2+J_{YH}-J_{YL}}{\sigma^2}\right] \# \frac{J_{YH}}{J_{YL}}\exp-\left[\frac{J_{YH}^2-J_{YL}^2}{\sigma^2}\right] \tag{4.46}$$

from which it can be seen that the high-spin state is favoured. The value of $\sigma^2 = \zeta\tau/\hbar$ has a considerable influence since it may enhance or attenuate the ratio J_{YH}/J_{YL} appreciably, depending on whether σ^2 is small or large. From the measurement of the isomer ratio, it might be possible to derive information about σ^2 and the moment of inertia ζ. Some tentative efforts have been made in this direction, in most cases at rather high excitation energies, where an effective nuclear moment of inertia approximately 0.5–0.7 as large as the rigid body value has been found.

4.4.11 Deviations from the statistical theory. The pre-equilibrium decay model

Predictions of the statistical theory have been verified in many experiments. However, some deviations appear for high excitation energies; these may be summarised essentially as being the high energy tails of particle spectra and of excitation functions. A typical example[49] is shown on Figure 4.10 for the reaction $^{197}Au(\alpha,2n)^{199}Tl$. At a bombarding energy larger than 40 MeV, the excitation of the compound nucleus ^{201}Tl is high enough for the emission of three neutrons, and the probability of evaporation of only two neutrons leaving ^{199}Tl excited to bound states should be very small. However, there is still a large fraction of (α,2n) cross-section. Other similar results have been found on (α,xn) reactions involving lead and vanadium, and also for p,xn and (p,p′,xn) involving cobalt, thulium and bismuth.

It is very striking to observe, for example, that only half the reaction cross-section arises through the pure compound nucleus mechanism when 70 MeV α-particles bombard lead[50], or when 40 MeV protons bombard bismuth[51].

A possible explanation has been suggested by Griffin[52] and developed by Blann[53]. It is assumed in Griffin's pre-equilibrium decay model that the energy is initially shared between some small number of particles and holes in the compound state. The initial state then progresses towards an equilibrium distribution of 'particle–hole' pairs, called excitons, through a series of successive interactions. The decay probabilities are calculated for each intermediate state and a particular level density, appropriate to the particular number of excitons, is considered. The pre-equilibrium spectral contributions are summed and it becomes possible to describe the energy spectra of the particles emitted before complete equilibrium. The yield for high-energy particles is enhanced, and consequently for a given number of emitted particles the excitation energy becomes larger. The application of this description of what happens before complete equilibrium leads to fairly good agreement with the experimental data, as is shown in Figure 4.10, in

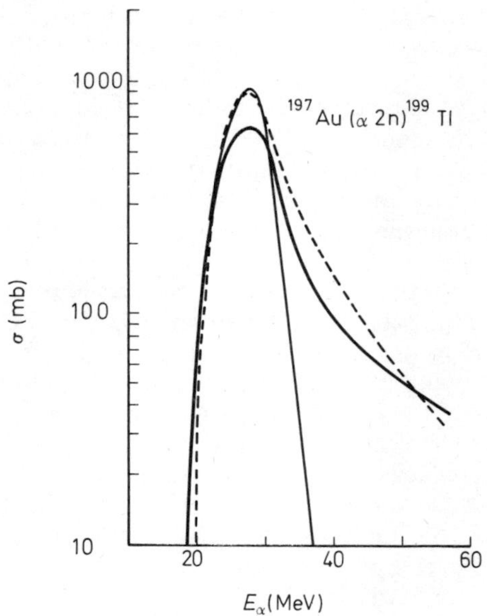

Figure 4.10 Experimental (heavy solid curve) and calculated $^{197}Au(\alpha,2n)$ excitation function. The dashed curved is the result of a 35% pre-equilibrium calculation. The thin solid line is the result of an equilibrium calculation. (From Blann and Lanzafame[49], by courtesy of North-Holland Publ. Coy.)

the energy range between 30 and 60 MeV. At higher energies, direct reactions have to be taken into account.

4.5 DIRECT REACTIONS. TRANSFER OF NUCLEONS

When prompt emission of particles follows a single nucleon–nucleon collision, or when one or more nucleons are transferred from the bombarding particle to the target or vice-versa, the reaction process is defined as a direct reaction. Both phenomena might occur at low and medium energies, although at higher energies all reactions proceed as a first step through one or several collisions resulting in the direct ejection of one or more interacting particles.

4.5.1 Quasi-elastic collisions at medium energies. Knock-out reactions

The mean free path of an incident proton in a nucleus of nuclear density ρ is $\Lambda = 1/\rho<\sigma_{ef}>$, where $<\sigma_{ef}>$ is the average effective proton–nucleon cross-section. It differs from the proton–nucleon cross-section for free nucleons in that the Pauli exclusion principle forbids collisions leading to occupied states. Λ is of the order of 3 Fermis for protons between 10 and 40 MeV, a value approximately three-times the wavelength. This suggests that a quasi-elastic collision may occur at some depth from the surface of the target nucleus. However, a small impact parameter will lead to compound nucleus formation after several interactions, since a central collision will preferentially scatter the incident proton into the core of the nucleus and Λ is only a fraction of the nuclear diameter. In contrast, a collision with large impact parameters, i.e. in the diffuse surface of the nucleus where ρ is smaller, will permit the direct re-emission of a scattered nucleon. Thus, direct processess will first occur in surface reactions and will extend to volume reactions only at higher energies.

There has been a large variety of observations of (p,p′), (p,2p) reactions which have been explained by a knock-on mechanism. In addition, the assumption of compound nucleus formation under-estimates the cross-sections of the (p,2n) and (p,n) reactions. Read and Miller[54] have calculated the collision probability for the incident proton with a loosely bound neutron in the diffuse nuclear edge. According to these estimates, the observed quasi-elastic processes are explicable. It has also been shown that in some cases the knock-out process occurs with an α-particle. Strong evidence for α-particle structure in the nuclear surface has led to the inclusion[55] of α-particle knock-out even in the energy range 50–100 MeV. This explains the emission of α-particles in the forward direction.

4.5.2 Transfer reactions. Pick-up and stripping for light bombarding particles

The commonest example is the (d,p) reaction. The interaction with the target nucleus is restricted to a particular excited level in which the neutron from the deuteron is captured. This is a one-step process. It was first recognised by Oppenheimer and Philips[56] and the theoretical work of Butler[57] has successfully demonstrated that the angular distribution of the forward peak of the proton is related to the angular momentum $l\hbar$ of the state in which the neutron is captured. Thus, stripping reactions have been widely used to study the characteristics of discrete excited levels. A large number of reactions such as $^{31}P(\alpha,p)^{34}S$, $^{23}Na(d,p)^{24}Na$, $^{7}Li(p,d)^{6}Li$ and $^{13}C(^{3}He,\alpha)^{12}C$ have shown, by forward peaking, the characteristics of 'stripping' or the inverse 'pick-up' reaction. In the latter, the incident proton approaches very closely to the target nucleus. A strong interaction occurs between the proton and an outer neutron, resulting in the formation of a deuteron. The theory shows that only very specific trajectories are responsible for such nucleon transfer reactions. Pick-up processes like (^{3}He, ^{4}He) are very exo-energetic reactions because of the large binding energy of ^{4}He. This is of practical interest since low-energy beams of ^{3}He might be used for the activation of many targets. For example, reactions like $^{16}O(^{3}He,^{4}He)^{15}O$ or $^{12}C(^{3}He,^{4}He)^{11}C$ produce radioactive species which are extremely useful in activation analysis for oxygen and carbon atoms.

Reactions of type (p,p′), (α,α′) can also be interpreted in terms of direct reaction processes. When the incident particle enters the nucleus it is subjected to the nuclear potential, loses some of its energy by interaction with a single surface nucleon (hole–particle interaction), and then escapes from the nucleus with smaller kinetic energy. The direct-reaction theories, based on the distorted-wave Born approximation (DWBA), give accurate descriptions of the direct reactions. They are widely used for nuclear spectroscopic studies.

4.5.3 Heavy ion surface reactions

It has been stated above that the compound nucleus mechanism has found very strong support since heavy ion beams became available. It was first believed that the total reaction cross-section was simply available for compound nucleus formation and that direct interactions could not occur since the complex bombarding particle had a large volume. However, it should be remembered that nuclear matter is not uniformly distributed in a nucleus and it is now well established that a halo of surface neutrons, protons and clusters of nucleons, like α-particles, surround the nucleus. The diffuseness of the nuclear surface has a striking effect when two complex nuclei travel on grazing trajectories. For large orbital angular momenta, a number of processes occur which do not proceed through a compound nucleus mechanism. Part of one (or both) of the interacting nuclei is transferred to the other without perturbing the bulk of the two nuclear masses. The number of different transfer reactions is large and the complexities increase with increasing energy. For the simplest nucleon transfers, the reactions resemble the stripping and pick-up processes observed in studies with lighter projectiles, and they might be used for nuclear spectroscopic investigations. Single proton or single neutron transfer reactions have been thoroughly studied on ^{14}N, ^{16}O, ^{12}C, ^{15}N, ^{13}C, either by the measurement of the radioactivity of ^{13}N or ^{15}O, or with more sophisticated techniques such as the identification of the light product with a telescopic arrangement of ΔE, E counters. For peripherical surface collisions, with l very close to l_{max}, the projectiles travel on classical Rutherford orbits before and after the reaction. For neutron transfer below the Coulomb barrier, the neutron tunnels from one nucleus to the other without the two nuclear surfaces ever coming in contact[58]. The distance of closest approach is related to the angle of observation $\bar{\theta}$ by means of the formula $R_{min} = Z_1Z_2e^2/2\bar{\varepsilon})(1+\text{cosec}\bar{\theta}/2)$. For small detection angles, i.e. large distances of approach, the probability of transfer is low. As θ increases, the maximum in the distribution occurs when R_{min} has decreased to an optimum value for which the two nuclei are more or less tangential (Figure 4.11). If the distance for maximum transfer probability remains the same, it is clear that as $\bar{\varepsilon}$ increases the peak shifts to smaller angles. Now, as

θ becomes larger R_{min} becomes smaller and smaller, the two nuclei coalesce and the compound nucleus reaction begins to predominate. Such behaviour has often been observed for single nuclear transfer reactions, following the first experiment of McIntyre *et al.*[59].

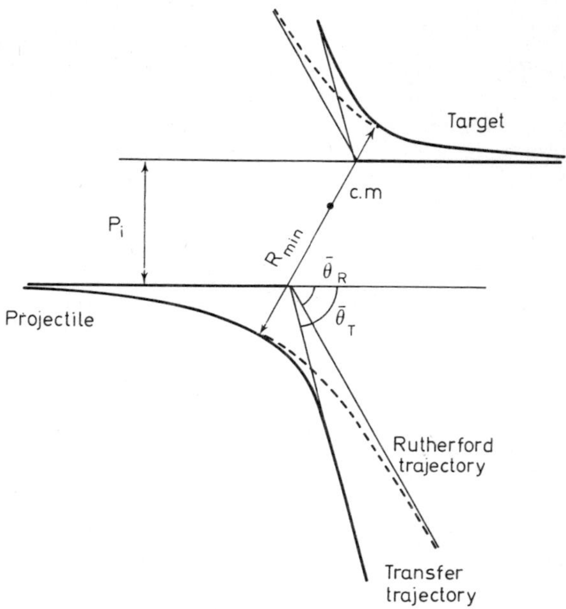

Figure 4.11 Treatment of the Coulomb scattering in the centre-of-mass system. θ_r is the experimental scattering angle. θ_R is the scattering angle for a pure Rutherford trajectory

A quantum mechanical approach to this process has been developed which is valuable so long as the Sommerfeld parameter $\eta = Z_1 Z_2 e^2/\hbar v$ is larger than 1. However, when the energy increases above the Coulomb barrier, another process occurs which yields transfer reaction products mainly in the forward direction. It has been shown by Volkov *et al.*[60] and by Galin *et al.*[61] that multi-nucleon transfer as well as single events occur with increasing probability as the energy increases. Moreover, the mechanism is not quasi-elastic any more, and a large fraction of the total kinetic energy appears as excitation energy. The light product is excited to a number of known levels and the heavy partner has such a large excitation energy that unbound levels are reached and particles emitted, as has been shown[62] for astatine nuclei which result from an α-particle transfer to Bi targets bombarded by ^{12}C ions. This is indeed an unexpected type of reaction, which occurs for values of the angular momentum $l\hbar$ in a region between large $l\hbar$ values corresponding to the quasi-elastic transfer ($l_\tau \# l_{max}$) and the critical value $l_c\hbar$ below which the compound nucleus is formed, as has been described above in Section 4.3 (equation (4.34)).

This region of inelastic processes has been a little explored[63] and many products may be synthesised by the nuclear chemist when more is known. In this region one may observe the disintegration of the incident particle into lighter fragments (like ^{4}He, ^{6}Li or ^{6}He etc.), the knock-out of one or more nucleons from the target nucleus and the capture of one or more nucleons from the target by the projectile. It has been observed, for example, that the incident particle may be enriched by two to six neutrons. For this reason, new very neutron-rich nuclei may be produced[64] such as ^{20}O, ^{22}O, ^{20}N, ^{18}C, ^{24}F, ^{26}Ne, ^{30}Mg, ^{36}Si, ^{38}P, ^{40}S and ^{42}Cl.

On the other hand, it is also possible to measure rather large cross-sections for neutron deficient nuclei. A very promising but relatively unstudied field is the study of these grazing reactions induced by ^{14}N, ^{16}O, ^{20}N and ^{40}Ar heavy ions on medium A and heavy targets at energies between 7 and 15 MeV per nucleon. Large fragments are exchanged between the two nuclei and there is certainly a connection between this process and nuclear fission. New theoretical ideas are required to describe the nuclear potential of the two colliding complex nuclei, most of them being inspired by atomic and molecular chemistry[65, 66] (molecular potentials, valency nucleons, etc).

4.6 NUCLEAR CHEMISTRY AND THE SYNTHESIS OF KNOWN AND UNKNOWN NUCLEI

As mentioned in Section 4.1, the principle aim of nuclear chemistry is the artificial production of new isotopes (usually radioactive) of known and, eventually, new elements. The study of nuclear reactions is directed towards an understanding of the processes by which natural isotopes have been synthesised. It is very different from the purpose of nuclear physics which is the understanding of nuclear forces. This is why the nuclear chemist and nuclear physicist are not always interest in the same type of nuclear reaction and why they do not try to extract the same kind of information.

For many years, most of the new artificial isotopes were synthesised by (n,γ) reactions, or as products from fission processes in nuclear reactors. However, it is well known that technecium was discovered after a (d,n) reaction on molybdenum, astatine after a (α,2n) reaction on bismuth and neptunium and plutonium after a (d,2n) reaction on uranium. All these new elements were synthesised using a cyclotron beam. At the present time new isotopes are sought mainly in three regions; light nuclei on both sides of β stability region up to the limits of particle instability, neutron-deficient nuclei in the rare earth region and above, and trans-plutonium elements.

A large number of new light nuclei have been produced by heavy ion transfer reactions. The bombardment of very heavy targets like Au or Th by projectiles like ^{16}O, ^{20}Ne leads to the pick-up of a number of neutrons, as described above. Although high-energy reactions (fragmentation and spallation) have also been used since 1966 to produce a large number of new species[67], it seems that at the present time the limit of particle instability has been reached for light nuclei on the neutron excess side since Volkov *et al.*[68] have shown that ^{13}Be and ^{14}Be are not bound.

The lower part of the chart of nuclides has been nearly entirely filled up on the right side of the β stability region (Figure 4.12). On the left side, reactions like (p,t), (^{3}He,n), (^{3}He,t) have been used to produce neutron-deficient nuclei, which are in some cases delayed proton emitters like ^{9}C, ^{13}O, ^{17}Ne, ^{21}Mg, ^{33}Ar, ^{40}Ti and ^{108}Te, or delayed α emitters like ^{20}Na.

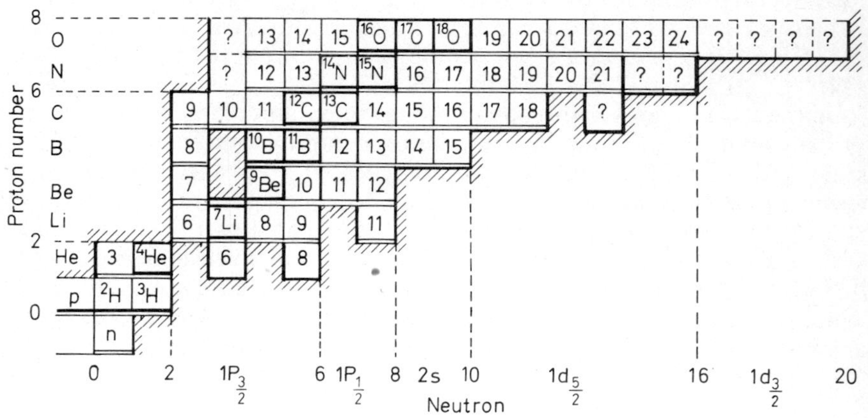

Figure 4.12 Chart of light isotopes. Theoretical boundaries are shown on the neutral rich side and on the proton rich side. Stable isotopes are represented by the symbol of the element. Neutron shells are given on the neutron scale

The bombardment of rare earth elements by ^{4}He and ^{3}He particles accelerated between 80 and 120 MeV yields compound nuclei. De-excitation by a series of neutron emissions produces residual nuclei which are neutron deficient and α-particle emitters. In many cases, it is more convenient to use heavy ions like ^{14}N, ^{16}O or ^{12}C on lighter targets. The effect of the 82 neutron closed-shell on α decay has been studied[69] for many rare earth elements such as $^{151-153}$Er, $^{150-154}$Ho, $^{156-157}$Yb. Light isotopes of Pt, Hg and Pb, most of them α-emitters, have been discovered with the help of the helium jet technique. Finally, α-decay chains initiated by $^{222-225}$Pa, $^{221-224}$Th and $^{213-222}$Ac have been produced by bombardments of Tl, Pb and Bi targets with beams of Ne, F, O, N and C. A large number of light isotopes of Ra, Fr, Rn, At and Po have been identified and their α decay characteristics have been determined[70, 71].

The search for new trans-uranium isotopes, and for super-heavy elements, is of course one of the most exciting purposes of nuclear chemistry at the present time. It is not within the scope of this Section to describe in detail the various attempts which have been made to identify elements 102, 104, 105 and very recently 106. Some of the results are still the subject of controversy. Seaborg[72] has given a comprehensive review of this subject (See Chapter 3.)

Intensive slow-neutron irradiations, in high-flux reactors or in nuclear explosives are now available for adding neutrons to heavy nuclei. By cumulative capture, 15 to 20 neutrons may be added successively without destroying the intermediate products through competitive fission reactions. It should be recalled that einsteinium and fermium were first isolated from the debris of thermonuclear explosion. The heaviest element produced in high flux reactors is ^{257}Fm.

^{238}Pu, ^{242}Cm, ^{243}Bk, ^{245}Cf and ^{256}Md were first synthesised by (α,xn) reactions on appropriate targets. The reaction ^{253}Es(α,n)^{256}Md has a cross-section of the order of 1 mb because of large fission competition. In recent years, a great variety of heavy ion nuclear reactions have been used in order to produce trans-mendelevium isotopes. The problem of the identification of a new nucleus has been resolved by a very ingenious fast chemical separation and by a systematic study of (HI,xn) reactions. After a long series of experiments in Sweden, Dubna and Berkeley, it now seems clear that isotopes 251, 254, 255, 257 and 258 of element 102 have been produced by bombardments of curium with ^{12}C and ^{13}C, and of plutonium by neon. Lawrencium (103) has been produced by bombardment of ^{249}Cf with ^{11}B, isotope 256 being the best characterised. Element 104, called rutherfordium in the United States, has been synthesised at Berkeley[73] by bombarding ^{249}Cf targets with ^{12}C and ^{13}C (^{257}104, ^{258}104 and ^{259}104). Flerov has announced the discovery of 104, and proposed the name of Kurchatovium, after an experiment in which ^{242}Pu was bombarded by ^{22}Ne. The assignment for the spontaneously fissioning product was ^{260}104. The Berkeley group has been unable to obtain such a nucleus by bombardment of ^{248}Cm with ^{16}O and ^{18}O.

An isotope of element 105 with mass number 260 has been identified[74] after bombardment of ^{249}Cf with ^{15}N. The name of hahnium was proposed. At Dubna, the bombardment of ^{243}Am by ^{22}Ne was assumed to have induced the formation of element 105 but there is disagreement with the conclusions of the Berkeley group[75].

The search for trans-mendelevium elements and the attempts to find super-heavy elements of Z values as large as 114 or 126 by using beams of krypton or higher masses, on very active areas of nuclear chemistry[27]. A knowledge of the mechanism of compound nuclear reactions, and of the neutron and fission widths, is very important for this kind of research in which the cross-sections produced are of the order of 10^{-30}–10^{-32} cm^2.

4.7 CONCLUSIONS. APPLICATIONS OF NUCLEAR CHEMISTRY

Nuclear chemistry has found application to a wide variety of research in many laboratories and there have been several books and conferences devoted specifically to the subject[76]. Activation analysis, a technique of growing importance is fundamentally dependent on nuclear chemistry.

Perhaps one of the most interesting applications of nuclear chemistry is in astrophysics. The origin of the elements is indeed one of the most fascinating questions, and it seems probable that the elements have all evolved from hydrogen. Man inhabits a universe composed of a great variety of elements and their isotopes and it is believed that the stars are the sites of synthesis of the elements. The general features of stellar synthesis are governed by the laws of nuclear chemistry. It is sufficient to read the table of contents of the well-known article by Burbidge *et al.*[77] to be convinced that (a) hydrogen burning which synthesises helium and particular isotopes of carbon, nitrogen (occurs) through radiative capture proceeding by the compound nucleus mechanism, (b) helium burning, (is) responsible for the synthesis of ^{12}C, ^{16}O, ^{20}Ne, ^{24}Mg by α-induced reactions, (c) *e* processes occur, in which the elements around the iron peak in the abundance curve are produced*, (d) the *s* process, which is the well-known process of neutron capture, takes place on a long time-scale in stars where a high flux of neutrons has been created by (α,n) and other reactions† and (e) the *r* process, i.e. neutron capture on a very short time-scale also occurs.

Neutron capture occurs at a rapid rate compared to β decay, and the synthesis of uranium, thorium and perhaps even heavier nuclei like californium is due to this mechanism, comparable to the artificial synthesis of trans-uranium elements. It is possibly the most noteworthy achievement of nuclear chemistry to have contributed so much to the explanation of the origin of the elements which are the building blocks for all chemistry.

---

* This latter process corresponds to a large set of reactions, either radiative captures, (p,γ), (α,γ), or photonuclear reactions, (γ,α), (γ,p), (γ,n), which have the effect of transforming a nucleus (A,Z) into (A',Z') even when the Coulomb barrier is large.

† This mode of synthesis, in which the product Z is neutron-rich and β-decays into a daughter (Z+1), is responsible for the majority of isotopes in the range $63 < A < 209$ and produces the abundance peaks at $A = 90$, 138 and 208.

References

1. Blatt, J. M. and Weisskopf, V. F. (1952). *Theoretical Nuclear Physics.* (New York: Wiley)
2. Bernas, R. (1969). *Adv. in Mass Spectrometry,* **4,** 919
3. Klapisch, R. (1969). *Ann. Rev. Nucl. Science,* **19,** 33
4. MacFarlane, R. D. and Griffioen, R. D. (1963). *Nucl. Instr. Meth.,* **24,** 461; MacFarlane, R. D., Gough, R. A., Oakey, N. S. and Torgerson, D. F. (1969). *Nucl. Instr. Meth.,* **73,** 285
5. Valli, K. and Hyde, E. K. (1968). *Phys. Rev.,* **176,** 1377
6. Ghiorso, A., Sikkeland, T. and Nurmia, M. J. (1967). *Phys. Rev. Lett.,* **18,** 401
7. Valli, K., Treytl, W. J. and Hyde, E. K. (1968). *Phys. Rev.,* **167,** 1094
8. Moseley, H. G. (1913). *Phil. Mag.,* **26,** 1024
9. Peed, W. F. and Spitzer, E. J. (1949). *Phys. Rev.,* **75,** 86
10. Alexander, J. M. (1968). *Nuclear Chemistry,* Yaffé, L. (Ed.), pp. 273–355. (New York and London: Academic Press)
11. Harvey, B. (1960). *Ann. Rev. Nucl. Sci.,* **10,** 235
12. Butler, G. W., Poskanzer, A. and Landis, D. A. (1970). *Nucl. Instr. Meth.,* **89,** 189
13. Thomas, T. D. (1968). *Ann. Rev. Nucl. Sci.,* **18,** 343
14. Halpern, I., Shepherd, B. J. and Williamson, C. F. (1968). *Phys. Rev.,* **169,** 805
15. Motz., H. T. (1970). *Ann. Rev. Nucl. Sci.,* **20,** 1
16. Smither, R. K. and Bollinger, L. M. (1969). *Proc. Int. Symp. Neutron Capture,* Studsvik, 601–605
17. Chrien, R. E. (1969). *Proc. Int. Symp. Neutron Capture,* Studsvik, 627–649
18. Ghoshal, S. N. (1950). *Phys. Rev.,* **80,** 939
19. D'Auria, J. M., Fluss, M. J., Herzog, G., Kowalski, L., Miller, J. M. and Reedy, R. C. (1968). *Phys. Rev.,* **174,** 1409
20. Fluss, M. J., Miller, J. M., D'Auria, J. M., Dudey, N., Foreman, B. M. Jr., Kowalski, L. and Reedy, R. C. (1969). *Phys. Rev.,* **187,** 1449
21. Porile, N. T., Tanaka, S., Amano, H., Furukawa, M., Iwaka, S. and Yagi, M. (1963). *Nucl. Phys.,* **43,** 500
22. Ahrens, S. T., Simon, W. G. and Eldrige, H. B. (1970). *Phys. Rev.,* **C2,** 1433
23. Grover, J. R. (1962). *Phys. Rev.,* **127,** 2142
24. Sperber, D. (1963). *Phys. Rev.* **130,** 468

25. Beringer, R. and Knox, W. J. (1961). *Phys. Rev.*, **121,** 1195
26. Natowitz, J. B. (1970). *Phys. Rev.*, **C1,** 623, 2157
27. Lefort, M. (1971). *Ann. Physique,* **5,** 355
28. Kalinkin, B. N. and Petkov, I. Z. (1964). *Acta Phys. Polon.*, **25,** 265
29. Sikkeland, T., Haines, E. L. and Viola, V. (1962). *Phys. Rev.*, **125,** 1350
30. Galin, J., Guerreau, D., Lefort, M. and Tarrago, X. (1971). *J. Phys.*, **32,** 7
31. Jodogne, J. C., Kowalski, L. and Miller, J. M. (1968). *Phys. Rev.*, **169,** 894
32. Basile, R., Lefort, M., Galin, J., Guerreau, D. and Tarrago, X. (1971), in press
33. Ericson, T. (1960). *Adv. Phys.*, **9,** 425
34. Sherr, R. and Brady, F. P. (1961). *Phys. Rev.*, **124,** 1928
35. Katsanos, A. A. and Huizenga, J. R. (1967). *Phys. Rev.*, **159,** 931
36. Benveniste, J., Merkel, G. and Mitchel, A. (1966). *Phys. Rev.*, **141,** 980
37. Vonach, H. K. and Huizenga, J. R. (1965). *Phys. Rev.*, **B138,** 1372
38. Lindhard, J. (1965). *Mat. Fys. Medd. Dan. Vid. Selsk.*, **34,** 14
39. Gibson, W. M. and Nielsen, K. O. (1970). *Phys. Lett.*, **24,** 114; Galin, J., Guerreau, D., Lefort, M., Oganecian, I. and Tarrago, X. (1971), in press
40. Clark, G. J., Poate, J. M., Fuschini, E., Maroni, C., Massa, I. G., Uguzzoni, A. and Verondini, E. (1971). *A.E.R.E. Report,* 6756; (1971). *Nucl. Phys.*, **A173,** 73
41. Benveniste, J., Merkel, G. and Mitchell, A. (1968). *Phys. Rev.*, **174,** 1357
42. Ericson, T. and Strutinski, V. (1958). *Nucl. Phys.*, **8,** 284
43. Grover, J. R. (1961). *Phys. Rev.*, **123,** 267; Grover, J. R. and Nagle, R. J. (1964). *Phys. Rev.*, **B134,** 1248
44. Gilat, J. and Grover, J. R. (1971). *Phys. Rev.*, **C3,** 734
45. Mollenauer, J. F. (1962). *Phys. Rev.*, **127,** 867
46. Alexander, J. M. and Simonoff, G. N. (1964). *Phys. Rev.*, **B133,** 93, 104
47. Le Beyec, Y., Lefort, M. and Vigny, A. (1971). *Phys. Rev.*, **C3,** 1268
48. Williams, D. C. and Thomas, T. D. (1967). *Nucl. Phys.*, **A92,** 1
49. Blann, M. and Lanzafame, F. M. (1970). *Nucl. Phys.*, **A142,** 559
50. Bimbot, R., Jaffrezic, H., Le Beyec, Y., Lefort, M. and Vigny, A. (1969). *J. Phys.*, **30,** 33
51. Birattari, G., Gadioli, E., Grassi Strini, A. M., Strini, G., Tagliaferri, G. and Zelta, L. *Nucl. Phys.*, **A166,** 605
52. Griffin, J. J. (1966). *Phys. Rev. Lett.*, **17,** 478
53. Blann, M. (1968). *Phys. Rev. Lett.*, **21,** 1357
54. Read, J. B. J. and Miller, J. M. (1965). *Phys. Rev.*, **B140,** 623
55. Lefort, M. (1964). *Ann. Physique,* **9,** 249
56. Oppenheimer, J. R. and Philips, M. (1935). *Phys. Rev.*, **48,** 500
57. Butler, S. T. (1950). *Phys. Rev.*, **80,** 1095
58. Breit, G. and Ebel, M. E. (1956). *Phys. Rev*
59. McIntyre, J. A., Jobes, F. J. and Watts, T. L. (1960). *Phys. Rev.*, **119,** 331
60. Volkov, V. V., Gridnev, G. F., Zorin, G. N. and Chelnokov, L. P. (1969). *Nucl. Phys.*, **126,** 1
61. Galin, J., Guerreau, D., Lefort, M., Pèter, J., Tarrago, X. and Basile, R. (1969). *Phys. Rev.*, **182,** 1267
62. Bimbot, R. and Gardès, D. (1971). *J. Phys.*, in press
63. Galin, J., Guerreau, D., Lefort, M. and Tarrago, X. (1971). *J. Phys.*, **32,** 7
64. Artukh, A. G., Gridnev, G. F., Mikheev, V. L. and Volkov, V. V. (1969). *Nucl. Phys.*, **A137,** 348; (1969). *JINR Reprint,* Dubna, **E7,** 4800
65. Demeur, M. Reidemeister, G. (1966). *Ann. Physique,* **1,** 181
66. Scheid, W. and Greiner, W. (1969). *Z. Physik.*, **226,** 364; (1970). *Phys. Rev. Lett.*, **25,** 176
67. Poskanzer, A. M., Butler, G. W., Hyde, E. K., Cerny, J., Lindis, D. A. and Goulding, F. S. (1968). *Phys. Lett.*, **27B,** 414
68. Artukh, A. G., Avdeichikov, V. V., Gridnev, G. F., Mikheev, V. L., Volkov, V. V. and Wilczynski, J. (1970). *Phys. Lett.*, **31B,** 129
69. Toth, K. S., Hahn, R. L., Roche, M. F. and Brenner, D. S. (1969). *Phys. Rev.*, **183,** 1004; (1970). **C2,** 1480
70. Borggreen, J., Valli, K. and Hyde, E. K. (1970). *Phys. Rev.*, **C2,** 1841; Valli, K., Hyde, E. K. and Borggreen, J. (1970). *Phys. Rev.*, **C1,** 2115
71. Torgerson, D. F., Gough, R. A. and MacFarlane, R. D. (1968). *Phys. Rev.*, **174,** 1494
72. Seaborg, G. T. (1968). *Ann. Rev. Nucl. Sci.*, **18,** 53
73. Ghiorso, A., Nurmia, M., Harris, J., Eskola, K. and Eskola, P. (1969). *UCRL* 18819

74. Ghiorso, A., Nurmia, M., Eskola, K., Harris, J. and Eskola, P. (1970). *UCRL* 19577
75. Flerov, G. N., Oganecian, Y. T., Lobanov, Yu. V., Lasarev, Yu. A., Tretiakova, J. P., Kolesov, I. V. and Plotko, V. M. (1970). *JINR Preprint,* Dubua, **E7,** 5252
76. Sayre, E. V. (1963). *Ann. Rev. Nucl. Sci.,* **13,** 145; Koch, R. C. (1966). *Activation Analysis Handbook.* (London: Academic Press); Bowen, J. M. and Gibbon, D. (1963). *Radioactivation Analysis.* (Oxford: Clarendon Press)
77. Burbridge, E. M., Burbridge, G. R., Fowler, W. A. and Hoyle, F. (1957). *Rev. Mod. Phys.,* **29,** 547
78. Dittner, P. F., Bemis, C. E., Hensley, D. C., Silva, R. J. and Goodman, C. D. (1971). *Phys. Rev. Lett.,* **26,** 1037

5
Chemical Reactions of Nuclear Recoil Particles in Gases and Liquids

D. S. URCH
Queen Mary College, University of London

5.1 INTRODUCTION

Nuclear reactions liberate large amounts of energy which are shared with conservation of momentum amongst the products of the reaction. The energy which each particle receives in this way is its recoil energy. Such energies are usually very much higher than the energies that are normally encountered in chemical reactions, i.e. up to *c.* 200 kcal mol^{-1} or *c.* 8 eV per molecule. The products of nuclear reactions are excited energetically in all possible ways, translationally and electronically, and they are usually ionised. These excitations are degraded by collision and cause radiation damage to the system through which the particle passes. There will come a time, however, when the particle is in such a state that it can enter into a chemical reaction with a collision partner to become bound chemically.

In this review I discuss studies of such chemical reactions initiated by recoil particles. In all cases the atoms or ions enter the energetic range in which chemical reactions (leading to stable products), are possible in an entirely different way from atoms, ions or molecules in conventional thermally initiated chemical reactions. Thermal excitation of a collection of molecules will produce some intermolecular collisions in which there is just sufficient energy to crawl over an activation energy barrier, associated with a particular energetically favourable orientation of the colliding species. In collisions involving atoms or molecules with recoil particles the opposite is the case. In many events the energy brought by the recoil particle will be too great and fragmentation will result. At somewhat lower energies chemical reactions associated with high activation energies and unfavourable geometries will be possible. Should such encounters be sometimes unsuccessful the recoil particle, with a somewhat reduced energy, may well be able to undergo other reactions.

There is a variety of chemical reactions possible between any two particles, characterised by different activation energies. The only reaction that would normally be studied is that characterised by the lowest activation energy whereas in the study of recoil particle reactions the whole gamut of possible reactions can be investigated.

For clarity it is of course most necessary to distinguish between the different possible types of excitation which the recoil particles may bring into their collisions with other atoms or molecules. Thus atoms which are neutral and in a ground electronic state, where the only energy of excitation is translational, can be compared in their reactions with thermally initiated chemical reactions. If the particle is ionised, then a comparison with ion–molecule reactions is relevant and, if it is suspected that the recoil particle may be reacting in an electronically excited state, photochemical comparisons will be in order.

The unfortunate truth is that in most cases the charge state, or the electronic state of the recoil particle, at 'chemical' energies is not always well known; it is the subject of conjecture and surmise. A second most unfortunate feature of the study of the chemical reactions initiated by recoil particles is that the energy of the particle, when it undergoes chemical reaction, is not known. In some cases the recoil energy will not be clearly defined, as in n, γ reactions, or in other cases, so enormous by chemical standards as to be wholly irrelevant. In an attempt to gain some knowledge concerning the energy of the atom or ion at the time of reaction, considerable interest has been aroused in other more quantitative attempts to study the chemistry of 'hot' particles which are produced with a clearly defined energy. Reference will therefore be made in this review to such investigations directly related to the study of recoil particle chemistry.

Much of the work in the field of recoil particle chemistry has been directed towards an understanding of the role of excess energy in a chemical reaction. This review will therefore be confined (with only minor exceptions) to experiments in the gaseous and liquid phases. Collisions between recoil particles and atoms or molecules in the gas phase will be unaffected by the presence of other molecules, so that the role of the energy brought to the collision by the recoil particle, can be studied in isolation. Comparison can then be made with the liquid phase and the effect of the close proximity of other molecules and the short time between intermolecular collisions studied. The special effects that occur in crystalline solids are the subject of another review (Chap. 6).

The first investigation of a chemical reaction initiated by a nuclear recoil was made by Szilard and Chalmers[1] in 1934 but detailed studies of such reactions have only been made during the past 10–15 years. During this time two important conferences on the Chemical Effects of Nuclear Transformations[2,3] have been organised by the International Atomic Energy Agency and a number of reviews[4–11] have appeared. In order to present as coherent a picture as possible some of the earlier theories and experimental results will be discussed but the main emphasis of the review will be on work carried out during the past 5 years.

The recoil particles selected for study have been determined largely by the caprice of Nature. By far the easiest way of detecting the products that result from recoil particle reactions is by a radioactive tracer technique; if the recoil particle is itself radioactive, it acts as its own tracer. Since the products of most nuclear reactions are radioactive this requirement is by no means stringent. However, the half-lives of radioactive atoms can vary from microseconds to megayears. A suitable half-life for detection can therefore limit the number of recoil particles whose reactions can be conveniently studied by the auto-tracer method. If an isotope of long half-life is to be used, then very many more recoil particles must be produced than if a short-lived isotope is selected. This in turn will affect the total energy deposited, in the sample being investigated, by *all* the products of the nuclear reaction. It might well be that in order to detect molecules labelled by a long-lived isotope so many nuclear reactions would be required that the radiation damage in the original sample would completely alter its character. An example of such a situation is ^{14}C produced by ^{14}N (n,p) ^{14}C.

Nuclear reactions are of many types but those initiated by thermal neutrons have been most widely used in recoil particle chemical studies. Neutron reactions with nuclei are broadly of two types; (n, photon) and (n, particle). In the former case the photon removes most of the energy of the reaction as hard γ-rays which usually cause little radiation damage in the system under investigation. The recoil energy of the atom is relatively small, typically a few hundred electron volts. In the latter example the particle carries away very much less of the energy of the reaction than the photon so that the recoil energy of the atom is very much greater, i.e. many kilo electron volts, sometimes mega electron volts. The short ranges of the recoil atom and particle both ensure that most of the recoil energy is deposited in the sample itself. These considerations have resulted in a rather short list of recoil particles whose chemical reactions have been investigated with any degree of thoroughness (Table 5.1).

Table 5.1

Recoil nucleus	*Half-life*	*Decay mode*	*Nuclear reaction* (n^* = *fast neutron*)
^{3}H	12.3 y	β^-	$^{3}He(n, p)^{3}H$.
			$^{6}Li(n, \alpha)^{3}H$
^{11}C	20 m	β^+	$^{12}C(n^*, 2n)^{11}C$ etc.
^{14}C	5400 y	β^-	$^{14}N(n, p)^{14}C$
^{13}N	10 m	β^+	$^{12}C(d, n)^{13}N$
^{15}O	2 m	β^+	$^{14}N(d, p)^{15}O$
^{18}F	112 m	β^+	$^{19}F(n^*, 2n)^{18}F$ etc.
^{31}Si	2.6 h	β^-	$^{30}Si(n, \gamma)^{31}Si$
			$^{31}P(n^*, p)^{31}Si$
^{32}P	14.3 d	β^-	$^{31}P(n, \gamma)^{32}P$
			$^{32}S(n, p)^{32}P$
^{35}S	86.7 d	β^-	$^{34}S(n, \gamma)^{35}S$
			$^{35}Cl(n, p)^{35}S$
^{36}Cl	3×10^{5} y	β^-	$^{35}Cl(n, \gamma)^{36}Cl$
^{38}Cl	37.3 m	β^-	$^{37}Cl(n, \gamma)^{38}Cl$
^{80}Br; $^{80}Br^{m}$ ^{82}Br, $^{82}Br^{m}$ ^{128}I	thermal neutrons initiate complex nuclear reactions with ^{79}Br, ^{81}Br and ^{127}I which give rise to both ground state and metastable excited nuclei.		

The recoil chemistry associated with many of these particles will be considered in separate reactions of this review. General problems associated with experimental methods and theoretical ideas of recoil particle chemistry will be discussed briefly first.

5.2 EXPERIMENTAL TECHNIQUES

5.2.1 Gas phase

5.2.1.1 Sample preparation

Samples for the study of recoil chemistry in the gas phase are usually encapsulated in quartz[12] or 1720 glass[13] (10–20 mm dia) ampoules of *c.* 10–20 cm^3 volume. Quartz has the advantage of being transparent to thermal neutrons but the disadvantage of containing many labile hydrogen atoms which can react with recoil particles that enter the ampoule walls[12, 14–16]. In the case of recoil tritium, the HT molecules so formed can sometimes diffuse through the quartz and re-enter the gas phase, distorting the observed yield of this compound. Corrections for this 'wall-effect' of HT have been suggested by measuring HT yields in systems with wholly deuterated samples[15]. In

1720 glass the advantages and disadvantages are reversed. The boron of the glass attenuates the neutron flux entering the ampoule but those recoil particles that collide with the walls remain there, and do not re-enter the gas phase[13].

The order in which gases are admitted to the ampoule during filling is also important; those with lowest vapour pressures (at the filling temperature, usually 77 K) should enter first. Special care is needed, if two or more components have appreciable vapour pressures, to ensure proper equilibration before the ampoule is sealed off from the vacuum line. Alternatively, non-condensable gases can be swept into the ampoule in such a way that the one with the largest partial pressure is added last[17]. Another way of determining the actual composition of the sample is by chromatographic analysis after irradiation[18, 19]. In many systems (e.g. ^{3}H, ^{18}F), the recoil particles are generated within the ampoule itself; in other cases the particles enter the vessel from outside (e.g. ^{11}C from a nuclear-stripping reaction).

5.2.1.2 Analysis

The labelled products produced by the recoil particle reactions are usually analysed by gas liquid and/or gas solid chromatography. The chromatographic effluent is then passed directly through a proportional counter[20–22] or a scintillation counter. A completely automated version of this type of apparatus has recently been described[23]. The products are identified by their retention times and the amount of activity determined by conventional counting equipment. If small ampoules are used, the whole sample may be broken in the gas stream before the chromatographic column; aliquots are withdrawn from larger samples. Special taps have been described to permit clean sample injection[24]. A wide variety of chromatographic columns and substrates have been used. Extremely long columns have, for example, been used to separate olefines labelled at paraffinic and at olefinic sites[25]. The use of long columns may be avoided with a pair of short columns and a column-switching technique[26, 27].

5.2.1.3 Proportional counting

The widespread use of proportional counters has produced many critical investigations of the method.

(a) *Counters*—For recoil atoms that emit 'hard' radiation a 'window' counter may be used in which the gas stream from the chromatographic column is separated from the actual counter(s) by a very thin film of material (e.g. mylar)[28]. An improved design for such a counter has been described by Welch, Withnell and Wolf[29]. New simple designs for conventional flow counters have also been given[30, 31].

(b) *Flow rates*—The number of disintegrations recorded by a flow counter will depend not only on the activity of the sample but also the time spent in the counter. It is therefore essential that the flow rate be constant and known during an analysis. When large amounts of sample are passed through the counter, the flow rate will be altered and the flow rate must be monitored constantly[22, 23].

(c) *Quenching*—With open flow counters of the type that must be used for the efficient detection of weak β-emitters, such as tritium, the passage of macroscopic amounts of sample through the counter may affect not only flow rates but also the characteristics of the counting gas[22]. Oxygen[32], bromine, alkyl halides, etc. can cause quenching of the counter and so also can noble gases. Hydrocarbons too can cause deviations in counting characteristics[21]. It is therefore essential, when windowless counters are used, that the effect of all components of the sample, in their proper amounts, upon the performance of the counter, be tested.

(d) *Counting efficiency*—A further factor must be taken into account, namely the actual efficiency of the counting system. It is usually expedient to set a particular discriminator level which will more or less eliminate millivolt noise in the counting electronics. In so doing, a certain portion of the energy spectrum of a β-emitter will also be lost. This can be estimated by comparing the energy spectrum of radiation that is actually counted with the complete spectrum[33].

5.2.2 Liquid phase

5.2.2.1 Sample preparation

Samples are sealed in quartz or Pyrex ampoules. After irradiation the ampoules may be broken under the surface of a liquid mixture, usually of suitable organic and inorganic solvents, or if sufficiently small the ampoule may be broken into the gas stream entering a chromatographic column.

5.2.2.2 Analysis

Many of the general comments in Section 5.2.1 also apply for the analysis of products made by recoil particles in the liquid phase. In those cases where a solvent extraction has been performed, scintillation counting is usually the simplest method of determining the activity. Care needs to be taken that labelled gaseous compounds are not lost when the solvent extraction technique is used.

5.3 THEORY

Theoretical considerations have played a large part in the interpretation of recoil particle reactions and also in determining the types of experiments that have been performed. Three main aspects will be considered, (a) slowing down of the recoil particle, (b) calculations aimed at describing the actual mechanics of the interaction of a low-energy particle with a molecule, and (c) more general overall approaches in which specific reaction mechanisms are not invoked.

5.3.1 Slowing down of recoil particles

Recoil particles fall broadly into two types, those with high (>1 keV) recoil energy, and those with low recoil energy (<1 keV).

5.3.1.1 High-energy recoil particles

(a) *Charge exchange*—Particles not produced by an (n, photon) reaction have, in general, very high recoil energies. At their moment of birth they are highly charged and they will lose energy in ionising collisions with other atoms and molecules. The probability that such a particle will pick up an electron will depend on the ionisation energy of the particle and the atom or molecule it collides with, and their relative velocities. The adiabatic principle[39] suggests that if the difference in ionisation energy is large (i.e. many eV) then the most probable relative energy of the particles for electron transfer will also be large (many keV), conversely, if there is a small difference in ionisation energies, electron transfer will take place at lower relative kinetic energies. The probability of electron transfer also depends upon the difference in ionisation energies directly; it is greatest when the difference is least.

Arguments based on the adiabatic principle have been used to suggest that recoil tritons probably are[12], and rather later on, probably are not[40], completely neutralised in samples that contain a great deal of helium. A warning that triton neutralisation might not be complete in excess helium was however made considerably earlier[41]. The differences in ionisation energies between helium and hydrogen and neon and hydrogen are both considerable and the inefficiency of high-energy neutralisation suggested for helium should also obtain but to a somewhat lesser extent for neon. Indeed, a gradation in neutralising efficiency might well be observed throughout the rare gases, helium to xenon.

The actual energy region of interest is probably 0.5–50 eV, and detailed experimental determinations of triton–tritium atom populations as a function of collision gas in this range have not been made. A great deal of work is being done[42–48] to study collisions between a wide variety of particles, charged and uncharged, at higher energy ranges, e.g. 5 keV; a few experiments are now being performed at lower energies[49]. Fleischman[50] has studied the reaction $H^* + X \rightarrow H^- + X^+$ where X is a rare gas and H^* is a translational excited hydrogen atom, 50–1000 eV. Quite large reaction cross-sections (10^{-1}–10^{-2} Å^2) were found in neon and argon at 50 eV. which increased as the ionisation energy of the rare gas decreased. With recoil tritium in the presence of excess xenon, it is just possible that T^- might have an appreciable population at chemically important energies.

(b) *Moderation of neutral atoms*—Early models about the way in which recoil particles lost energy upon collision were developed by Libby *et al.*[51,52], Miller, Gryder and Dodson[53] and also Capron and Oshima[54]. By and large these theories assumed hard-sphere elastic collisions ('billiard ball'). More detailed calculations about the way in which hot atoms lose energy in collisions with other atoms or molecules have been made during the past decade. The simplest approach was that made by Estrup and Wolfgang[55] who assumed that the the theories developed by Placzek[56], to describe the cooling of neutrons, could be applied to the cooling of hot atoms. Thus the number of collisions experienced by a hot atom in the energy range E to $(E+\mathrm{d}E)$ was $\mathrm{n}(E) = (\alpha E)^{-1}$: α is the average logarithmic energy loss per collision.

$$\alpha = <\ln\left(\frac{\text{energy before collision}}{\text{energy after collision}}\right)>_{\text{av}}$$

and was first calculated assuming hard-sphere elastic collisions between the hot atom and the atoms or molecules. Experimentally it was shown that the effective collision was with just one atom in a molecule rather than the molecule as a whole[57]. Later calculations by Estrup[58] by use of more realistic collision-interaction potentials showed that the collisions were probably quite 'soft' giving lower values of α. Hsiung and Gordus[59] have given a detailed and critical account of the applicability of neutron cooling theories to systems involving hot atoms. In particular, they discuss the anisotropic nature of the collision between a hot atom and another atom or a molecule, that is, there is a much larger cross-section (or probability) for collisions involving a small scattering angle than for collisions in which the hot particle is deflected through a very large angle. This in turn ensures that small energy losses characterise the most frequent collisions (i.e. the most probable energy loss will be less than the average energy loss upon collision). Hsiung and Gordus conclude that equations of the Estrup–Wolfgang type (Section 5.3.3.1) work best when applied to systems of low total reactivity. Felder and Kostin[60] have related functions for $\mathrm{n}(E)$ directly to collision interaction potentials without any assumptions from neutron-cooling theory and have obtained similar results. Often it is possible to express the interaction potential in the form $V(r) = A.r^{-s}$ where A and s are constants. Mason and co-workers[61, 62] by use of particles with initial energies of 1–2 keV have determined A and s for a variety of collision partners (e.g. H–He [63] $A = 3.7\times10^{-12}$, $s = 3.29$, $V(r)$ in ergs). Then,

$$\mathrm{n}(E) = \mathrm{c}.E^{-m}$$

where $m = 1-(2/s)+(s-1)^{-1}$ and c is a constant

thus $$\alpha = (E^{m-1}/\mathrm{c}).$$

Since s is related to the anisotropy of the collision, the possible energy dependence of α can also be related to this factor and to the nature of the interaction potential ($s = \infty$, corresponds to 'hard-sphere' and $s = 1$ 'soft-sphere', coulombic scattering). For values of $s>6$ the energy dependence of α for a tenfold change in E is not great (i.e. less than 50%).

Felder[64] has extended the calculations to consider mixed systems in which one component can undergo chemical reactions with the recoil particle. The general conclusions as to the applicability and limitations of the Estrup–Wolfgang theory are similar to those reached by Hsiung and Gordus. Baer and Amiel[65] and also Thomson[66] have also considered the problem of the moderation of recoil particles in reactive media and also the related problem of the moderation of excited species produced by the chemical reactions of the recoil particles. Despite the care and thoroughness of all these theoretical approaches two general omissions must be emphasised. The first is that only elastic collisions are considered, however, there is no reason to believe that the collisions will necessarily be completely elastic. At energies >20–30 eV many 'hot' particles can initiate ionisation, or receive electrons from their collision partners. At lower energies and in collisions with molecules, there seems every reason to expect highly inelastic collisions in which much translational energy will be transferred to vibrational modes of the molecule, leading often to bond breaking and chemical reactions. The capture of the recoil particle in such a collision to form a labelled molecule is just a special example of this type of inelastic collision. Secondly there is an unfortunate paucity of experimental data on n(E) functions in the energy range of chemical interest namely 0.5–50 eV [67]. Such data as have been used for collision potentials have been derived from experiments where the particles have much higher initial energies (*c.* 1 keV).

A more direct theoretical approach to the problem is to be found in the trajectory-type calculation of the actual collision process between a recoil particle and a molecule (see Section 5.3.2.2).

5.3.1.2 *Low-energy recoil particles*

Such particles are usually produced as a result of an n,γ reaction or by an isomeric transition (I.T.) taking place in an excited nucleus. In both cases photons will be emitted, usually more energetic in the former than the latter case, which will confer recoil energy upon the nucleus. Even if γ-rays with energy in the MeV range are emitted the recoil energies produced are only in the range 100–1000 eV. In many cases de-excitation of the nucleus following neutron capture will proceed by rapid multiple emission of γ-rays[68]. The final recoil energy will be the vector sum of the individual recoil energies imparted by the different γ-rays[69–71]. Considerable cancellation may result giving rise to the possibility of very low recoil energies indeed. The passage of γ-rays from either the (n, γ) or (I.T.) reaction, through the extra-nuclear electron cloud will often cause the ejection of an inner electron. Subsequent relaxation can give rise to the emission of Auger electrons so that the final result of the nuclear reaction will be the production of a particle with a low recoil energy charged, maybe highly charged, and probably electronically excited[72]. The number of collisions such a particle will make before reaching 'chemical energies' will be very much less than for a high-energy recoil particle. It is therefore much more likely that features such as charge and electronic excitation will be more important in controlling the chemical reactions of low-energy recoil particles than for high-energy recoil particles.

5.3.2 Reaction mechanics. Calculation of reaction cross-sections

When a recoil particle collides with a molecule two possibilities exist, (a) a chemical reaction will take place or (b) it will not. In the first case the product molecule may be stable or excited and if it is excited it may decompose in a variety of ways; it may even regenerate the recoil particle. In the second case, the recoil particle will just lose energy to the molecule possibly causing it to decompose. There seems no reason to suppose any fundamental difference in these two types of collision. At high energies, collision times will be so short as to preclude any type of collision other than atom–atom. If one of the atoms were chemically bound, chemical bond-strengths would be energetically irrelevant. At lower energies, say <100 eV, the possibility of quite inelastic collisions between recoil particles and molecules must be considered. This will be especially true in the energy range where chemical reactions can take place, as a chemical reaction can be quite an inelastic process.

5.3.2.1 *Hard-sphere models*

Libby and his co-workers[51, 52], using a 'billiard-ball' model to describe the collisions of the recoil particles, were the first to attempt a detailed mechanistic description of how a 'hot' particle actually participates in a chemical reaction. Miller, Gryder and Dodson[53] developed these ideas, not without difficulty. Cross and Wolfgang[73] also assumed the same type of basic collision in their attempts to describe the reactions of recoil tritium. If a chemically bound atom received greater energy than the sum of its bond strengths, then it was deemed to have been displaced. If the kinetic energies of any two of final particles were less than the bond strength between them, they were regarded as chemically bound. This model was found wanting by comparison with experiment. Undeterred, Suplinkas[74] has recently carried out a complete kinematic analysis of the atom–diatomic molecule collision using hard spheres. The results are in excellent agreement both with experiment (for $T+H_2$, $T+D_2$ etc.) and with more sophisticated calculations (Section 5.3.2.2).

5.3.2.2 *Calculations using trajectory methods*

(a) $H+H_2$—Within the past 5 years there have been many more attempts to try and calculate what happens to the excess energy of a recoil particle when it collides with a molecule, whether it reacts or not. One of the first such calculations was by Karplus, Porter and Sharma[75–77] who studied the $H+H_2 \rightarrow H_2+H$ reaction for a variety of isotopic possibilities using a modified London–Eyring–Polyani–Sato (LEPS) method[78]. Reaction cross-sections as a function of energy are shown in Figure 5.1.

A somewhat simplified version of the Karplus–Porter–Sharma method has been developed by Baer and Amiel[79], which give results in very good agreement with more sophisticated calculations. Porter and Kunt[80] have extended the theoretical treatment of $T+H_2$ and $T+D_2$ systems and have considered a variety of models for the energy loss in non-reactive collisions. They find that an energy-loss function in which most collisions are grazing (i.e. the recoil particle is scattered through only a small angle), with small energy losses, corresponds best with quantum mechanical calculations.

Another way of describing such a function is to say that the average energy loss upon collision is greater than the most probable energy loss. In all the models described by Porter and Kunt α is independent of energy. This contrasts with the earlier calculations of Estrup which showed appreciable energy dependence for α at least for He–T collisions.

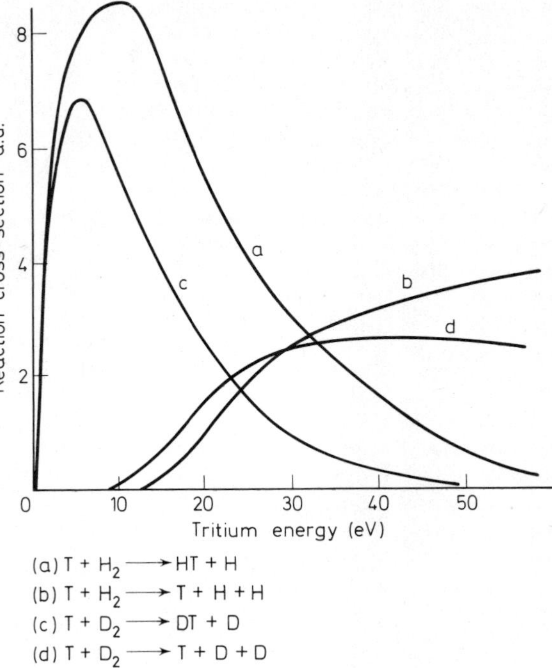

Figure 5.1 Reaction cross-sections as a function of energy (From Karplus *et al.*[77], by courtesy of The American Institute of Physics and the authors)

(b) T+HR—Polyani and co-workers[81] have also used and extended the LEPS potential function technique in order to calculate the potential energy hypersurface for reactions of the type:

$$\begin{aligned} A+BC &\rightarrow AB+C \quad &\text{(abstraction)} \\ &\rightarrow AC+B \quad &\text{(displacement)} \\ &\rightarrow A+B+C \quad &\text{(fragmentation)} \end{aligned}$$

If B is hydrogen and C an alkyl radical, the calculation can be used to describe the main hot-atom reactions with a hydrocarbon, when B is deuterium isotope effects can also be determined. The basic parameters in the calculations are determined by the use of functions for the various possible particle interactions. The equation also contains adjustable parameters related to overlap integrals between pairs of atoms.

Application of this method to the system: A = T*, B = H, C = alkyl, have shown three main types of reaction to predominate. At lowest energies (*c.* 2 eV), abstraction must be associated with T+H collisions, competing at higher energies (*c.* 4 eV) with displacement, brought about by T+R collisions. At these higher energies, the hot-atom trajectories that lead to abstraction are increasingly oblique to the H—R bond, i.e. interaction with R somewhat moderates T so that TH is not ruptured as it forms. At higher energies still (*c.* 12 eV) the cross-section for displacement falls off but a third type of reaction takes over. This is a new abstraction mode in which the T approaches from the R end of the molecule and is severely moderated by collision with R but continues past R to react with H and form HT. Whilst a high-energy abstraction mode, the mechanism is diametrically opposed in its concept to the stripping mechanism proposed by Wolfgang[7,8]. Abstraction by a high-energy atom via a spectator-stripping process was found not to be important.

The calculations show the importance of the time of the collision. Slower hot-atoms spend longer in the collision volume but are more easily deflected by repulsive forces. Higher energy atoms interact more quickly and can overcome higher energy barriers. During the collision the speed with which the various partners to the reaction can move (inversely proportional to their masses) determines the outcome. Thus heavy atoms in a molecule will tend to reduce reactivity because they cannot move quickly and high-energy hot atoms will not become chemically bound in any way since their time of interaction is too short for much energy transfer; slightly moderated and leaving an excited molecule in its wake the hot atom will hurtle on.

The main objection to these calculations is the approximation whereby a complex group is considered as a single atom.

(c) $T+CH_4$—Bunker and Pattengill[82] have taken the necessary next step and considered the interaction between a hot-atom and methane as a six-atom system. All the hydrogens and the carbon in the methyl group are considered explicitly. Abstraction is found to be a lower-energy process (most probable energy *c.* 2.8 eV) than substitution (probable range 4–10 eV) and at energies >7 eV fragmentation is an increasingly important process. The high-energy abstraction process found by Polyani *et al.* in the HR system is not observed for $H—CH_3$. The presence of the C—H bonds and the structure of the methyl group is clearly important in determining the course of possible reaction of a recoil atom with the remaining H atom of methane. One most interesting implication of Bunker's calculations is the lack of any isotope effect between CH_4 and CD_4. Such effects have been observed and it is therefore concluded that the sharp drop in CD_3T when compared with CH_3T is due to a sharp drop in the efficiency of a Walden inversion[83] type substitution-reaction in going from CH_4 to CD_4. Since considerable atomic motion is required and time is of the essence in a hot collision reaction, CH_4 will produce much more CH_3T by inversion than will CD_4.

Angular distributions of the products are also of great interest. As the incident-particle energy, increases, so the angles deviate more and more from those expected intuitively for the lowest activation energy routes. This is entirely reasonable; if the geometry of a collision corresponds to a low activation energy, then a high-energy atom will not be slowed down sufficiently to react. Thus Bunker and Pattengill find that abstraction proceeds at more and more oblique angles as the hot atomic energy increases, corresponding finally to a quasi-stripping reaction (7–10 eV).

5.3.3 General description of recoil particle reactivity

The specific calculations described in the previous section attempt a detailed account of the actual hot-atom collision with a molecule. A much more modest approach is to admit that hot processes may exist and to try and formulate a general description of the way in which product yields may vary with experimental conditions. Calculations of this type were first attempted by Magee and Hamill[84] who compared hot atoms with thermal reactions in the light of absolute rate theory. A more general theory was developed by Estrup[52, 82] which has had a profound effect upon the type of experiments that have been performed in the last decade.

5.3.3.1 Estrup–Wolfgang kinetic theory of hot reactions

(a) *Basic theory*—As the development of the theory has been given by the authors[55, 85], only an outline will be included here. It is assumed that a hot atom (energy E) can react with a molecule to produce a product i with probability $p_i(E)$ or a reaction cross-section $\sigma_i(E)$ which depends upon whether the collision cross-section (S_j between hot atom and component j) is regarded as known or not. If there is a variety of possible hot products produced by hot atoms at different mean energies, then the yield of the product characteristic of the highest energy hot atoms will deplete the hot atom flux (as a function of energy) so that fewer hot-atoms are available to produce those products made by hot atoms with lower energies. The yield of 'high energy' products will therefore appear enhanced. The relative attenuation of low-energy hot atoms can be reduced by the addition of a chemically inert moderator. The yield of both high-energy and low-energy products will be reduced but the *relative* yield of the lower energy product will increase.

If hot-atom reactions are occurring, the the effect of addition of an inert moderator upon the total yield of all hot reactions (P) may be described by:

$$P = 1 - \exp(-fI/\alpha)$$

f = mole fraction of reactive component corrected for collision cross-section

($f_j = S_j X_j / \Sigma S_j X_j$; X_j = mole fraction of component j)

$$I = \int_{E_1}^{E_2} \frac{\Sigma p_i(E)\mathrm{d}E}{E} = \text{reactivity integral}$$

$E_{2,1}$ = upper, lower, energy limits for hot atom reactions

α = logarithmic energy loss parameter

(it is assumed that in a mixture $\alpha = \Sigma f_j \alpha_j$)

If a single reactive substance (R) has been allowed to react with recoil particles in the presence of increasing amounts of an inert moderator (M), then a graph of $[-\ln(1-P)]^{-1}$ *v.* $([1-f_R]/f_R)$ should be a straight line with a slope $[\alpha(M)/I]$ and an intercept $[\alpha(R)/I]$. Such a graph has been called a 'first-kind' plot. From the values of slope and intercept it is possible to express I and α(R) in terms of α(M). Their values are experimentally determined rather than calculated (Section 5.3.1.1 (b)]. By use of these ratios it is then possible to construct 'second-kind' plots for the individual products[86].

$$\left(\frac{\alpha}{f}\right)P_i = A_{i0} - \left(\frac{f}{\alpha}\right)A_{i1} + \frac{1}{2!}\left(\frac{f}{\alpha}\right)^2 A_{i2} - \frac{1}{3!}\left(\frac{f}{\alpha}\right)^3 A_{i3}\ldots$$

$$A_{in} = \int_{E_1}^{E_2} \frac{p_i(E)}{E}\left[\int_{E}^{E_2} \frac{\Sigma p_i(E).\mathrm{d}E}{E}\right]^n \mathrm{d}E$$

$$A_{i0} = I_i,\ A_{i1} = K_i,\ A_{i2} = L_i,\ \text{etc.}$$

Provided the range of (f/α) happens to be small (i.e. <0.5) and the values of I_i and K_i are not too large, then the higher terms involving L_i and M_i etc. will be negligible and $(\alpha/f)\ P_i$ *v.* (f/α) will give a straight line. From the slope the parameter K_i can be determined. This is the 'energy shadowing' term which measures how much hot reaction has taken place at energies above, and in, the range for the production of i. K_i therefore permits the position of the mean energy for the production of i to be determined relative to the mean energy for other products. It is possible to calculate maximum and minimum values for K_i for each possible product and to compare these with the experimental values using a parameter R.

$$R_i = \{K_i(\text{observed}) - \tfrac{1}{2}[I_i]^2\}/I_i[\Sigma_{k\neq i} I_k + \tfrac{1}{2}I_i]$$

If R_i is zero, the product is formed in a reaction of the highest-energy hot atoms, if $R_i \rightarrow 1$, then this product is formed by the lowest energy hot atoms. (Wolfgang[35] uses $F_i = 1 - R_i$.)

Extensions of the theory to mixtures of reactive substance are also possible. The first attempt[86] assumed a direct proportionality between the probability of reaction functions for the two components. This assumption is really too restrictive and later modifications have attempted to elucidate the relative energies of tritium atoms reacting with different components. Milman[87] and Wolfgang[88] have also described modifications of the basic theory to consider reactions in mixtures of components. A general extension of the theory to two reactive components has been given by Johnston[89].

(b)*Limitations of the theory*—Assumptions are inherent both in the theory and in the conclusions that have been drawn from its use. The theory implicitly assumes that the number of collisions in a reactive zone will be large. This justifies the use of an integral form and also the assumption that the population at energy E can be determined by considering reactive attenuation between the limits E_2 and E itself. Calculations by use of a computer model[90] in which only a few collisions took place when the average particle traversed a reactive zone showed that the general form of the theory could still be retained. What does alter, of course, is the recoil-atom population in the reaction zone itself. It will not be so attenuated as the theory would suggest. This will give rise to values of K_i which could be less than the 'minimum' theoretical value (R negative)[91].

It is assumed that the value of α for the various components will be constant in the energy range considered. Estrup[58] has shown that at least for helium this may not be so. If it is desired to try and compute reactivity integrals in terms of probability functions rather than reaction cross-sections, then collision cross-sections must be introduced into the calculations. Again it must be assumed that these are invariant with energy and indeed that values determined from thermal experiments are relevant to collisions with hot atoms.

Despite these qualifications the data derived from experiments, especially with recoil tritium, have been shown to fit the theory very well; i.e. anticipated straight-line relations have been found. However, it has recently been shown[92] that the functions used in the kinetic theory are not critical and that it is possible to violate many of the conditions of the theory and still produce linear graphs. To observe a straight line on an Estrup–Wolfgang graph does not therefore signify much, certainly not as much as was once thought.

The actual value of α that should be used in a mixture, even with the assumption that the component values of α have constant energy, has been the subject of some discussion. First it must be admitted that α(mixture) = $\Sigma f_j \alpha_j$ is only an approximation but its use does not lead to much error unless the values of α_j are very disparate. If one of the components is reactive, then the value of α derived from the equation will be distorted in favour of the α for the reactive substance[93]. Thus the kinetic theory will be most accurate when it is applied to systems of low overall activity.

(c) *Computer models*—With the ready availability of large digital computers, it is now possible to simulate recoil-atom reactions in a simple way[90, 93]. Such calculations were first attempted by Rowland and Coulter[93] who used their results to test the Estrup–Wolfgang[55] kinetic theory. In more recent and detailed calculations[90], functions have been introduced which describe loss of energy upon collision in a random way, controlled by the requirement of an average energy loss over a wide number of collisions. Probability functions have also been devised and the chance of an atom reacting, or merely being moderated, at a particular energy, determined. In this way it is possible to simulate the actual slowing down and possible reactions of recoil atoms. The results can then be compared with kinetic-theory analyses of actual experiments, and the usefulness and limitations of the kinetic theory can be more clearly defined.

(d) *Conclusion*—The Estrup–Wolfgang theory provides a useful method for the analysis of recoil-atom chemistry provided that particles of a clearly defined type are reacting. If ions or electronically excited species are also present, as well as translationally excited atoms, whose concentrations may vary as the experimental conditions in a typical series of samples varies, then only misleading results will be obtained. It is a pity that the analytic functions the theory uses are so uncritical of deviation from the 'ideal' system as defined by the assumptions made in the derivation of the theory.

5.3.3.2 Other theoretical methods

Brodski and Temkin[94, 95] have produced a series of most interesting papers in which yields from hot-atom reactions are calculated. Their theories are based on an adaptation of the Bright–Wigner theory of nuclear reactions, based on the concept of rather narrow (on an energy scale) reaction resonances. Other interesting approaches[96, 97] to hot-atom reactions have been discussions concerning the nature of the parameters in the Arrhenius equation, the relationships between these parameters and their connections with yields from hot-atom reactions.

5.4 REACTIONS OF RECOIL TRITIUM

5.4.1 Introduction

The radioactive isotope of hydrogen, tritium, can be easily produced by two thermal neutron-induced reactions ^{3}He (n,p) ^{3}H and ^{6}Li (n,α) ^{3}H. The former lends itself naturally to gas phase, the latter to condensed phase, experiments. The total recoil-energy per tritium atom from the two reactions is 0.8 and 5 MeV respectively. When the radio-gas chromatographic analysis technique described in Section 5.2 is used, sample activities of about 10^{5} dps are convenient. This requires the production of about 10^{13} tritons so that in a sample of about 10^{20} molecules the average radiation dose is 10^{-1} eV per molecule when helium-3 is used as the source atom. Whilst this dose is small some attempt should be made to control the ion–molecule reaction chains that might alter chemically the nature of the starting material. This is usually done by the addition of small amounts of other molecules of low ionisation potential, e.g. bromine, iodine, etc. The additives are chosen to have a second property, that of reacting readily with free radicals and thermalised tritium atoms. Such compounds are referred to as scavengers. Early experiments[12, 98, 99] demonstrated that the addition of a few mole per cent of oxygen or even iodine vapour reduced the yield of HT from paraffins by *c.* 30 %, this result may be taken as evidence of the efficacy of the scavenger molecules in removing tritium atoms once they have reached thermal energies. A typical sample bulb for a gas-phase experiment would therefore contain a convenient amount of helium-3 (say 1–2 cmHg), a suitable pressure of a scavenger molecule (usually *c.* 5 cmHg oxygen, bromine, etc.) and the actual compound or mixture of substances to be studied. It is assumed that the scavenger removes all thermal atoms and suppresses radical reactions and that therefore the labelled products that are observed are due to chemical reactions initiated by hot species.

In the liquid phase lithium compounds are used. It is most desirable that a homogeneous system be used so that the range of the recoil triton can be estimated and the total anticipated activity calculated. When this is not possible slurries have been used. Such systems suffer from the great disadvantage that the amount of recoil tritium emerging from each solid particle cannot be known and its energy cannot even be estimated. Many liquid systems have been studied and gas-phase experiments extended to the liquid phase by working at low temperatures. The results and conclusions from gas- and liquid-phase experiments will be discussed below.

It should be remembered that whilst many authors talk freely of hot *atom* reactions and whilst indeed the phrase tritium *atom* may creep into the subsequent sections of this review this assumption is by no means proven. The charge state, indeed the electronic state of the tritium species in the chemically reactive hot zone has not been determined experimentally. Also it is most reasonable to believe that the state of the tritium will be a function of environment. Thus although it would be nice and convenient if recoil-particle chemistry were synonymous with hot-atom chemistry, experimentally we have no conclusive reasons for assuming this to be so.

5.4.2 Basic concepts and results

5.4.2.1 Reactions

(a) *Alkanes*—With alkanes (RH) it has been established that recoil tritium reacts 'hot' very efficiently (i.e. *c*. 60% of the tritium atoms that are produced and do not recoil into the walls) to produce HT, labelled hydrogen and RT, labelled parent molecule[41, 100–102]. About 90% of the tritium observed in chemical combination is found in these two products. The remaining 10% is divided amongst alkanes and alkenes with fewer carbon atoms than R. The following points may be significant: (i) the HT:RT ratio varies from compound to compound increasing as the number of CH_3 carbon sites decreases; (ii) isomeric alkane RT molecules are not formed; (iii) the yields of the olefinic counterparts of RT are very small; (iv) the relative yields of fragment alkanes decrease as the number of carbon atoms increases and (v) labelled radicals with fewer carbons than R can be detected.

(b) *Alkenes*—With unsaturated compounds the same general reactions are observed as for alkanes. It is found that the HT:RT ratio is low for ethylene (0.25) but increases rapidly as the relative degree of saturation increases in larger olefines[103]. A new reaction[104] is also observed which can be rationalised by assuming that the hot atom adds on to the double bond. As a result, an excited radical is formed which decomposes by breaking, if possible, a carbon–carbon bond to form a radical and an olefin. The mechanism is such that the olefin carries the label,

$$RCH_2{\cdot}CH{=}CH{\cdot}CH_2R' + T^* \rightarrow [RCH_2{\cdot}CHT{-}\dot{C}H{\cdot}CH_2R']^* \rightarrow \begin{cases} RCH_2{\cdot}CHTCH{=}CH_2 + R'{\cdot} \\ \text{or} \\ RCH_2{\cdot} + CHT{=}CHCH_2R' \end{cases}$$

or

$$RCH_2{\cdot}CH{=}CH{\cdot}CH_2R' + T^* \rightarrow [RCH_2{\cdot}\dot{C}H{-}CHT{\cdot}CH_2R']^* \rightarrow \begin{cases} RCH_2{\cdot}CH{=}CHT + CH_2R'{\cdot} \\ \text{or} \\ R{\cdot} + CH_2{=}CH{-}CHT{\cdot}CH_2R \end{cases}$$

Since an excited radical intermediate is proposed it would be expected that it should show pressure dependence, with an increased chance of decomposition at lower pressures. This has been confirmed by Rowland *et al.* (see Section 5.4.4.4).

(c) *Aromatic compounds*—Benzene[105–107] has been studied, mostly in the liquid phase. A low HT:RT ratio is observed as for ethylene. Some labelled acetylene has also been observed in the gas phase[108]. Toluene has been studied by Ache[109] who found little evidence for selectivity amongst ring hydrogens but also showed that displacement of H by T was more likely at a ring site than in the methyl group.

(d) *Silicon derivatives*—Silane[110] and alkyl silanes[111,112] are more reactive than corresponding alkanes but only conventional products were observed, i.e. those that would easily be rationalised as due to abstraction or replacement of hydrogen. Reactions at Si—C positions were also observed. It appears that Si—H bonds are more susceptible to the abstraction reaction than corresponding C—H bonds, but that the reactivity of CH_3 groups is little affected by being bound to silicon rather than carbon[113].

(e) *Alkyl halides*—Reactions which involve both H and X replacement are observed; the former are characterised by relatively larger yields. The total efficiency of all reactions decreases as halogen substitution increases[114]. Alkyl halides are more liable to unimolecular decomposition[115], by cleavage of an HX molecule, than are alkanes. The spectrum of products from T* + RHX suggests that such reactions take place as a result of the formation of an excited RTX molecule. This type of reaction is observed even for halogenomethanes[116], in particular $T^* + CH_2F_2 \rightarrow [CHTF_2)^* \rightarrow :CTF + HF$, to give rise to tritium labelled derivatives of methylene.

(f) *Other molecules*—Preliminary studies of alcohols, ketones[117] and substituted aromatic molecules[118] have been made in liquid and solid phases.

5.4.2.2 *Stereochemistry*

(a) C sp^3-*sites*—The efficient replacement reaction of hydrogen by tritium could proceed by the direct expulsion of the atom or by the rather more subtle approach from the carbon end of the C—H bond to give rise to Walden-type inversion. Experiments in the solid phase with optically active compounds showed that the reaction proceeded with the retention of configuration[119–121]. In condensed phases, however, a stereochemical constraint may be placed on a radical or intermediate thus preventing inversion. Henchman and Wolfgang[122] showed conclusively in their gas-phase study of s-butanol that the replacement reaction did proceed with retention of configuration. This result has been confirmed more recently by Rowland and co-workers[123,225] by use of compounds with two asymmetric carbon atoms. Gas-phase chromatography was then used to separate DL- and *meso*-forms rather than the more delicate procedures of Henchman. The only evidence for a possible Walden inversion type mechanism has come from the calculation of Bunker [Section 5.3.2.2 (c)] for the CH_3 group in methane. Heavier groups (even CD_3) would probably invert too slowly for this route to be a possible replacement-reaction mechanism, for a rapidly moving hot atom.

(b) C sp^2 *sites*—A study of *cis*- and *trans*-dichloroethylenes[124] showed that the labelled parent compound was formed with substantial retention of configuration.

(c) *Conclusion*—It seems reasonable to conclude from these experiments that the replacement reaction takes place without Walden inversion in all cases, except perhaps methane.

5.4.2.3 *Duration of a reactive collision*

The stereochemistry of the replacement reaction, the speed of a tritium atom with what one imagines to be the correct amount of energy to initiate an 'hot' reaction, and the theoretical results of trajectory calculation all suggest a hot particle-molecule collision in which time is of the essence. There is no time for complicated group migrations, no time for heavy atoms to lumber out of the way, the reaction will take place quickly or it will not take place at all.

Such arguments form the basis for Wolfgang's[125] inertial model by which he seeks to rationalise the less efficient replacement of heavy atoms. The most direct piece of evidence as to the time-scale of a hot reaction comes from Rowland's[126] work with propylene. Rowland[127] has established a relationship between the yield of HT formed by abstraction, from a variety of compounds under controlled conditions, and bond strength. In conventional experiments the measurement of the *D*(C—H) in the methyl group of propylene gives a low value (77 kcal mol^{-1}) because as the C—H bond is weakened the C—C bond lengths in the proto-allylic radical are simultaneously changing. Some of the energy necessary to break the C—H bond is off-set by the resonance stabilisation of the lone electron in the allylic system. Such a relaxation would probably be complete in the time taken for a few C—C bond vibrations, namely 10^{-12} s at most. Rowland[126] found that the yield of HT from the CH_3 of propylene was comparable with that from a normal methyl group and gave no indication of coming from a weak C—H bond. One can therefore conclude that the hot abstraction process is effectively complete before any C—C relaxation has taken place. This indicates a reactive collision time of round *c.* 10^{-13} s or less. There seems no reason to suppose that there will be any fundamental difference in collisions that happen to lead to different labelled products.

5.4.2.4 *Scavengers*

A few mole per cent of reactive molecules, scavengers, are added to experimental systems in order to eliminate thermal (and ion–molecule) reactions. Whilst such molecules may be added with this intention there is no guarantee that the molecules themselves will be so disciplined as to avoid reactions with the hot atoms; that scavengers do eliminate thermal reactions for tritium atoms has been effectively demonstrated, even though in heavily moderated systems more scavenger will be needed than in more reactive systems. That typical scavenger molecules (oxygen, bromine, iodine) may also react with hot tritium atoms has been suggested by Urch and Welch[128]. The evidence is suggestive but not conclusive. The HT:RT ratio from ethane or neopentane decreases steadily as the mole fraction of bromine is increased[129]. This can be due either to competition between bromine and hydrocarbon for some hot tritium species that would otherwise make HT or it may be due to specific efficient moderation by bromine of the tritium atoms that make HT. It should be remembered that very efficient moderation and a chemical reaction are really indistinguishable since both result in the removal of a tritium atom from the 'hot' energy zone.

Whatever the cause, the experimental result is a dramatic intervention in 'hot' yields by a scavenger molecule. Oxygen also affects HT:RT yields but to a lesser degree[130]. It was observed that a few mole per cent considerably enhanced the rather small yield of ethylene from ethane, by comparison with experiments in the absence of scavenger[128]. A rather complicated reaction involving triplet states was first proposed but the suggestion of Baker and Wolfgang[131] that a macroscopic amount of oxygen protects the minute carrier-free amounts of labelled unsaturated molecules from radical attack during the irradiation seems more reasonable.

In order to explain the differences in alkyl halide yields from alkane–tritium systems, with either bromine or iodine scavenger, Urch and Welch[128] suggested that the low alkyl iodide yields are due to the lower vapour pressure of iodine at room temperature. This was demonstrated by Baker and Wolfgang[131].

Seewald and Wolfgang[40] have expressed doubts about the efficiency of oxygen as a scavenger and have favoured bromine. Bromine can however react with hydrocarbons. Samples must therefore be shielded from the light (usually by wrapping in aluminium), however, it is not so easy to stop radical chains from being started by the recoil reactions within the ampoule itself. Samples with methyl groups only seem to be stable, but the presence of methylene groups can lead to short radical chains which produce secondary bromides in macroscopic yields (e.g. isopropyl bromide from propane, s-butyl bromide from butane[132]). The actual degree of macroscopic decomposition seems to vary from laboratory to laboratory[35] and may be related to background γ-radiation during irradiation.

Iodine chloride[133] has also been suggested as a scavenger but it seems prone to decomposition to yield iodine and chlorine readily.

The choice of scavenger and the amount used will have experimental implications which must be taken into account when assessing the results of hot atom experiments; euphoria and faith are not sufficient.

5.4.2.5 *Relative reactivities*

The actual yield of a particular labelled product depends on two factors, the intrinsic reactivity of a particular site to recoil tritium particles of a particular energy and the number of tritium atoms that reach that energy. This second factor can be affected by the reactions that the tritium atoms can undergo at higher energies and also by the moderating properties of the medium. To isolate the first factor it is therefore necessary to operate under standard conditions and to ensure that the substance studied is present at a very low concentration. A rather more precise technique is to determine the yield of the particular product as the concentration is decreased and to extrapolate to 0%. With the symbolism of the Estrup–Wolfgang kinetic theory the actual yield P_i is related to the reactivity integral by, $P_i = I_i/\alpha_{xi}$ where α_{xi} is α for the substance X which is used in excess to create standard conditions of tritium flux (as a function of energy).

A comparison of yields, obtained under standardised conditions, of the same reaction from a variety of different compounds should therefore show how the reactivity integral varies for that reaction from compound to compound. This work has been done by Rowland and co-workers[127, 134–136] for the abstraction reaction that produces HT from a wide variety of substances. Implicitly it is assumed that α_{xi} will be the same for all compounds, i.e. the energy range of the tritium atoms initiating abstraction is constant. Since the 'constant conditions' are a constant small mole fraction of compound being studied, and since the observed yield is then divided by the number of C—H bonds in the molecule, an implicit relationship is assumed between collision cross-section and number of C—H bonds[137]. This is not a very good approximation, especially for unsaturated systems, but even so the range of yields observed far transcends any that could be explained merely by cross-section corrections. It may therefore be concluded that a relationship has been established between bond strength and overall reactivity in the hot atom abstraction process, the yield increases as the strength decreases. Indeed Root and Rowland have suggested that yields of HT may be used to determine bond strengths[138]. The correlations have been extended to alkyl halides and to systems containing N—H bonds[139].

Rowland has suggested that weaker bonds cound react successfully with lower-energy tritium-atoms so that the probability for reaction upon collision, as a function of energy would extend to lower energies.

Since all schemes for energy loss suggest an increase of n(E) as E decreases, such changes at the lower energy end of a reactivity curve will have a much greater effect upon the total yield than a corresponding effect at the upper energy end.

Wolfgang[40] having concluded from neon moderated experiments that the HT was formed by, on average, rather high-energy atoms, and conceding the correctness of Rowland's bond-energy relationship, sought to rationalise it by postulating a high energy sideways stripping reaction[7, 8] in which only a little energy was transferred to RH; thus slight changes in a C—H bond strength could be important. Detailed trajectory calculations of tritium atom–hydrocarbon molecule collisions have shown no evidence for stripping reactions of this type. Baker and Wolfgang[140] have however also shown that the average energy of the tritium atoms that produce HT *increases* in the series CH_4, C_2H_6, C_3H_8 which would seem to contradict Rowland's proposal that weaker bonds can react with lower energy atoms. Baker and Wolfgang suggest that their result can only be reasonably explained by the high-energy stripping reaction. Other explanations are of course always possible. Energy transfer to the group R in RH may be of some importance in determining the stability of the HT formed. The larger is R the more effectively can high-energy atoms transfer energy to it leaving a relatively unexcited tritium atom to complete the abstraction reaction. The bigger R the better an energy sink it is subject to the limitations imposed by the efficiency of energy transfer during the collision. Arguments similar to this have been proposed by Arthy[141] to explain variations in HT yield from primary and tertiary sites in isobutane.

5.4.3 Relative energies of 'hot' tritium reactions

5.4.3.1 Determination of relative energy ranges for particular products

The kinetic theory developed by Estrup and Wolfgang[55] (Section 5.3.3) holds out the possibility of characterising particular labelled products as being formed by hot tritium particles in a particular energy range. If A were formed by cooler tritium atoms, on average, than B then the ratio A:B would increase if an inert moderator were added. In order for this conclusion to be valid the rate at which tritium atoms lose energy in the particular medium must be known accurately; deviations can then be ascribed to chemical reactivity.

Much early work with a variety of hydrocarbons, methane[99], ethane[142], propane[17], butane[17,143], isobutane[91], neopentane[143], n-pentane[143], ethylene[144], propylene[144], but-1-ene[17] and but-2-ene[144], was carried out using helium as a moderator. In general, HT:RT increased as the helium concentration rose (whether oxygen or bromine was used as scavenger) and so it was concluded that on average the abstraction reaction was brought about by cooler hot atoms than the replacement reaction. Experiments with other moderators, krypton[142], xenon[142] and neon[35,40], did not give results in agreement with those from helium. In particular, more HT seemed to be formed when helium moderator was used than with other moderators. Seewald and Wolfgang[40] suggested that in an excess of helium the tritons might not be neutralised [Section 5.3.1.1 (a)]. Another explanation is that the assumption made about the rate of loss of recoil energy may not be quite correct, i.e. α may be a function of energy.

Seewald and Wolfgang[40] and also Baker, Silbert and Wolfgang[35] have shown that if neon moderator is used (with bromine scavenger), the HT:RT ratio falls slightly for methane, ethane, propane and butane, suggesting that HT is produced by slightly higher energy tritium atoms than RT. Westhead and Urch[145] have moderated a mixture of ethane (20 cmHg) and oxygen (5 cmHg) with the rare gases from helium to xenon in turn and shown that the limiting value, in 100% moderator, of HT:RT falls as the atomic number of the moderator increases. It really is rather difficult to say now which is a 'good' moderator and which is not. It may be that ions are only really successfully and completely neutralised by an inert gas such as xenon or it may be that the change in yields of HT and RT in various 'inert' gases reflects the changes in n(E), (the probability of a tritium atom having a particular energy) as a function of energy for the different gases. It might also be that the stability of labelled products HT, or RT is affected by collisions with different inert gases.

Certainly the rather definitive conclusions that are to be found in the literature can no longer attract credence.

5.4.3.2 Competitive reactions; mixtures of reactive components

Competitive reactions between hydrocarbon molecules have been attempted. The interpretation of the results is complicated by the variety of factors that must be considered. The moderating powers of the components may be different and their reactivities as a function of recoil-atom energy may be different. Two main types of experiment have been conducted; variation of concentrations of the two components and moderation of a mixture of fixed composition. In the former case the relative moderating powers must be known (e.g. from experiments in which the components are individually moderated with a rare gas) if the deviations in yields from those anticipated purely on a collision basis are to be interpreted as due to relative energy effects. When a mixture of fixed composition is moderated, the situation is a little more simple. The mixture can be regarded as a single component and an ordinary kinetic theory analysis attempted. Wolfgang[86] has suggested an extension of the kinetic theory to mixtures but the assumptions made are quite drastic. Urch and Welch[146] used this theory to interpret their results from ethane–butane and ethane–neopentane mixtures and concluded that the larger molecules were more reactive with high-energy atoms than ethane. Later experiments by Johnston[89] used helium as a moderator for olefin mixtures (ethylene, propylene, ethylene, *trans*-but-1-ene, etc.) and a more sophisticated form of kinetic theory analysis. In general, quite good agreement was found between competitive experiments and moderated mixtures. It was found that but-2-ene and propylene could react better with higher energy tritium atoms in the displacement reaction than could ethylene[147]. Work with isotopically labelled mixtures (H_2,D_2 [148]; CD_4, H_2 [149]; CH_4,D_2 [150]; CH_4,CD_4 [151]) has produced some confusing results (see Section 5.4.5). Competitive experiments between methyl silanes and the corresponding alkanes have established that the reactivity of these molecules to substitution is comparable but that the abstraction reaction increases in efficiency with the number of Si—H bonds[113].

5.4.3.3 *Hot* H, D *and* T *atoms by photolysis*

Despite the ease with which recoil tritium experiments can be performed, quantitative information about the actual energies of the reacting atoms is wholly lacking. Even much of the information thought to indicate relative energies is now seen to be uncertain. Other means of producing hot-atoms have therefore been sought; in particular, methods which will generate atoms of a known energy. Results can then be compared with those from recoil studies and the reactive energy ranges of the recoil atoms deduced. Photolysis of hydrogen halides HI and HBr is the simplest method of making such hot atoms, because 1849 Å light can produce 3.6 eV H atoms and 2.8 eV T atoms from HI and TI respectively. The enhanced reactivity of hydrogen atoms produced photolytically was observed many years ago[152–154]. More recently, Martin and Willard[155,156] have investigated the reactions of H and D atoms with hydrogen, methane and ethane and Dzantiev *et al.*[157] with cyclopropane.

(a) *Hot* T *from* TBr *and* TI—Chou and Rowland[158,159] showed that 2.8 eV T atoms would react with methane causing both abstraction (to be expected) and also replacement. The HT:CH_3T ratio was about 3.5, much higher than for recoil tritium suggesting that the reaction probability functions below 2.8 eV were much more in favour of the abstraction reaction, with a lower threshold energy, than the replacement reaction. As the wavelength of the light is increased, this ratio also increases, showing that the CH_3T yield is falling more rapidly than HT as the T atom's energy decreases[160]. At *c.* 1.5 eV the yield of CH_3T is effectively zero. 2.8 eV tritium atoms react with HD to produce HT and DT in the ratio 0.7:1.0 whereas in H_2 and D_2 mixtures there seems to be no isotope effect[161]. Chou, Smail and Rowland[162,223] have studied the relative yields of DT from perdeutero-hydrocarbons in the presence of excess CH_4 (cf. Section 5.4.2.5). A rough parallelism with labelled hydrogen yields using recoil tritium under comparable conditions was observed suggesting that the bond-energy effect is due to the reactions of low energy 'hot' atoms, (even so this should not be construed as proof since photolytic atoms can only probe part of the reaction probability function).

(b) *Hot hydrogen atoms in other photolytic reactions*—Photolysis (2138 also 1859 Å) of hydrogen sulphide produces hot hydrogen atoms with energies in the range 1.65–2.05 eV[163,164]. Their reactions with olefins have been studied by Dzantiev[165,166] and also Cvetanovic[167]. Tritriated water when photolysed at 1849 Å produces 1.4 eV T atoms. Chou and Rowland[168] have studied the reactions of these atoms with D_2, CD_4 and ethylene. Dubrin and co-workers[169,170] have used a variety of different sources to generate hot atoms of different energies and so study the energy dependence of the hot hydrogen abstraction reaction.

5.4.3.4 Hot tritium atoms by other techniques

Tritium ions can be produced relatively easily, accelerated and decelerated to a specific low energy. Menzinger and Wolfgang[171,172] have argued that when such ions are near (i.e. a few ångstrom units away) from a solid hydrocarbon surface they will be neutralised. The actual chemistry is therefore the chemistry of hot atoms, albeit in the solid phase. In this way these authors determined a threshold energy for the hot tritium replacement reaction of *c.* 1 eV with solid cyclohexane. Firzova and her co-workers[173,174] have used a similar technique to make extensive studies of the reactions of tritium 'atoms' of known energy with solid aromatic compounds.

5.4.4 Excited species produced by recoil tritium reactions

5.4.4.1 Introduction

The first major discovery in the field of hot tritium chemistry was the demonstration that the reactive collision between the recoil atom and a paraffin molecule was of short duration, leading to either HT or RT as main products, and with relatively little energy transfer to the rest of the molecule, so that subsequent decomposition of the labelled molecule was not of major importance. In alkenes evidence was obtained for an excited radical intermediate formed by addition[104]. More detailed work did, however, show that some products were formed by the decomposition of excited intermediates, in alkane systems also.

5.4.4.2 Decomposition to radicals

(a) *Alkanes*—The reaction of recoil tritium with ethane in the presence of bromine showed that labelled methyl and ethyl radicals were formed, as well as HT, CH_3T and C_2H_5T. Since the yield of labelled methyl bromide increased as the pressure of the parent hydrocarbon, ethane, decreased, it was concluded that some of the labelled ethane molecules were undergoing unimolecular decomposition[175]. A kinetic analysis suggested that only about one-third of molecules labelled acquired enough energy of excitation in the replacement to have a chance of undergoing decomposition, i.e. in two-thirds of the labelling reaction less than 3.7 eV excess energy was left in the molecule. Of the other third it was concluded that the excess energy was *c.* 5 eV.

Neopentane has also been studied[176]. Labelled methyl radical is the major radical despite careful searches for t-butyl and neopentyl radicals. A slight pressure dependence was observed for methyl radical formation but it would seem that the half-life of the excited molecule is less that in the case of ethane. This result seems to suggest that decomposition of the excited neopentane takes place at the C—C bond adjacent to the labelling site and not as a result of the equilibration of excitation energy throughout the molecule.

The rupture of C—H bonds seems to be much less likely that of C—C bonds. Thus the yield of ethyl radicals from ethane, and neopentyl radicals from neopentane, is very much smaller that the yield of methyl radicals in either system. Only a very slight pressure dependence was found by Baker and Wolfgang[177] for ethyl radicals produced from ethane. This suggests that those excited C_2H_5T that decompose by C—H rupture have a shorter half-life and a higher average amount of excitation energy than those that give rise to methyl radicals. The actual life-time of such excited ethane molecules may be comparable with the time of the displacement collision itself. This limiting case of the 'excited molecule' is the double-knock reaction mechanism first proposed by Urch and Wolfgang[102]. In such a reaction it is proposed that two hydrogen atoms are simultaneously ejected by one tritium atom to give a labelled radical. Rowland's[178] arguments based on studies of propyl chloride that the double-knock reaction is of insignificant importance have been criticised by Wolfgang[177].

(b) *Alkyl compounds*—The reaction of recoil tritium with alkyl halides gives rise to excited molecules which are more prone to decomposition than alkanes, e.g. methyl chloride[179].

$$CH_3Cl + T \begin{cases} \rightarrow HT + CH_2Cl\cdot \\ \rightarrow CH_3T + Cl\cdot \\ \rightarrow [CH_2TCl]^* \rightarrow CH_2T\cdot + Cl\cdot \end{cases}$$

Cipollini and Stöcklin[180] have related the ease of decomposition of $[CH_2TX]^*$ with the C—X bond strength for X, fluoride, hydroxyl, chloride, amide, thiol. Ethyl chloride[115] yields a wider variety of excited products,

$$[C_2H_4TCl]^* \begin{cases} \rightarrow C_2H_4T\cdot + Cl\cdot \\ \rightarrow CH_2T\cdot + CH_2Cl\cdot \\ \rightarrow CHTCl\cdot + CH_3\cdot \\ \rightarrow C_2H_3T + HCl \text{ [see Section 5.4.4.3 (b)]} \end{cases}$$

Yields derived from such reactions are *c.* 70% of those of direct substitution reactions in samples at atmospheric pressure, increasing to about 100% at 100 Torr.

5.4.4.3 *Decomposition and isomerisation to molecular products*

(a) *Hydrocarbons*—Rowland *et al.* have studied a variety of small ring hydrocarbons which readily undergo isomerisation (e.g. cyclopropane→ propylene) or fragmentation (e.g. cyclobutane → 2 ethylene), to molecular and not radical products. Pressure dependence has been demonstrated for cyclopropane[181], cyclobutane[182] and substituted cyclobutanes[183]. In all cases the average energy deposited in those molecules that react further seems to be *c.* 5 eV. When a methyl group rather than a hydrogen atom is replaced, however, a greater amount of energy seems to be left behind since the yield of monomethylcyclobutane from dimethylcyclobutane is quite small and the corresponding yields of ethylene and propylene are enhanced.

There is also evidence that molecular fragmentation may also occur in recently-labelled alkane molecules. Rowland[185] has investigated the position of the tritium atom in the isobutane formed when recoil tritium reacts with neopentane. If an excited neopentyl radical of the type $[(CH_3)_3C{\cdot}CHT{\cdot}]^*$ were formed, either by a double-knock process or by C—H cleavage rapidly following replacement, the isobutene formed would be exclusively $(CH_3)_2C = CHT$. Yet tritium is also found in the methyl positions which suggests molecular cleavage of a rather random kind to produce methane and isobutene.

A molecular fragmentation reaction might also be invoked to explain the quite small yield of the alkene molecule that corresponds to the parent alkane, i.e. ethylene from ethane, or propylene from propane[102]. In this case molecular hydrogen would be eliminated. The yields are only a few per cent of the labelled parent (except in the case of ethane, methyl chloride, iodine scavenger system where ethylene:ethane ratios in the range 3:10–1:10 were observed[180]), and diminish with increasing molecular complexity.

This is in keeping with the suggestion of an average excitation energy of *c.* 5 eV remaining in the molecule. Most molecules will decompose by C—C bond cleavage but a few will breakdown by the higher activation energy routes of molecular hydrogen elimination or C—H bond rupture. Experiments with ethane $(CH_3{\cdot}CD_3)$ show[184] that hydrogen elimination is a 1,2- and not a 1,1-process.

An indication that similar reactions may occur in recently labelled fragment alkanes has been found by Avdonina *et al.*[186] by an examination of the ratios of alkanes and alkenes with fewer carbons than the initial paraffin molecule. The ethylene: ethane and propylene:propane ratios were found to be constant even when derived from different larger alkanes, suggesting a common origin for both members of the saturated–unsaturated pair. Such an origin could be an excited fragment alkane produced by direct tritium attack on the original alkane. That such a species might have considerable excitation energy is suggested by Rowland's work with 1,3-dimethylcyclobutane[183]. The excited fragment alkane could then undergo a variety of reactions, e.g.

$[CH_3{\cdot}CH_2{\cdot}CH_3\text{-t}]^*$	$\rightarrow CH_3{\cdot}CH_2{\cdot}CH_3\text{-t}$	moderation
	$\rightarrow CH_3{\cdot}CH_2{\cdot} + CH_3{\cdot}$	C—C cleavage
	$\rightarrow CH_3{\cdot}CH_2{\cdot}CH_2{\cdot} + H{\cdot}$ (n- or iso-)	C—H cleavage
	$\rightarrow CH_3{\cdot}CH = CH_2 + H_2$	molecular fragmentation

of which the latter routes are characterised by higher activation energies than C—C cleavage but probably not by more than 10–20 kcal mol^{-1}. A molecule with say *c.* 6 eV. of excitation energy will not discriminate greatly between reaction routes whose different activation energies are considerably less than the excitation energy it possesses. Rowland[185] has proposed a similar but more direct mechanism for the production of fragment molecules from an excited parent species,

$$[RT]^* (C_n) \rightarrow \text{alkane}\,(C_{n-x}) + \text{olefin}\,(C_x)$$

so that either the alkane or olefin would be labelled. Such a mechanism would not easily explain the basic ratios noticed by Avdonina *et al.*

C—H cleavage is also characteristic of electronically excited hydrocarbons so that it is possible that such species may be involved in these reactions. This in turn would require either electronically excited tritium atoms (e.g. in a $2s^1$ configuration) or that a hot tritium atom (10–15 eV) could, upon collision, effectively convert translational energy to electronic excitation. At the present state of knowledge these are possibilities rather than probabilities.

In ethylene the C—H bond is weaker that the carbon–carbon double bond and so a thermally excited molecule would be expected to decompose by a route involving C—H rupture. It is therefore of interest to note the observation of a small yield of acetylene from ethylene by Johnston[147]. This is presumably formed by a direct molecular cleavage $[C_2H_3T]^* \rightarrow C_2HT + H_2$. If any $C_2H_2T\cdot$ radicals were also formed they would not have been detected.

Acetylene (*c.* 1% of labelled benzene yield) has also been observed[108] in the labelled products that result from the reaction of recoil tritium atoms with benzene, suggestive of molecular fragmentation $[C_6H_5T]^* \rightarrow C_2HT + 2C_2H_2$.

(b) *Alkyl halides*—Recently labelled alkyl halides are much more prone to decomposition following the reaction in which tritium replaces hydrogen, e.g.

$$[C_2H_4TCl]^* \rightarrow C_2H_3T + HCl$$

Tang and Rowland[187] have also established that the halogen hydride is formed by a 1,2, rather than a 1,1-elimination. Similarly, a study of the propylene formed from isopropyl chloride[184] (where half of the tritium is found in the methyl group and the other half at olefinic sites) suggests that HCl is eliminated in a random fashion from $[CH_3CHClCH_3\text{-t}]^*$.

(c) *Methyl cyanide and isocyanide*—These compounds have been studied by Ting and Rowland[188, 189], who showed that the isocyanide isomerised completely (in the gas phase) to the cyanide by the tritium labelling reaction in the methyl group; addition of argon helped to stabilise the methyl cyanide-t. No labelled methyl isocyanide was observed as a result of recoil tritium reactions with methyl cyanide in the gas phase. With the assumption of equilibrium conditions, Ting and Rowland attempted to calculate the ratio CH_2TCN:CH_2TNC. They were, however, unable to rationalise the complete lack of labelled methyl isocyanide. Perhaps this was because of the assumptions made in the calculations or perhaps because the excess energy has not equilibrated before recently-labelled methyl isocyanide isomerises.

Non-equilibration of excess energy is also indicated by the way in which neopentane-t preferentially decomposes by breaking the C—C bond adjacent to the labelling site[175]. Certainly the estimate of the average energy left in the methyl isocyanide molecule (*c.* 4 eV) is not in agreement with other experiments with ethane[175], cyclobutane[182] etc.

5.4.4.4 Decomposition of excited radicals

The decomposition reactions of excited radicals formed by hot-atom addition to a double bond have been studied in great detail by Rowland and his co-workers[190]. The anticipated pressure dependence has been confirmed, indeed it has been found that the reaction proceeds even in the liquid phase[191, 192]. There is suggestive evidence that the excited radical can decompose not only by the specific C—C bond rupture outlined in Section 5.4.2.1 (b) but also by breaking a C—H bond. This mechanism provides a convenient explanation for the formation of but-1-ene from but-2-ene[25], i.e. $[CH_3{\cdot}\dot{C}H{\cdot}CHT{\cdot}CH_3]^* \rightarrow CH_2{=}CH{-}CHT{\cdot}CH_3 + H{\cdot}$ (and also for but-2-ene from but-1-ene) although a small amount of but-1-ene labelled at the olefinic site is also found. This may be formed by a 1,3 hydrogen shift with a rather high activation energy in a but-2-ene molecule excited by a recent labelling event.

If C—H bond cleavage can occur, then it should also take place in radicals which have no C—C bonds that can be easily broken, i.e. $C_2H_4T{\cdot}$, $CH_2T{\cdot}\dot{C}H{\cdot}CH_3$ and $(CH_3)_2\dot{C}{-}CH_2T$. Evidence that such radicals do decompose to give an enhanced yield of labelled parent olefin has been obtained by Garland[193] (this in turn depresses the HT:RT ratio, as is found in small olefins). A lack of selectivity in hot-atom attack on a double bond has been observed but cool hot atoms show a marked preference for terminal addition to but-1-ene forming the more stable s-butyl radical[194]. Temperature effects have also been found[195]; the intrinsic energy of the olefin affects the ratio of fragment olefins formed after tritium atom addition. Labelled radicals are also formed by tritium attack on small ring molecules (e.g. cyclopropane[181]). The radicals so formed decompose by both C—C and C—H cleavage.

5.4.4.5 Conclusion

It can be concluded that when recoil tritium reacts to produce labelled molecules, the molecules will be vibrationally excited to a certain extent. The average amount of excitation will probably vary from molecule to molecule depending upon the nature of the reactive collision (e.g., T for H or T for CH_3, etc.) and upon the type of molecule. The main reactions are comparable with thermally induced decompositions or isomerisations indicating that the excitation energy is primarily vibrational in character. Only minor yields are observed from reactions paths which have somewhat higher activation-energies than the minimum for reaction. This indicates that the percentage of molecules produced with a lot of excess energy is rather small. Indeed from ethane there is the complementary evidence that two-thirds of the molecules have less energy deposited in them, by the replacement reaction, than is required to break a C—C bond (i.e. < 3.7 eV).

5.4.5 Isotope effects

When recoil tritium atoms react with perdeutero-molecules the yields of labelled products are not the same as from their hydrogen-containing counterparts. Very roughly it is found that the deutero-compounds give smaller yields. The isotope effect upon yields can be a function[185] of (a) a change in reactivity of the molecule associated with the H—D change and (b) a change in the moderating power of the molecules so that the tritium atom flux, as a function of energy (i.e. n(*E*)), is different for the two systems, per-H molecule and per-D molecule. The reactivity isotope effect can be subdivided into a 'first-order' effect and a second-order effect depending upon whether the tritium reaction is at an H or D site or at another position (e.g. a C—C or C—X bond).

Isotopically differentiated molecules have been studied by Lee, Musgrave and Rowland[196] and by Lee and Rowland[151], (H_2, D_2 and CH_4,CD_4). In order to avoid the changes in moderating power that might accompany mixtures of varying isotopic composition, Root and Rowland[197] investigated CH_2D_2 and found an isotope effect in favour of tritium replacing H of 1.35 (oxygen scavenger). Jurgeleit and Wolfgang[57] studied CH_3F and CD_3F and found the former more reactive. These authors concluded that the diminished yield in CD_3F was due to its greater moderating power. However, in the results from early experiments it is difficult to distinguish between moderator and reactivity effects. Smail and Rowland[198] have investigated CHF_3 and CDF_3 and Lee, Miller and Rowland[199] have studied CH_3F and CD_3F in more detail; they concluded, as a result of competitive and moderator experiments, that the second-order effect favours T for F substitution in CH_3F and also that CH_3F is a better moderator than CD_3F [α (CH_3F): α(CD_3F) = 1.2]. The results also indicated that T for H substitution was more efficient than T for D. These conclusions are in accord with current opinion as to the nature of the tritium-molecule collision. The lighter hydrogen atoms move more quickly, either out of the way of the tritium to permit reaction, or more quickly to absorb excess energy as a vibration, than their deutero-counterparts. That energy transfer should be more efficient between T and D than T and H is anticipated only in models based on the idea of an isolated hard sphere in elastic collision. It would seem that in colliding with either an H or a D in a molecule a tritium atom is neither engaging in an elastic hard-sphere collision nor can it be regarded as colliding with an isolated atom[57].

In a detailed study of isobutane $(CH_3)_3CD$, $(CD_3)_3CH$ and $(CD_3)_3CD$, Smail and Rowland[200] have demonstrated an isotope effect for the abstraction reaction at the tertiary site of *c.* 1.6 and argue that the low value at the primary site (1.15) is a combination of moderator and reactivity isotope effects. An isotope effect of 1.25:1 was found for T for H:T for D at the tertiary position[201].

Direct measurements[148] of reactivity integrals for H_2 and D_2 with argon moderator gave $I(H_2) = 6.9 \pm 0.7\ \alpha(Ar)$ and $I(D_2) = 7.1 \pm 0.7\ \alpha(Ar)$ but the moderation of a H_2, D_2 mixture suggested that the overall reactivity of H_2 was slightly greater than D_2, $I(H_2):I(D_2) = 1.15 \pm 0.04$. Other experiments have studied the isotopic pairs CH_4, D_2 [150, 202], CD_4, H_2 [149] and CH_4, CD_4 [151]. The interpretation of results from reactive mixtures is beset with difficulties due to the different moderating powers and different reactivities of the components. Even so, their interpretation has not been aided by the insistence of some authors in presenting results as a function of mole fraction of the components and then attempting to draw conclusions about molecular reactivities. Hsiung and Gordus[203] pointed out the fallacy of such arguments 8 years ago seemingly without avail. (Mole fractions may be related to reaction cross-sections; molecular reactivities can only be deduced if the mole fraction is corrected by the collision cross-section.) The application of 'pseudo-kinetic theory' arguments to reactive mixtures of this type has recently been criticised by Wolfgang[204]. Detailed conclusions as to the relative reactivities and reactivity ranges for the molecules CH_4, CD_4, H_2 and D_2 cannot yet be made[204, 205].

5.4.6 Phase effects

Whilst most experiments with recoil tritium have been conducted in the gas phase some have been carried out in the liquid and also the solid phase. Lithium compounds provide a convenient source of recoil tritium. The most important general conclusion is that liquid phase results are quite similar to those in the gas phase indicating that the nature of the reactive collision is essentially the same. The fate of excited species can, however, be modified as can the outcome of reactions involving radicals.

5.4.6.1 Aliphatic hydrocarbons

The reaction of recoil tritium with alkanes gives the same spectrum of labelled products, in the same yields, in both gaseous and liquid phases[12, 192]. The only specific exception to this generalisation has been found in isobutane[200] where the tritium replacement reaction proceeds relatively less favourably at the tertiary position than at the primary position when the reaction takes place in the liquid phase. In the gas phase little or no selectivity was observed.

When alkenes are studied, the yields of products which arrive from the decomposition of an excited radical are reduced; this has been observed for hex-2-ene, but-2-ene and propylene[194]. Excited radical products are not eliminated in the liquid phase (usually reduced to *c.* 30–50% of gas-phase yield) which suggests that many radicals are so highly excited that intramolecular energy rearrangements compete quite successfully with intermolecular energy transfer. This in turn suggests that radicals with such a high degree of excitation energy should be able to break C—H and C—C bonds with almost equal ease. Thus it is interesting to note that the relative yield of but-1-ene-t from but-2-ene is almost unaffected by change from gas to liquid phase, whilst the yield of propylene-t is reduced by *c.* 60%. Mahon and Garland[193] suggest a similar effect in propylene but also propose that the reaction $C_3H_6+C_3H_6T\cdot \rightarrow C_3H_5T+C_3H_7\cdot$ can augment the yield of labelled propylene in the liquid phase.

Decomposition of recently-labelled molecules is also observed in the liquid phase although on a somewhat diminished scale. Thus labelled ethylene is still formed from liquid cyclobutane[182].

5.4.6.2 *Aromatic systems*

Benzene[206], other aromatic compounds[207, 208] and mixtures containing benzene have been studied extensively for a number of years. In the gas phase $HT:C_6H_5T$ are in the ratio 1:5; this increases to 2:5 in liquid benzene. Garland and Rowland[105, 106] suggest that it is the yield of labelled benzene which falls on going to the condensed phase and propose that $C_6H_6T\cdot$ radicals are involved. These may either decompose to yield benzene-t in the gas phase or indulge in chemical reactions in the liquid (forming 1,3- and 1,4-cyclohexadiene-t and polymeric products[209]). In mixtures of benzene with aliphatic compounds, e.g. benzene and cyclohexane, it has been suggested by Avdonina[210, 211] that enhanced yields of cyclohexane-t are produced and this is due to very rapid collision-stabilisation of excited cyclic $C_6H_{11}T$ molecules by benzene. The benzene yield is depressed by an approximately corresponding amount. These conclusions are, however, not based upon an independent assessment of the hot tritium available for reactions but merely from variations in total observed activity (liquid and gaseous). By taking the activity observed in cyclohexane as a standard (and assuming some activity to be lost in benzene due to polymer formation) Sokolowska[212] demonstrated linear relationships between benzene-t, cyclohexane-t and the corresponding mole fractions. Studies of this type have been extended to mixtures of aniline or benzoic acid with ethanol and acetone[213], benzene with alkane[214] and toluene with cyclohexane[214]. These conclusions have been disputed by Avdonina and Nesmeyanov[215–217] who claim to observe exaltation of the cyclohexane-t yield in cyclohexane–benzene mixtures even when polymer yield is taken into account. These workers have also studied the effect of iodine; the HT yield remains constant as does that of cyclohexane-t but that of benzene-t is increased[218]. This may be due to I_2–benzene complexes forming a structured liquid which can more easily deactivate $C_6H_5T^*$, or it may be due to specific reactions involving $C_6H_6T\cdot$ radicals and iodine.

5.4.6.3 *Alkyl compounds*

Phase changes in methyl chloride[179] and ethyl chloride[115] only affect the yields of those products thought to be made via an excited intermediate. As in similar cases discussed in Section 5.4.6.1, such yields are *not* eliminated but reduced by 50–70% indicating that some excited species have very short half-lives. In studies with methyl cyanide and methyl isocyanide[189, 190] labelled isocyanide produced by the reaction of recoil tritium with methyl cyanide, by isomerisation, is only observed in the liquid phase.

5.4.7 Conclusion; models and mechanisms

The data which have been reviewed above have invited many interpretations; as always the more scant the evidence the easier it is to theorise.

5.4.7.1 Steric model

After observation that the HT:RT ratio increased as the complexity of a paraffin increased, Urch and Wolfgang[102] proposed the steric model for hot tritium reactions. The abstraction reaction was assumed to be associated with axial (along C—H axes) or near axial approach by the hot atom whilst the labelling reaction was thought to be more characteristic of atoms attacking the C—H bond region from a more perpendicular direction; clearly the latter but not the former would suffer some steric hindrance in the progression CH_4, CH_3R, CH_2R_2, CHR_3. Indeed in a preliminary study of isobutane, Odell, Rosenberg, Fink and Wolfgang[219] claimed to show relatively less efficient tritium labelling at the tertiary site than at the primary sites in isobutane. Later work by Rowland[200] suggested a rather less dramatic effect (how the results are to be interpreted depends rather more on personal prejudice than anything else, thus primary:tertiary labelling efficiencies of 1:0.87 can either be regarded as more or less equal, clearly showing the steric model to be wrong or clearly not equal, showing labelling at the primary site to be more efficient than at the tertiary, more or less in accord with steric model ideas!)

Competitive experiments between butane and methane[220] established that C_4H_9T was formed about two and half times more efficiently than CH_3T, i.e. more or less in the ratio of the number of C—H bond sites, or approximately the ratio of the collision cross-sections. This evidence can be used to refute the steric model or to establish that butane is larger than methane.

5.4.7.2 The abstraction reaction and C—H bond strengths

Rowland[127] has succeeded in establishing a correlation between C—H bond strength and HT yield which would seem to supersede the steric model as the true reason why the HT:RT ratio increases with molecular complexity. Root[221] has also suggested that C—C bond strength may also be important in determining the yields of labelled compounds in recoil tritium–alkyl replacement reactions. Since the C—H bond strength is a function of the s/p character of the hybrid orbital used by the carbon atom an exactly parallel correlation can be established from n.m.r. coupling constants[224]. This in turn suggests the idea of a hot tritium atom, bent on an abstraction reaction, as an electrophilic reagent[222]. All these ideas have been used to support the proposal that the increased HT yield at weaker C—H bond sites is due to a lower cut-off energy for the reaction. Data from the reactions of 2.8 eV tritium atoms are also in accord with this mechanism[162]. This view has been criticised by Wolfgang[7,8] who suggests that the increase in total reactivity that could be brought about by lowering the lower energy limit for reaction by a few kilocalories is not nearly sufficient to explain the really considerable changes in HT yield.

Wolfgang[7,8] has therefore proposed the stripping mode, a high energy abstraction reaction that would be sensitive to bond strength. In support of this view, Baker and Wolfgang[140] have shown that the average energy of tritium atoms abstracting hydrogen from weaker bonds is *greater* than those abstracting hydrogen from strong bonds.

Of course it could be that bond strength has little to do with the observed results. In a series of molecules CH_4, CH_3—Z, CH_2=Z*, CH≡Z*, the C—H bond strength decreases as the complexity of Z increases. It could be that when Z is complex, representing many groups, it is better able to absorb excess energy brought to a collision by a hot tritium atom, than when it is simple. Thus higher energy atoms would abstract hydrogen more efficiently and give higher yields of HT and the basic reason would not be the nature of the C—H bond but of the other atoms or groups attached to the carbon atom!

5.4.7.3 *Inertial effects*

Odum and Wolfgang[114] observed that as halogen substitution of methane increased so overall reactivity decreased. They chose to interpret this as due to an inertial effect[125], which can be thought of in two stages. At first the hot atom collides with an atom of a molecule and attempts to displace it. The lighter this atom is the more quickly it will move away. The second step concerns the remainder of the molecule; if it is light it could rotate more quickly to capture the newly arrived hot atom than if it were heavy (although exactly why it should rotate has never been explained). These two effects will clearly favour an efficient reaction in methane, namely $CH_4 + T^* \rightarrow CH_3T + H\cdot$, and an inefficient reaction in, for example, perfluoromethane namely $CF_4 + T^* \rightarrow CF_3T + F\cdot$.

This idea has also been applied to the reaction of hot tritium atoms at C—C bonds in neopentane and $CH_3\cdot CF_3$ where it is found that the yield of CH_3T far outweighs that of $(CH_3)_3CT$ or CF_3T. In both cases this could be explained as due to the relative ease with which the lighter group can rotate to capture the tritium atom. The result with $CH_3\cdot CF_3$ disproves the possibility of a bond strength factor being important.

By contrast with the results from $(CH_3)_4C$ the ratio of yields of CH_3T and $TSi(CH_3)_3$ from $(CH_3)_4Si$ [112] is only 2.5:1 even though the trimethyl silicon radical is heavier that trimethyl carbon. This suggests that inertial factors may not be fundamental in determining such product ratios. Clearly more detailed work is required, preferably with substituents other than halogens.

5.4.7.4 *Hot atom energies*

Both the steric and inertial models are implicitly competitive models, i.e. they depend for their effect upon a tritium atom with a given energy being

---

* = and ≡ indicate two or three σ bonds, not double or triple bonds

able to perform one or another reaction. There is no reason for this necessarily to be so. It may be that the potential reactions of a hot atom are closely controlled by its energy, that at one energy a hot atom has *only* the possibility of forming say RT, and at another energy can *only* form HT. In this case the arguments upon which the steric model is based and the arguments given above about the preferential formation of CH_3T from neopentane or 1,1,1-trifluoroethane, are null and void. Experiments with a helium moderator indeed have suggested that the average energy of tritium atoms forming small labelled molecules is less than that of tritium atoms that form large labelled molecules[147]. This observation in turn promoted the suggestion that the size of the group with which the tritium atom combined was of importance. The larger it was the more successfully could a hot atom remain chemically bound because of the greater number of degrees of freedom. Such arguments, whilst plausible, are unfortunately without experimental foundation since it does not now seem possible to draw meaningful conclusions as to the relative energies of hot atom reactions from moderator experiments with inert gases[145].

5.4.7.5 *Conclusions*

What then remains of the past decades labour in this particular vineyard? It must be admitted it is really not known if HT is formed at higher or lower average energies than RT, or whether the range of energies in which these two products are formed is more, or, less the same or widely different. It must be admitted that although the kinetic theory can be applied most successfully to experimental results it contains many assumptions which are not in accord with reality. Consequently the conclusions drawn from experiments analysed in this way will often be in error.

It can be regarded as established that the collision between a hot tritium atom and a molecule is a rapid event, 10^{-13}–10^{-14} s, and that, therefore, very little motion by atoms in the molecule other than the one or two directly hit will take place. Since a simple molecule vibration takes about the same period of time, the conclusion is rather obvious, yet it has been honoured with the appellation of 'Golden Rule'. The nature of the collision has important consequences: (a) retention of configuration, (b) transference of relatively little excitation energy to the rest of the molecule, (c) reaction with atoms on the outside of the molecule (e.g., H rather than C in hydrocarbon), and (d) a hot atom must not bring too much excess energy into its collision with a molecule. The atom will not have time to divest itself of much energy so that success in a hot collision (in as much as success is the formation of a stable labelled molecule) will go to those atoms that have the right amount of energy, or a little more, to go over the activation energy barrier that characterises their particular trajectory. Too much will cause fragmentation or just non-capture of the hot atom.

These general conclusions will be modified for other hot atoms because of mass differences or because they are not univalent.

5.5 RECOIL FLUORINE

The chemistry of recoil fluorine species is most conveniently studied by following the reactions of the radioactive isotope ^{18}F (half-life 109.7 min).

This species is produced by a variety of nuclear reactions, e.g. ^{16}O (α, p+n or d) ^{18}F, $^{16}O(^{3}H, n)$ ^{18}F, $^{19}F(n, 2n)$ ^{18}F, ^{19}F (γ, n) ^{18}F and even the double reaction ^{6}Li (n, α) ^{3}H–$^{16}O(^{3}H, n)$ ^{18}F [226]. In all cases the fluorine is produced with a high recoil energy (i.e. many keV), so that it probably reacts chemically as an atom in its ground electronic state (ref. 227, p. 159). The chemical reactions of recoil fluorine with hydrocarbons and fluorocarbons have been investigated in solid, liquid and gas phases.

5.5.1 Experimental methods

5.5.1.1 General techniques

The (γ, n) reaction is the most widely used to produce ^{18}F, the source of high-energy x-rays being the bremsstrahlung from an electron accelerator (e.g. 25 MeV electrons falling on a tungsten or tantalum target). Fast neutrons to initiate ^{19}F (n, 2n) ^{18}F come either from the D, T reaction (14 MeV) or from the bombardment of beryllium with 25 MeV deuterons. Samples have been encapsulated in Pyrex, stainless steel or gold-plated copper. The latter types of vessel permit yields of $H^{18}F$ and $F^{18}F$ to be estimated. Analysis of volatile compounds has been by radio-gas chromatography using a 'window' proportional counter (Section 5.2). Aten[228] has introduced a preliminary organic–inorganic separation.

5.5.1.2 Estimation of total ^{18}F production

It is quite difficult to estimate the total amount of ^{18}F induced in a sample because two very probable products, HF and F_2, are so reactive. Wolfgang and co-workers[88, 227] have used pairs of sample vessels, one containing the system to be studied and the other an identical amount of source material (e.g. CF_4), neon moderator, ethylene and iodine (see Section 5.5.1.3). The latter two compounds act as a scavenger for thermalised fluorine atoms. It is hoped that all the fluorine that is stopped in the gas phase will be present as fairly volatile organic compounds. The total activity could be estimated directly by counting the ampoule before analysis. The amount of ^{18}F that recoiled into the wall can be determined after the volatile products have been analysed. Aten[228] has determined inorganic ^{18}F and wall ^{18}F by washing with sodium fluoride and hydrochloric acid.

5.5.1.3 Scavengers

It is always of interest in recoil chemistry to know if the atom utilises any of its excess energy to initiate chemical reactions, i.e. does it react hot? To this end some method must be devised to remove thermalised atoms whose reactions would otherwise interfere with the results. Colebourne and Wolfgang[230], have shown that a mixture of ethylene and iodine (at room temperature and vapour pressure) is a satisfactory scavenger for thermal fluorine atoms. Later experiments with methane[227] showed that C—H bonds are almost as effective for reaction with thermalised ^{18}F atoms. Rowland and co-workers[231] have also used oxygen and tetramethylethylene as scavengers. It seems that usually *c.* 5–20% of the recoil fluorine reacts 'hot', to give organic products. It is not known how much of the HF yield results from a hydrogen-abstraction reaction initiated by 'hot' atoms.

5.5.2 Experimental results; the chemistry of hot ^{18}F

5.5.2.1 Reactions with perfluorocarbons

The simplest reaction, with CF_4, showed that a hot ^{18}F-for-fluorine-replacement reaction does exist[227, 229] but that it has a much lower collision efficiency than the corresponding T for H reaction (only *c.* 4% of the recoil fluorine available for reaction in the gas phase was finally detected as $CF_3{}^{18}F$). Labelled CF_3I was also observed ($\sim$2% yield). This compound presumably arises via the decomposition of vibrationally excited $CF_3^{18}F$. Later work by Tang, Smail and Rowland[232] showed that $:CF^{18}F$ is also produced in a yield comparable to that of $CF_3^{18}F$. If it is assumed that the difluorocarbene is also produced by the decomposition of excited $CF_3^{18}F$, then the total initial yield of this compound is *c.* 12–14% of the available hot fluorine compared with a figure of *c.* 30% for tritium. With tetrafluoroethylene[234], recoil fluorine gives rise to highly excited $CF_2{\cdot}CF^{18}F$ by displacement and to $\cdot CF{\cdot}CF_2^{18}F$ by addition. Both molecules are prone to decomposition and the yield of labelled parent molecule is quite small. Excited tetrafluoroethylene decomposes to perfluorocarbene which reacts with the hydrogen iodide present as a scavenger to give CF_2HI. The excited radical also splits the carbon–carbon bond and the labelled trifluoromethyl radical reacts with HI to give trifluoromethane. Thermalised C_2F_5 radicals also react with HI to produce C_2F_5—H as the major product. Sulphur hexafluoride has been used as a moderator and the experiments indicate that fluorine atoms of decreasing energies give rise to $:CF_2$, $\cdot CF_3$ and $C_2F_5\cdot$ as might be expected. Dulman and Aten[228, 229] have investigated a variety of perfluorocarbons of much higher molecular weight (e.g. perfluoro-benzene, -toluene, -hexane, -hexene, -methylcyclohexane). In all cases a wide variety of labelled products was observed but often only the parent fluorocarbon could be identified. The compounds were studied in the liquid phase in the absence of scavenger. Typically *c.* 10–15% of the total available activity was found as labelled parent and *c.* 50% as high boiling and polymeric material. The inorganic yield was *c.* 20–30%. It seems probable that the complexity of products is due to extensive recoil fluorine reactions with carbon–carbon bonds with the production of labelled radicals and diradicals. A comparison of yields between toluene and perfluoro(methylcyclohexane) demonstrated that recoil fluorine reacts with comparable readiness at C—C and C—F positions. When recoil fluorine reacts with cyclic perfluorocarbon molecules, excited species are also produced[233]. The favoured decomposition route is by the fission of two carbon–carbon bonds to produce $:CF_2$ and C_2F_4 from c-C_3F_6 and $:CF_2$ and c-C_3F_6 from c-C_4F_8. [c-C_4F_8]* also cleaves to produce $2C_2F_4$ as might be expected. The products *not* found are those olefins that would be formed by fluorine migration, e.g. $CF_3{\cdot}CF{=}CF_2$ from c-C_3F_6. Recoil fluorine also attacks the C—C bonds in these cyclic molecules producing perfluoro radicals. If hydrogen sulphide is added to provide a source of hydrogen atoms, molecules of $R_nF_{2n}^{18}FH$ are produced. Thus it can be shown that recoil fluorine attack at C—C sites is about twice as efficient as at C—F sites, in striking contrast to the preference shown by recoil tritium for C—H positions.

5.5.2.2 *Reactions with hydrocarbons*

Todd, Colebourne and Wolfgang[227] have compared the reactivity of methane and perfluoromethane to recoil fluorine. By use of an extension of the kinetic theory, they showed that the reactivity integral for the reaction: $^{18}F+CH_4 \rightarrow CH_3^{18}F$ was three times higher than that for $^{18}F+CF_4 \rightarrow CF_3^{18}F$. On the other hand, decomposition of the recently labelled molecule was much less marked in $CH_3^{18}F$ than $CF_3^{18}F$ [$I(CH_3^{18}F){:}I(CH_2^{18}F) = 3.2{:}1$]. The possible decomposition of $CH_3^{18}F$ to $:CH_2+H^{18}F$ could not, of course, be determined. However, in the reactions of recoil fluorine with partially fluorinated methanes extensive decomposition to carbenes and HF has been noted[232]. Thus $:CF^{18}F{:}CHF_2\ ^{18}F$ from fluoroform is 5.2:1. In CH_2F_2 it is interesting to note that if HF expulsion only is assumed then comparable decomposition (75%) of the recently labelled molecule is observed whether the reaction has been F for F or F for H, (but since it would appear that two fluorines or molecular fluorine can also be expelled from molecules excited by fluorine labelling this conclusion may not be valid). The fate of excited $CH_2F^{18}F$ can be compared with excited $CHTF_2$ made by the T for H process. The former is much more prone to decomposition, indicating that much more energy is deposited in a molecule by the reaction with a hot fluorine atom than with a hot tritium atom. Indeed it would seem that recoil fluorine can deposit as much as 11–12 eV in a molecule (e.g. $CH_3{\cdot}CF_3$) by the F for F reaction[235, 236]. Such extensive decomposition reactions detract from mechanistic conclusions drawn from an incomplete observation of all the labelled products. Spicer, Todd and Wolfgang[237] have attempted to draw such conclusions from yieids of labelled compounds and labelled mono-radicals that combined with iodine. These data, which show a diversity of possible decomposition routes, (e.g. C—H or C—F cleavage both probably take place in molecules what have suffered ^{18}F for F or ^{18}F for H replacement), when combined with the results of Tang, Smail and Rowland[232], suggest that recently-labelled molecules can have a very wide range of vibrational excitation energies; to suggest detailed reaction mechanisms is, at the present state of knowledge, premature.

When recoil fluorine reacts with cyclic hydrocarbons[231] about half the labelled fluorocyclopropane isomerises to the various isomeric fluoropropylenes and two-thirds of the fluorocyclobutane decomposes to vinyl fluoride. In both cases the parent hydrocarbons were at *c.* 100 Torr pressure and the relative yields of isomerisation and decomposition products decrease as pressure increases. Vinyl fluoride is also a prominent product from cyclopropane, possibly indicating C—C attack followed by rearrangement and decomposition or methylene elimination from c-$C_3H_5F^*$. Vinyl fluoride is the major product from the reaction of recoil fluorine with 1,3-dimethylcyclobutane. It is probably formed by the decomposition of a molecule excited by the F for CH_3 reaction (cf. tritium, Section 5.4).

The reaction of recoil fluorine in an aromatic system was first studied by Aten *et al.*[238] by the fast neutron irradiation of fluorobenzene. They found 36% of the ^{18}F activity in organic combination. Brinkman and Pauvels[239] have recently studied the fluorobenzene system in more detail. About 10% of the total available activity was found in difluorobenzenes in both the liquid and the gas phase. The distribution of activity between *o*-, *m*- and *p*-isomers in the absence of iodine indicated fairly equal reactivity on the part of the C—H sites but when iodine was added the yields reflected an apparent change in reactivity $o:m:p \approx 35:20:45$. Ledrut[240] has investigated the reactions of recoil ^{18}F with $CF_3 \cdot C_6H_5$ in a variety of solvents. Elias and Ndiokwere[241] have studied the reactions of recoil fluorine in perfluorobenzene–hydrocarbon mixtures and have concluded that the yield of inorganic ^{18}F-labelled material can be related to the ratio of the number of H and F atoms in the mixture, up to about $H:F \sim 2:1$ after which the inorganic yield is fairly constant at about 60% (cf. Aten *et al.* above).

5.5.3 Conclusions

The mass of fluorine is such that the velocity of a 10 eV atom is *c.* 1 Å per 10^{-14} s. Thus even quite high energy (from a chemical point of view) atoms will be moving relatively slowly by comparison with nuclear vibrational frequencies. A collision between a hot fluorine atom and a molecule will therefore be a rather protracted affair in which much translational energy can be transferred to vibrational modes of the molecule. Molecules labelled by recoil fluorine will thus have the possibility of being quite highly excited and this is borne out by the experimental evidence presented above. This type of collision also provides an explanation for comparable reactivity of C—C and C—F or C—H sites. Whilst the actual contact will be, as with tritium, with atoms on the outside of the molecule, the duration of the collision with recoil fluorine permits energy to be transferred to the weaker carbon–carbon bonds and to extend them. The fluorine atom can therefore enter deeper into the molecule in the time of the collision than is possible in the case of recoil tritium. Since the decomposition of recently-labelled molecules is such a prominent feature of recoil fluorine chemistry, the change from gas to condensed phase will have quite a profound effect upon the nature and yields of the labelled products as has been demonstrated by Richards and Wolfgang[242]. The short duration between collisions that characterises liquids will reduce unimolecular decomposition and the inhibition to motion which the condensed phase presents will enhance the yields of products formed by radical recombination reactions (cage effect).

5.6 RECOIL CHLORINE

Both isotopes of chlorine react with thermal neutrons by the n, γ-process to produce recoil species. These reactions have been used to study the chemistry of recoil chlorine but suffer from the disadvantages inherent in any particle that results from an n, γ reaction. Other nuclear reactions also produce recoil chlorine with much higher recoil-energy, e.g. ^{35}Cl (γ, n) $^{34}Cl^m$, ^{40}Ar (γ, p) ^{39}Cl. Both $^{34}Cl^m$ and ^{39}Cl, as well as ^{38}Cl, have half-lives of between 30 and 60 min and are therefore very suitable for study. A variety of isotopes also permits comparison to be made in the yields of labelled products. It is found that yields are constant[243], although the recoil atoms are made by diverse processes; this is good evidence for a single type of reactant species in all cases assumed to be a ground-state translationally excited atom.

5.6.1 Reactions of recoil chlorine

5.6.1.1 Reactions with alkanes and alkyl compounds

(a) *Alkanes*—Recoil chlorine reacts with methane and ethane in the gas phase[244]. As with fluorine, the efficiency of the hot reactions is much less than for tritium. About 8% of the total available activity is found in products formed by hot chlorine, methyl chloride (6.4%) and the chloromethyl radical (1.3% detected as CH_2ClI). Earlier studies had shown somewhat higher total yields[245] with methane[246] (17%) and ethane[246] (10%) (HCl was the source of recoil chlorine). The difference is probably due to ethylene and iodine (used by Spicer and Wolfgang) being a more efficient scavenger for thermal chlorine atoms than the C—H bonds of methane. A lower yield (7%) from butane[247] may reflect more efficient scavenging by the weaker C—H bonds of this molecule. Methyl chloride and n- and iso-propyl chlorides are the principal products of the reaction with propane[248]. A kinetic theory analysis[244] (with argon as a moderator) showed that the reactivity integrals for the Cl for H reaction were comparable (0.12) for methane and ethane but that ethane was a better moderator for hot chlorine than either argon or methane. This may be due to the presence of the C—C bond in ethane whose vibrational period is perhaps similar to the duration of the hot-chlorine–molecule collision. No pressure dependence for the yields of hot products was observed save that which could be accounted for as incomplete scavenging of radiation damage species. Since however, the range of pressures studied (6–50 cmHg of methane) covers intermolecular collision times of only $10^{-9}-10^{-11}$ s, it might well be that there are decomposition reactions of excited species with half-lives less than 10^{-11} s which have not yet been detected.

In the liquid phase much higher yields[245, 249, 250] ($\sim$20%) of labelled alkyl chloride are observed than in the gas phase. Carbon tetrachloride, alkyl[245, 249] and aryl[250] chlorides have been used as sources of recoil chlorine. The increase in yield is almost certainly due to 'cage effects'.

(b) *Alkyl halides*—Methyl chloride has also been studied in the gas phase and labelled CH_3Cl and CH_2Cl_2 (3.1 and 0.7% of available hot chlorine) are both formed. Moderator studies showed that moderating powers for recoil chlorine decrease in the order: methyl chloride, neon, helium. It is found that whilst the yield of labelled methyl and methylene chloride was susceptible to kinetic theory analysis, the ratio of the two products changed with moderation [CH_3Cl:CH_2Cl_2 = 4.4 (no neon), = 2.7 (95% neon)]. Since the total yields are so low this change cannot be explained as due to the competition for hot atoms from high-energy and low-energy reactions. It could be due to changes in the relative values of α for methyl chloride and neon (or helium) with energy or due to the omission from the analysis of HCl produced by hot atoms. However, Spicer and Wolfgang[244] have observed HCl in their work with methane and found the total hot yield to be low ($\sim 2\%$). In an earlier study in the absence of a scavenger, Willard observed much higher total organic yields (20.7 %) with increased yields of both methyl and methylene chlorides[245].

Other alkyl halides[251] (CH_2Cl_2, $CHCl_3$, CCl_4, C_2H_5Cl and CH_3F) have also been studied in the gas phase. Willard[245, 248] has investigated ethyl chloride and n- and iso-propyl chloride. In all cases low (2–10 %) total yields were observed. Replacement of Cl by Cl* is more efficient than replacement of H. Also in the case of ethyl chloride[251] attack at the C—C bond produced CH_2Cl_2 and CH_3Cl in the ratio 1:1.6 (cf. Willard[245] who found 1:1). The corresponding process with recoil tritium gave a ratio CH_2ClT:CH_3T of 1:6. This may be explained by assuming that inertial effects are important in tritium reactions. The longer collision time of the hot chlorine with a molecule permits the $CH_2Cl\cdot$ group to rotate and so compete with the methyl group for the recoil atom.

Alkyl chlorides have been more extensively studied in the liquid phase where 'organic yields' of *c.* 20 % are usually observed. Detailed analysis by Willard[245] showed that the labelled parent molecule was always the largest single product but by no means the major product; a wide variety of labelled molecules was also formed. It is interesting to note that n- and iso-propyl chloride when labelled did not isomerise. Black and Morgan[252] have described the preparation of labelled chlorinated hydrocarbons by use of recoil ^{38}Cl.

Wai and Rowland[253, 254] have studied the stereochemistry of the Cl for Cl replacement reaction and found that in the gas phase it proceeds with retention of configuration. However, in the liquid phase[255, 256] much larger yields of labelled material were observed but the increase in yield was associated with randomisation of configuration. This clearly indicates that radical cage processes are dominant in liquid-phase recoil chlorine chemistry (cf. ref. 242).

The reaction of recoil chlorine with carbon tetrachloride was one of the earliest reactions studied. In the liquid phase an organic yield of 38 % was found; the remainder was shown to be labelled chlorine by subsequent reactions with pent-2-ene[257]. The organic fraction was shown[258] to be mostly $CCl_3^{38}Cl$ with a few per cent of C_2Cl_4, C_2Cl_6, etc. Aten and co-workers have studied other perchlorocarbon compounds[229, 259]. The addition of a few mole per cent of compounds containing aliphatic C—H bonds dramatically reduces the yield of $CCl_3^{38}Cl$ to *c.* 3–4 %[249, 260] which is approaching the gas-phase yield of 0.6 %[251].

(c) *Alcohols*—Reactions with alcohols have been studied by Vasaros[261] and more recently by Kontis and Urch[250]. The former investigated carbon tetrachloride–alcohol mixtures, the latter authors used chlorobenzene as a source of recoil chlorine. The high organic yield characteristic of chlorobenzene (61 %) rapidly drops when the concentration of aliphatic alcohol is only a few per cent. Thereafter a slow general decline in organic yields is observed to a limit of *c.* 10 %; this limit is lower for ethanol and almost zero for methanol. It has been suggested[250] that recently labelled molecules will be vibrationally excited and that α-chloroalcohols will, in particular, be subject to decomposition, e.g.

$$\begin{bmatrix} \mathrm{Cl} \\ | \\ \mathrm{RC{-}OH} \\ | \\ \mathrm{H} \end{bmatrix}^* \longrightarrow \begin{matrix} \mathrm{R} \\ \\ \mathrm{H} \end{matrix}\!\!>\!\mathrm{C{=}O} + \mathrm{HCl}$$

This accounts for the observed increase in organic yields as the number of carbons in the alcohol increases. Vasàros[261] also found evidence of excited molecules being formed; in particular, in the reaction with carbon tetrachloride, $[CCl_3^{38}Cl]^*$ decomposes to produce a labelled trichloromethyl radical which can abstract hydrogen from an alcohol to form chloroform.

5.6.1.2 *Reactions with alkenes*

Only the reactions with *cis*- and *trans*-dichloroethylene have been studied[262]. In the gas phase a very small yield of the Cl for H product is found but the major yield is of isomerised parent molecule. This is presumably formed by hot addition to the double bond followed by random C—Cl cleavage to form $CHCl{=}CH^{38}Cl$. In the liquid phase there is enhanced retention possibly due to the 'caging' of the radical recombination reaction, $CHCl{=}CH\cdot + \cdot{}^{38}Cl$.

5.6.1.3 *Reactions with aromatic compounds*

The reactions of recoil chlorine in aromatic systems has been investigated by Stöcklin and co-workers[263–265] using carbon tetrachloride or chlorine itself as a source of recoil chlorine. Benzene is much less efficient as a scavenger of thermalised chlorine atoms than a paraffin. Cl for H replacement products predominate in the organic yields from benzene[263], toluene[263], halobenzene[265], trifluoromethylbenzene[264] and nitrobenzene[264]. In general, dramatic orientation effects between yields of *o*-, *m*- or *p*-isomers are *not* observed in the limit of 100% aromatic species (such effects are found, however, at high carbon tetrachloride concentrations). Detailed investigations of the effects of scavengers such as iodine have led to the conclusion that the excited recoil products such as $[Cl^{38}Cl]^*$ and $[CCl_3{}^{38}Cl]^*$ might be reactant species as well as hot chlorine atoms. Iodine scavenger also brings about a considerable reduction in the organic yield (e.g. chlorobenzene 60% to *c.* 12%). To explain the high organic yield in chlorobenzene which *is* susceptible to scavenger, Urch and Kontis[250] proposed a π-complex intermediate of rather long half-life, from which a scavenger could extract ^{38}Cl, or which could eventually undergo chlorine exchange to give labelled chlorobenzene. Berei and Stöcklin[265] reached the complementary conclusion that such a complex was probably not important in the replacement of Cl in chlorobenzene by hot Cl.

5.6.2 Reaction mechanisms

The general features of recoil chlorine chemistry are low 'hot' yields in the gas phase but greatly increased yields in the liquid phase. These results are in keeping with a hot collision being of rather long duration, i.e. a few bond vibrations, which enables the molecule as a whole to adjust to the impact of the incoming atom. Since the excess energy of the recoil atom is absorbed by the molecule the chance of breaking particular bonds are reduced, hence the low vapour-phase yields. Spicer and Wolfgang[251] have rationalised the fact that Cl for Cl processes seem to be more efficient than Cl for H as due to translational–inertial factors. If recoil chlorine strikes the methyl group of methyl chloride normal to the C—Cl bond, then the translational inertia of the bound Cl, *relative* to the recoil chlorine, will be such that it will continue to move in the direction opposite to that of the recoil atom. The C—Cl bond will be broken and the recoil atom has a chance to combine with the methyl radical.

The rather efficient transfer of recoil energy to an impacted molecule suggests that break up into radicals either during, or shortly after, the collision might well take place as with recoil fluorine. It is not surprising, therefore, that the change from gas to liquid phase is accompanied by a large increase in labelled molecule formation, since radical movement is inhibited in the condensed medium.

5.7 RECOIL BROMINE

The two naturally occurring isotopes of bromine both react with thermal neutrons by the n, γ process. In both cases metastable nuclei can be produced (see reactions) which decay by isomeric transition (I.T.) and the emission (or internal conversion) of x-rays, half-lives are indicated in square brackets.

$$^{79}\text{Br}(n, \gamma) \xrightarrow{\sigma = 2.9} {}^{80}\text{Br}^{m}\ [4.4\ \text{h}] \xrightarrow{(\text{I.T})} {}^{80}\text{Br}^{m}\ [7.4\ \text{ns}] \xrightarrow{\text{I.T}} {}^{80}\text{Br}^{g}\ [17.6\ \text{min}]$$

$$= 8.5$$

$$^{81}\text{Br}(n, \gamma) \xrightarrow{\sigma = 3{\cdot}0} {}^{82}\text{Br}^{m}\ [6\ \text{min}] \xrightarrow{(\text{I.T})} {}^{82}\text{Br}\ [36\ \text{h}]$$

$$= 0.2 \quad \sigma$$

Over the past 25 years or so it has proved delightfully easy to irradiate compounds of bromine with thermal neutrons but it has also proved extremely difficult to understand the conflicting results that have been reported. This is because of the late discovery of the isomeric transition in bromine-82 which remained undetected until 1965[266, 267]. Before that date it had been assumed that the chemistry of ^{82}Br was that of an n, γ recoil particle. The presence of a transitory state with a nano second half life also complicates the understanding of the chemical reaction of ^{80}Br^{m} [4.4 h] initiated by isomeric transition.

Carlson and White[268] have argued that if an atom, X, in a molecule, RX, undergoes an isomeric transition and much of the x-ray energy is converted internally then the emission of Auger electrons will cause X to become highly charged. Since X is bound to R the charge will spread throughout the molecule causing it to decompose violently (e.g. results of Wexler[269] for $CH_3\,^{80}Br^m$ decay). The charged fragments will repel each other and can acquire quite high translation energies (10–100 eV)[270]. Such particles are, in general principle, at least, similar to those formed directly by n, γ processes and so will be considered in this review.

Other nuclear reactions that yield bromine isotopes are $^{79}Br(n, 2n)^{78}Br$ [6.4 min] and $^{81}Br(n, 2n)^{80}Br^m$, both induced by 14 MeV neutrons. Recoil bromine atoms (^{84}Br, ^{86}Br, ^{87}Br) are also produced by uranium fission and have been found to react with methane to produce methyl bromide.

To distinguish (n, γ) recoil chemistry from that induced by isomeric transitions it is necessary to study those $^{80}Br^g$ atoms that are produced directly from ^{79}Br. This calls for intense irradiations and rapid analysis. The other (experimentally easier) method is to measure the $^{80}Br^m$ activity after all the initial $^{80}Br^g$ has died away. Results can be complicated by the $^{80}Br^g$ produced from the decay of the $^{80}Br^m$; in some cases the organic yield is found to vary with time[272] but even so the general assumption is made in the recent literature that the chemistry of $^{80}Br^m$ is the chemistry of a nuclear recoil species. The almost complete internal conversion of the γ-rays produced by the reaction: $^{80}Br^m$ (I.T.) $^{80}Br^g$ make direct detection of the $^{80}Br^m$ state almost impossible. The reactivity of ^{82}Br is now known to arise almost entirely from the isomeric transition process. Discussions of ^{82}Br recoil chemistry made before 1965 are based on the assumption that it was an n, γ-particle; conclusions based on this assumption must now be set aside.

5.7.1 Reactions of recoil bromine

5.7.1.1 Gas phase

(a) *Methane*—Recoil bromine reacts with methane in the gas phase to produce methyl and methylene bromide. Spicer and Gordus[273] found an increased organic yield for ^{80}Br activated by (n, γ) of 12% compared to the yield of 7% of that activated by isomeric transition. Smaller yields were observed for C_2H_6 (n, γ; 10%. I.T., 4.5%). Rack and Gordus[274] by use of moderators, concluded that all the (n, γ)-^{80}Br particles that reacted were hot atoms and not ions. I.T. species have been studied by Okamoto and Tachikawa[275, 276] who found that the yields of $CH_3^{82}Br$ and $CH_3^{80}Br$ both fell upon moderation to a common value of about 0.5%. The yields[277] of $CH_2Br^{80}Br$ and $CH_2Br^{82}Br$ 1.1% were indifferent to added moderator. The yields of $CH_3^{80}Br$ and $CH_3^{82}Br$ in the absence of moderator were 3.5% and 5.0% respectively, so that 4.5% of the ^{82}Br activity and 3.0% of the ^{80}Br activity was due to hot atom reactions. A comparison[278] of yields from (n, γ) ^{80}Br and (n, γ)-$^{80}Br^m$ with methane showed no difference in the two nuclear recoil species.

(b) CH_4, CD_4 *comparison*—Spicer and Gordus[273] found smaller organic yields with CD_4 than for CH_4 both when bromine was activated by (n, γ), 6.4%, and by (I.T.), 4.5%. Nicholas and Rack[278] however, were able to confirm these results only for the n, γ particles. Tachikawa and Kahara[279] have, however, observed an isotope effect in the reactivity of CH_4 and CD_4 to (I.T.) ^{82}Br, in agreement with the results of Spicer and Gordus. The yields of $CH_3^{82}Br$ and $CD_3^{82}Br$ were found to be in the ratio 2.7:1 for any amount of added moderator. Thus the isotope effect cannot be due to different moderating powers of CH_4 and CD_4 but must be due to a difference in reactivity between the two molecules. It would appear that in methane all the n, γ-particles that react are hot atoms but that in the I.T. process both hot atoms and ions can react. The yields of all products decrease with a drop in pressure suggesting that all products are somewhat excited and can decompose by C—Br cleavage.

(c) *Alkyl halides*—In the vapours of alkyl halides rather lower organic yields are found both for (n, γ) and (I.T.) reactions[273], thus CH_2F_2 (3.2%, 1.5%), CHF_3 (1.5%, 0.8%), CF_4 (0.4%, 0.3%), CF_3Br (1.3%, 1.4%) and CH_3Br (2.8%, 2.4%), but yields of the former process (n, γ) usually exceed those of the latter (I.T.). An earlier study[248] of n-propyl bromide vapour gave an organic yield of *c.* 6% for both $^{80}Br^m$ and ^{82}Br. The effect of added helium and xenon upon the products of the (I.T.) ^{82}Br reaction with methyl bromide has been investigated by Okamoto and Tachikawa[280]. No evidence was found for any ionic reaction; presumably the lower ionisation potential of methyl bromide facilitates their neutralisation. Kinetic analysis demonstrated that xenon and methyl bromide have comparable moderating powers i.e. *c.* 4.5 times better than helium. The products observed in methyl bromide were $CH_3^{82}Br$ (4.6%) and CH_2Br_2 (1.2%) which suggests, merely on a bond-count basis, that C—H and C—Br bonds are equally reactive.

5.7.1.2 Liquid phase

(a) *Hydrocarbons*—Increased organic yields[281] are observed in the liquid phase 40–55%. A comparison between the reactions of (I.T.) ^{82}Br and (n, γ) $^{80}Br^m$, $^{82}Br^m$ and ^{82}Br with the isomers of hexane has been made by Merrigan, Nicholas and Rack[272]. In all cases, and at all mole fractions of bromine, higher yields were found from (I.T.) than from the (n, γ) reaction. At about 5 mol % of bromine the (I.T.) yield, for all isomers was *c.* 20–25% and the (n, γ) *c.* 15%. Under these conditions *c.* 40% of the (I.T.) organic yield was labelled hexyl bromide. A more detailed analysis of product trends with scavenger concentration is required however, before mechanistic conclusions can be safely drawn [e.g. see (c) below].

(b) *Aromatic compounds*—Milman[87, 282] has studied the reactions of ^{80}Br with benzene [presumably the (n, γ) product] in the presence of varying amounts of bromine scavenger. The scavenger curve does not show the usual dramatic drop in organic yield associated with the addition of one or two mole per cent of bromine, presumably because the benzene system is self-scavenging for radicals. At zero bromine concentration an organic yield of *c.* 18% is observed. As with aliphatic hydrocarbons, higher organic yields are observed[283] from (I.T.) activation, 26% (of which 14% is bromobenzene). Gavoret and co-workers[284] have studied the reactions of ^{82}Br in benzene and derivatives of benzene. It seems as though these results should be interpreted as those of (I.T.) ^{82}Br. In competitive experiments reduced organic yields, relative to benzene (100), were found for toluene (89), nitrobenzene (26) and phenyl benzoate (39). Unfortunately, Stamouli and Katsanos[283] who studied phenyl–X reactions with (I.T.) ^{82}Br, where X = H, F, Cl, Br, I and NO_2, did not perform competitive reactions. Their results do, however, show variations in total organic yield (X = F, 32%; Cl, 42%; Br, 46%; I, 37%; NO_2, 24%). A relatively high yield of Br for X replacement is usually found which correlates well with the weakness of the C—X bond. It seems that Br^+ might play some role in the reactions. Berei and Stöcklin[265] also found that the organic yield of Br for X increases, X = F (1.4%), X = Cl (8.5%) and X = Br (18.5%). The *o-m-p* substitution reactions in fluorobenzene and chlorobenzene show some preference for *o-* and *p*-positions. For bromobenzene the ratios are reported[285] as 1:1.5:2. Naphthalene reacts more readily than benzene with recoil bromine[286]. This somewhat confused picture certainly does not present clear evidence for any one reactive species. Both Berei and Stöcklin and Stamouli and Katsanos propose intermediate complexes between the aromatic molecule and the recoil particle. Stamouli and Katsanos have rationalised results from the reactions of (I.T.) ^{82}Br in bromobenzene–methanol and bromobenzene–n-propyl bromide mixtures[287] and also in the methanol–benzene, methanol–iodobenzene and benzene–iodobenzene systems[288]. Although methanol and ethanol give no easily determinable bromine-labelled compound, Iyer and Willard[289] have shown for ethanol this is because the products are volatile or easily hydrolysed (CH_3Br, C_2H_5Br and C_2H_4BrOH); the actual yields are 20% for ^{82}Br and 13% for $^{80}Br^m$.

(c) *Alkyl halides*—Organic yields of *c.* 35% were found in many simple alkyl halides. This figure fell to *c.* 20% on the addition of about 2 mole per cent of bromine and then decreased more or less in proportion to the amount of bromine added. Much early work was devoted to attempts to establish[289–291] or to refute[292] the existence of isotope effects. It would seem that many of the apparently conflicting results were due to traces of impurities. The time between irradiation and analysis might also have given rise to discrepancies since the yields of labelled products from isomeric transitions would vary with time. In the past decade more exacting studies have been made and in many cases the actual labelled compounds have been identified by chromatographic analysis.

The addition of bromine to alkyl bromide systems not only helps to eliminate olefinic impurities but can also be used to help determine the nature of the chemical reactions that lead to labelled molecules. Thus products that depend for their formation upon a long diffusion path in the liquid can be intercepted by a scavenger such as bromine. The simplest and most common process of this type is the removal of a thermally diffusing recoil bromine atom: $Br^* + Br_2 \rightarrow BrBr^* + Br$, thus inhibiting the reaction: $R\cdot + Br^* \rightarrow RBr^*$. The radical R· in the bulk of a sample will be the most stable radical that can be derived from the parent molecule and $R\cdot + Br^*$ will therefore generate labelled parent bromide. This was clearly demonstrated by Chien and Willard[293] in their studies with n-propyl and iso-propyl bromide. In each case there was very little isomer interconversion and in each case the yield of labelled parent molecule was reduced by *c.* 50% by the addition of a few mole per cent of bromine. Products unaffected by bromine were assumed to be formed 'hot'. Such products could, of course, be stable molecules formed by a direct hot displacement reaction, or stable molecules formed by radical recombination at the hot collision-site or labelled radicals which will form dibromo products either by the reactions: $(RBr^*)\cdot + RBr \rightarrow R\cdot Br^*Br + R\cdot$ or $(R'Br^*)\cdot + Br_2 \rightarrow R'Br^*Br + Br$. Dibromopropane yields, found by Chien and Willard[293] to be unaffected by added bromine, are probably made by reactions of this latter type. Whilst bromine intercepts thermalised atoms and so increases the inorganic yield, 1,2-dibromoethylene also reacts rapidly with bromine atoms but in so doing increases the organic yield. Roy, Williams and Hamill[294] found that 40% of the recoil bromine could be scavenged one way or another; 12% of these scavengeable atoms would give rise to organic products and 28% (the fraction caught by 1,2-dibromethylene) to inorganic compounds. A small mole percentage of 1,2-dibromoethylene was much more effective than the same amount of bromine, showing that the longer the diffusion path of the bromine atom the more likely it was to make an inorganic product (probably $Br^* + RH \rightarrow R\cdot + HBr$). Shorter diffusion paths give rise to organic compounds ($R\cdot + Br^* \rightarrow RBr$) suggesting that radical concentration falls as the thermal recoil atom diffuses away from the region of its last 'hot' collision. This work also shows that *c.* 60% of the recoil atoms do not react at, or very near, the site of the last hot collision to give *c.* 20% organic material and 40% inorganic. Diffusive reactions also seem to be affected by the nature of the recoil process. Thus rather larger scavengeable yields are observed[295] when recoil reactions are initiated by the high recoil-energy (n, 2n) process rather than the (n, γ) reaction.

Harris[296, 297] has suggested that the results with bromine scavenger can be made to yield more information about reaction processes at the site of the last hot collision, the so-called hot zone or hot spike. When the yields of various compounds are corrected by a factor for the percentage of bromine present in a sample, it is found that such corrected yields for some products are constant (type A) whereas others decrease linearly to zero as the bromine concentration goes to 100% (type B). Harris proposes that type A products are formed by true hot reactions (in the gas-phase sense) but that type B products are produced by radical or other reactions in the hot zone itself, i.e. 'hot-spot diffusive reactions'. Those products whose corrected yields decreased linearly with increasing bromine concentration but were not zero at 100% bromine are formed by both processes. Thus the yield of labelled parent molecule in methyl bromide was found to be due to hot reactions (5%), hot-spot diffusive reactions (16%) and thermal-diffusion reactions (43%). In ethyl bromide the amount of labelled parent produced by the hot reaction was 3%. These 'hot' yields are quite similar to those observed in the gas phase. In this analysis labelled radicals made by a direct hot reaction will also show up as hot products. Thus in the methyl bromide system a 13% hot yield is found for the polybromomethanes CH_2Br_2, $CHBr_3$ and CBr_4. Whilst the first can be made by a direct Br for H reaction, the others probably originate from the reaction.

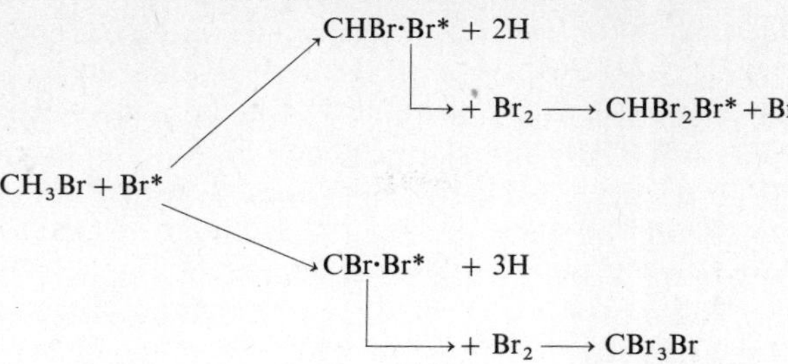

Unfortunately, Harris did not distinguish in this work between (n, γ) and (I.T.) processes.

Shaw and Milman[298–301] have made an extensive study of the reactions of recoil bromine with bromoethane with special emphasis on the diffusion-controlled processes. Further work by Shaw and co-workers[295, 302] has included a detailed study of the way in which various products labelled with either $^{80}Br^m$ or ^{82}Br vary with the concentration of added bromine. Some labelled radicals are prone to decomposition, e.g. $\dot{C}H_2CH_2Br^* \rightarrow CH_2{=}CH_2 + Br^*$. Work at high pressure[303] showed an increase in the yield of 1,2-dibromoethane suggesting that the $\dot{C}H_2{\cdot}CH_2Br$ radical was an intermediate in its formation. The yield of 1,1-dibromoethane was not affected, as would be expected since $\dot{C}HBr{\cdot}CH_3$ would not readily decompose. Such studies emphasise the importance of a thorough investigation of a system before sweeping mechanistic conclusions can be reached.

In perdeuterobromoethane[304] the yields of most compounds are greater than in C_2H_5Br (contrast with the gas phase) save for 1,1-dibromoethane. The general increase in yields may be due to slower decomposition reactions of recently-labelled species. The reduced production of $CD_3{\cdot}CDBr^*$ may be due to the slower C—D vibrations permitting energy transfer to the rest of the molecule during the collision rather than breaking as in the C—H case.

5.7.2 Reaction processes

It is clear that bromine atoms or ions with recoil properties can be made as a result of nuclear reactions and that their chemistry can be investigated in detail. The profusion of possible reactive species and their possible reaction pathways has stimulated research in recent years to find out exactly what the primary chemical reaction processes are.

Milman *et al.*[305, 306] have studied the effects of neutron irradiation of bromoethane with a view to elucidating chemical differences between $^{80}Br^m$ (assumed to be simply a nuclear recoil species) and ^{82}Br (assumed to be activated by an internally converted isomeric transition). Experimentally the results of Milman are in good agreement with those of Harris[296, 297], and the conclusions of Milman[305] concur with those of Gilroy, Miller and Shaw[307] that ^{82}Br seems to have a greater facility for H substitution whereas $^{80}Br^m$ favours C—C or C—Br bond attack.

Mia and Shaw[302] have discussed the probable reactions of $^{80}Br^m$ and ^{82}Br

in ethyl bromide in some detail. Neutron irradiation generates both $^{80}Br^{m}$ and $^{82}Br^{m}$ as recoil species and presumably there is no difference in their chemistry, as Merrigan, Ellgren and Rack[308] showed in liquid hexanes. The products labelled with $^{80}Br^{m}$ are detected directly but molecules labelled with $^{82}Br^{m}$ will be decomposed when the isomeric transition takes place. The internal conversion of the x-ray and the subsequent Auger processes lead to the formation of many charged fragments but mostly to charged hydrogen atoms. These protons, with high recoil energy, will move rapidly away leaving a site depleted in hydrogen. The (I.T.) ^{82}Br atom (or ion) will therefore often be found in molecules with more bromine than the parent.

It is true, however, that although variations are found in labelled product yields, the general spectrum of products is the same from both $^{80}Br^{m}$ and ^{82}Br. This similarity led Kazanjian and Libby[270] to conclude, from a study of n-propyl bromide that all the products could be accounted for as due to hot atoms, whether the original particle had been produced by an (n, γ) or an (I.T.) process. Conversely Merrigan, Nicholas and Rack[272] and also Gunter and Willard[309] have emphasised the importance of charge formation as a result of the internal conversion of x-rays.

The basic difference between these various ideas is exactly when the internal conversion of x-rays takes place. Kazanjian and Libby[270] propose that for $^{80}Br^{m}$ and for (I.T.) ^{82}Br it is completed, and that neutralisation has also taken place, in the time taken to lose recoil energy and before the particle can attempt to enter chemical combination. Willard's proposal is that some residual nuclear excitation might well remain in the recoil species after it has entered a chemical compound. Subsequent emission of internally converted x-rays would then lead to local charging and fragmentation which would be different for different nuclear species because of their different degrees of excitation and nuclear structure. The labelled chemical products to be finally observed would then be governed by reactions in these localised regions. Product patterns are generally similar because, for both isotopes the basic process is autoradiolysis; actual product yields show differences because of the different decay energies of $^{80}Br^{m}$ and ^{82}Br. It is known the γ-radiation can affect and indeed increase organic yields[310], but Fonseca, Fuller, Latham and Shaw[311], in a series of experiments to compare (n, γ), (I.T.) and radiolysis effects, concluded that autoradiolysis cannot explain the specific product ratios that are observed in recoil bromine reactions.

In the liquid phase a comparison of yields from the two isomeric transitions $^{80}Br^{m} \rightarrow {}^{80}Br$ and $^{82}Br^{m} \rightarrow {}^{82}Br$ in carbon tetrachloride showed no difference[312] (39%). This contrasts with the more detailed gas-phase study where higher yields were observed as a result of the $^{82}Br^{m}$ (I.T.) ^{82}Br activation process. This is probably because the decay of $^{80}Br^{m}$ is a two-stage process with a nanosecond intermediate step. The first stage will therefore form a variety of molecules and the internally converted second stage of decay will cause their disruption. It is quite probable[277] that the recoil energy of the ^{80}Br will be different from that of ^{82}Br since in the former case the partner of $^{80}Br^{m}$ might be H or R, whereas for ^{82}Br it will always be Br from Br—$^{82}Br^{m}$. This can account for higher yields often associated with ^{82}Br but it also requires that an important factor in determining the final chemical state of the bromine atom is recoil energy.

Milman[87] has successfully applied the Estrup–Wolfgang kinetic theory[55] to recoil-bromine chemistry which also suggests that hot atoms play an important role. Katsanos and Stamouli[313], on the other hand, have proposed that intermediate complexes are important in (I.T.) reactions.

The picture is still complicated and unresolved; hot atom processes and radical recombination reactions are clearly involved but the relative importance of ionic reactions and processes has not yet been unequivocally established.

5.8 RECOIL IODINE

Recoil iodine is produced by a wide variety of reactions, e.g. ^{127}I(n, γ) ^{128}I, ^{127}I(n, 2n)^{126}I, ^{127}I(d, p)^{126}I, ^{129}I(n, γ)^{130}I, U(n, fission)^{131}I, ^{132}Te $\xrightarrow{\beta}$ ^{132}I, $^{123 \text{ or } 125}$Xe $\xrightarrow{\text{electron capture}}$ $^{123 \text{ or } 125}$I with a wide variety of recoil energies. It is also possible that in many cases the recoil species will undergo an isomeric transition after it has dissipated its initial recoil-energy. If the x-ray from this process is internally converted a highly-charged iodine will result which will once again have recoil energy. In one case, such an isomeric state has a sufficiently long lifetime to enable its (I.T.) induced chemistry to be studied in isolation, namely, ^{130}I^{m} $\rightarrow$ ^{130}I. For the most part it is the chemistry of the ^{127}I(n, γ)^{128}I reaction that has been studied but many comparisons have been made with other processes.

5.8.1 Reactions of recoil iodine in the gas phase

5.8.1.1 Reactions with hydrocarbons

(a) ^{128}I—The reaction with methane is quite exceptional to give an organic yield of *c.* 54%. With other gaseous hydrocarbons (in the presence of iodine scavenger) yields of only a few per cent were found (C_2H_6 [315], 4%; C_3H_8 [246], 4%; n-C_4H_{10} [246], 5%; n-C_5H_{12} [318], 4%; n-C_6H_{14} [318], 3%; C_6H_6 [315], 7%). The higher alkanes gave iodine labelled parent molecule and also a rather larger yield of methyl iodide.

A preliminary study of the recoil–iodine–methane system was made by Willard and co-workers[314, 315] who, by judicious choice of additives, were able to establish the paramount importance of ions rather than hot atoms. Rack and Gordus[316, 317] made a very thorough investigation and showed that neon, argon and krypton caused the organic yield to drop from 54% to 36%, at the limit of 100% inert gas. Xenon caused the yield to drop to 11% at the limit; other additives were also investigated[317]. It was concluded that hot species contribute to only 18% of the total organic hot yield, that I^+ ions in the $^1D_2$ state (neutralised by Xe) form 25% and that I^+ ions in the $^3P_{0, 1 \text{ or } 2}$ states are responsible for 11% of the yield. Hydrocarbons and alkyl iodides in a few mole per cent are all equally effective in reducing the organic yield to *c.* 10%. This is due to their moderating powers and to their lower ionisation potentials. With CD_4 [318], very similar results are obtained except that the 'hot' yield is reduced (16%), indicating that CD_4 is a better moderator than CH_4 for hot iodine.

(b) ^{125}I—These results may be compared with those of Schroth and Adloff[319] who studied the reaction of ^{125}I with methane. Xenon-125 decays by electron capture to iodine-125. Auger electrons are emitted when the inner orbital vacancy is filled and a charged ^{125}I atom results. The organic yield observed was 58% and the addition of neon or argon in any amount had no effect whatsoever!; krypton and xenon both reduced the yield to 18%. These results indicate that 'hot' species play no part in this reaction, that 40% of the ^{125}I reacts as $(^1S_2)I^+$ and that 18% reacts either as an electronically excited atom or as $(^3P_0$ or $^3P)$ I^+.

(c) ^{123}I—Recently Welch[320] has shown that ^{123}I, also produced by electron capture by the corresponding xenon isotope, can react with methane. In the absence of any scavenger an organic yield of *c.* 80% was observed. Traces of iodine vapour reduced the yield to 48%, that this figure is similar to that in the (n, γ)^{128}I–methane system (which also contained traces of iodine and a very small amount of methyl iodide) is probably coincidental.

(d) ^{133}I–^{135}I—Recoil iodine can be produced[321,322] directly by nuclear fission and by β^--decay of fission fragments. Both species are found to react efficiently with methane but greater yields are associated with primary-fission iodine 'atoms'.

5.8.1.2 *Reactions with alkyl halides*

Horning, Levey and Willard[314] found the organic yield from neutron-irradiated ethyl iodide vapour to be only 1% but in a more detailed study Cross and Wolfgang[323] determined the yield of $CH_3^{126}I$ from high-energy γ-irradiation of methyl iodide to be 4.1%. These workers also confirmed the gas phase iodine exchange reaction: $CH_3I + II^* \rightarrow CH_3I^* + I_2$ studied by Laurence and Stranks[324]. By use of I_2 tagged with ^{129}I as a tracer, Cross and Wolfgang were able to determine the rate of this exchange and thus determine the amount of $CH_3^{126}I$ made by exchange and directly by a recoil species. Within experimental error Laurence and Stranks did not observe any hot reaction as a result of (n, γ) produced ^{128}I with methyl iodide. Rack (ref. 318 p. 69) found a hot yield of only 1.2%. These observations indicate a difference in reactivity between ^{128}I and ^{126}I. This is not unreasonable as their modes of production and their nuclear properties (in particular the possibility of internally converted isomeric transitions) are quite different. Fission iodine can also react with methyl iodide[321].

The reactions of a large number of alkyl halide vapours with (n, γ) ^{128}I have been studied by Rack[318]. The yields with alkyl iodides are all *c.* 1%. With chloro- and fluoro-derivatives, CH_nX_{4-n}(X = F or Cl, n = 0, 1, 2 or 3,) higher yields of 4–8%, depending on the structure, are observed. In contrast to the alkanes, the yield from perfluoromethane was only 4.5% but that from perfluoroethane was 9%. Gordus and Hsuing[325] neutron irradiated mixtures of *c.* 1% alkyl iodide vapour and 99% nitric oxide and found organic yields as follows: CH_3I (1.09%), CD_3I (0.68%), CF_3I (0.12%), C_2H_5I (0.08%), $CH_3{\cdot}CHI{\cdot}CH_3$ (0.3%) and $CH_3{\cdot}CH_2{\cdot}CHI$ (0.06%). The authors interpreted these results as due to non-rupture of the C—I bond after irradiation because of the great excess of nitric oxide.

5.8.2 Reactions of recoil iodine in the liquid phase

As with the other recoil halogens great increases in organic yield are observed upon going from the gas to the liquid phase.

5.8.2.1 Reactions with hydrocarbons

Reid[326] irradiated a solution of iodine in pentane with thermal neutrons and found that 38% of the activity entered organic combination. He also irradiated iodine in isolation and *then* added it to the pentane. The organic yield was >0.5%, which shows that there is no isomeric state of ^{128}I, produced by neutron capture in ^{127}I, with a half-life of more than a few minutes. Organic yields of 45% and 53% have been observed[281] for pentane and decane. These yields are reduced by *c.* 10% on the addition of 0.5 mole per cent of iodine and are also reduced by increasing temperature. These results show that some of the organic yield results from diffusion reactions of thermalised atoms and that a 'cage' mechanism is also important.

A detailed study of the reaction of recoil iodine, from a variety of alkyl iodides and iodobenzene, with cyclohexane has been made by Shaw[327].

The reaction of iodine–liquid methane is facilitated by air. Such solutions are difficult to control and the organic yields were found to vary wildly[314]. In one case an organic yield of 59% was found to consist of methyl iodide (42%), ethyl iodide (9%) and other labelled compounds (8%), indicating new mechanisms are operating in the condensed phase.

It is not clear, however, whether the species that react together within the solvent cage have been produced by hot atoms and ions from the primary nuclear events or whether there has been an isomeric transition leading to autoradiolysis. Geissler and Willard[309] have irradiated solutions of alkyl iodides (RI,) in pentane, with iodine scavenger. Even at low mole percentage of RI, the principal component of the organic yield is R^{128}I and not labelled amyl iodides. The authors[328] convincingly argue that the only reasonable mechanism is the combination of ^{128}I atoms with radicals such as would be produced by radiolysis in a RI–pentane mixture. The radiolysis would be brought about by the emission of Auger electrons from an ^{128}I atom undergoing an internally converted isomeric transition. Such a process would disrupt by virtue of large positive charge any molecule, organic or inorganic, which the recoil iodine had entered. This might have happened if the half-life of the I.T. process were anything longer than a few picoseconds. The ionic iodine would have a low recoil energy and would be quickly neutralised. It would then find itself amidst radicals produced by the Auger electrons it had so recently emitted.

To test this hypothesis solutions of methyl and ethyl iodides in pentane containing I^{131}I were subjected to γ-radiolysis[328]. The distribution of ^{131}I activity was the same as that for ^{128}I produced by radiative neutron capture in ^{127}I.

McCauley and Schuler[329] claim to have found larger organic yields for the ^{127}I(n, 2n)^{126}I reaction than from ^{127}I(n, γ)^{128}I in cyclohexane but not in alkyl iodides. The former reaction will cause more bulk damage to the solution because its recoil energy is greater; it will also have a longer (in time) recoil track and therefore more opportunity to relax to the ground state of ^{126}I. Some reactivity differences between isotopes are therefore to be expected.

Increases in organic yield in hydrocarbon solutions of alkyl iodides (no iodine scavenger) can be induced by increasing the γ dose. This has been explained by Vlatkovic and Willard[330] as due to the reaction of radiation-induced radicals with labelled inorganic compounds $H^{128}I$ and $I^{128}I$. Since these compounds are essentially carrier-free, they have a high specific activity; the labelled iodides on the other hand have a very low specific activity, hence the activity enters the organic yield to *c.* 35% no matter what γ dose was given.

5.8.2.2 *Reactions with alkyl halides*

Szilard and Chalmers[1] irradiated ethyl iodide with thermal neutrons and found that *c.* 40% of the activity was present as labelled organic compounds. Later analysis[331] showed these compounds to include methyl and methylene iodide. Traces of iodine added as scavenger reduced the yield of the labelled parent but did not affect the latter[332]. Scavenger insensitive products are formed in the 'hot zone' (but obviously not by direct reaction since the gas-phase yields are so low) and the remainder of the methyl iodide yield is formed by thermalised iodine atoms diffusing through the liquid.

Alkyl iodides are characterised by higher organic yields upon neutron irradiation than their chloro or bromo counterparts[333] (n-C_3H_7X organic yields, X = Cl, 20%; Br, 35%; I, 46%); a greater proportion of the activity is found as the labelled parent molecule[334] (e.g. 90% of organic yield is labelled n-propyl iodide). Macrae and Shaw[335] found the same effect with ethyl iodide. Iyer and Martin[336] irradiated mixtures of methyl and propyl iodides and concluded that the probability that recoil iodine would label methyl iodide was about three times greater than propyl iodide. This would seem to be a specific reactivity effect. Autoradiolysis of propyl iodide (the mechanism proposed by Willard for recoil iodine labelling) as a source of extra methyl radicals cannot be the reason since only a very low yield of labelled methyl iodide is found from pure n-propyl iodide.

Many workers have attempted to place an upper limit on failure to bond rupture as a result of the (n, γ) reaction[325], and figures of 4%[336] and 2–1%[327] etc. are bandied about. That this effect is probably quite insignificant in contributing to the observed organic yield is shown[337] by the great similarity in results of experiments with inorganic iodides as a source or with other nuclear reactions to produce recoil iodine.

Schuler and co-workers[329, 334, 338] established that the organic yields in alkyl iodides did not depend upon the nuclear reaction used to make the recoil iodine (as distinct from the result with alkanes). Svoboda[339] found that if the neutron irradiation of methyl iodide was carried out with the minimum of concomitant γ-radiation the organic yield approached 100%. Addition of iodine scavenger or the addition of γ-radiation to the normal reactor level both restored the organic yield to its usual value of 50%. Svoboda[339, 340] also found that if labelled inorganic compounds were continuously extracted during low γ-irradiation the organic yield again fell to 50%. It was suggested that this was due to an exchange reaction between carrier-free 'inorganic' iodine ($H^{128}I$ and $I^{128}I$) and methyl iodide. Taken in conjunction with the results of Willard the organic yield (in the absence of scavenger) is seen to be sensitive to gross changes in radiation dose; either very little or a great deal results in high organic yields. The results from pure alkyl iodides and hydrocarbon solutions which contain alkyl iodides may be compared since radiation-generated radicals in the latter system come predominantly from the alkyl iodide by fission of the C—I bond. The results of Stöcklin *et al.*[333] from propyl iodide are somewhat at variance with those of Willard. Stöcklin too found that the organic yield increased with γ dose but only to a limiting value (80%). Thereafter, more γ-radiation caused a decline interpreted as due to the bulk build up of scavenger iodine. Both Vlatkovic and Willard and Stöcklin and co-workers found the organic yields were sensitive to scavengers other than iodine, e.g. oxygen. In some cases the large changes in organic yields noted above may have been due to impurities or just to dissolved air.

The effect on organic yields in propyl iodide of a few mole per cent of amines was studied by Stöcklin, Schmidt Bleek and Herr[333]. The organic yield (unscavenged) of 46.2% was unaffected by aniline but increased to 100% upon the addition of *N*,*N*-dimethylaniline. The increased yield was almost entirely labelled n-propyl iodide. Quaternary ammonium salts based on aniline had a similar effect.

5.8.2.3 Reactions in aromatic systems

(a) *Benzene, toluene*—The organic yield of recoil iodine in liquid benzene (20%)[341] is rather more than double that observed in the gas phase. The yield is insensitive to the addition of small amounts of iodine showing that iodine acts as its own scavenger for radicals and thermalised iodine atoms. The variation of organic yields with varying concentration of alkyl iodides, iodobenzene and iodine in benzene have been determined by Macrae and Shaw[341]. The incorporation of iodine into organic molecules would seem to be due to a radical recombination reaction at the end of the recoil track.

An analysis[342] of the radio-iodine activity (total observed yield 17%) shows iodobenzene (15%), diphenyliodonium cation ((1.2%) and polymer (*c.* 1%). The organic yield does not seem to depend very much upon the initial nuclear process; McCauley and Schuler[343] observed yields of 18% and 21% from (n, γ) and (n, 2n) reactions. The organic yield increases again on solidification (*c.* 55%)[344], whether the iodine is produced by (n, γ) or by ^{123}Te decay[320]. Ormond and Rowland[345] have shown that iodobenzene (^{131}I) is produced when benzene reacts with fission fragments; ^{131}I is not produced directly but by the β^- decay of ^{131}Te. Denschlag *et al.*[322] have confirmed the high organic yield of *c.* 50–60% but found the yield was reduced to 20% by the addition of iodine scavenger.

The decay-induced reactions of ^{131}I have also been studied directly by Halpern and Sochacka[346,347] who used dibenzyltellurium-131 as a source.

Bond rupture in >98% of decay events was established but the yield of iodobenzene varied by only 1% in the presence or absence of benzene, or if benzene were accompanied by a radical scavenger such as allyl iodide. It would seem that only *c.* 1% of the iodine produced by β-decay in dibenzyltellurium can react with benzene to produce iodobenzene. The high yields from fission ^{131}I are presumably facilitated by the reactor radiation dose received by the sample during the reaction. Tellurium-132 also decays by β^- emission to the corresponding iodine isotope. In this case it is calculated that about 85% of the excited ^{132}I decays by internally converted γ-emission. This causes Auger charging and the production of charged translationally excited ^{132}I species. Their reactions have been studied by Llabador and Adloff[349–352] who used diphenyltellurium-132 and dibutyltellurium-132 as sources. With the latter molecule benzene gave iodobenzene (10%)[348, 349]. Halpern[353] compared the reactivities of ^{131}I and ^{132}I, produced by β-decay from $^{131 \text{ or } 132}TeCl_4$, with toluene. He found much higher organic yields (66% compared with 25%) from the latter isotope. Iodine scavenger had little effect but allyl iodide, which reacts with thermal iodine atoms, reduced the yields of iodobenzene by *c.* 50% in both cases.

Kontis and Katsanos[354] have studied the isomeric transition $^{130}I^m \rightarrow {}^{130}I$ in benzene and find an organic yield of 17%, two-thirds of which is iodobenzene. This result is very similar to that from other forms of nuclear activation and supports the view that the initial nuclear reaction is unimportant; the final yields of labelled products are determined as a result of internally converted isomeric transitions which generate charged translationally excited species.

(b) *Aryl halides*—Thermal neutron irradiation of iodobenzene gives an organic yield of *c.* 80%[341, 355] which is almost entirely labelled iodobenzene. Addition of iodine causes a dramatic fall in yield which then forms a smooth curve of declining yield with increasing iodine; extrapolation of this curve to zero iodine gives an organic yield of *c.* 25%. The possible reactions have been discussed by Shaw[341], Peacocke and Walton[356] and Wheeler and Lecumberry[342]. Much lower yields of labelled iodobenzene were found by Berei and Stöcklin[265], *viz.* 40% from (n, γ) and 32% from (n, 2n) reactions. With 1% scavenger, both nuclear processes gave iodobenzene yields of *c.* 29%. Lower yields were observed for the reaction of ^{128}I in C_6H_5X,X = Br (12%), Cl (8%), F (2%). In the reaction with chloro- and fluoro-benzene the yields of hydrogen-substituted products, $C_6H_4X\ {}^{128}I$, where X = Cl (7%) and F (6%). In both cases a preference for *para*-substitution was noticed; a π-complex mechanism was proposed. Direct halogen replacement has also been observed[357] in bromobenzene and α- and β-bromonaphthalenes by ^{131}I produced as a result of uranium fission. In the first compound 85% of the yield was iodobenzene and 15% was $C_6H_4Br^{131}I$.

5.8.3 Reactive species in recoil iodine reactions

The general conclusion would seem to be that most iodine isotopes will undergo isomeric transitions with half-lives long compared with the time taken for a recoil species to loose its energy and to enter chemical combination, i.e. nanoseconds. The internal conversion of the γ-rays causes labelled molecules just formed to be disrupted and the re-liberated iodine reacts with the radicals and ions it has just formed. The strongest confirmation of these ideas is that 'recoil reaction' can be simulated by γ-irradiation. The importance of hot species is demonstrated by the gas-phase reaction with methane.

5.9 CONCLUSION

Despite the considerable amount of work that has been done to study the chemical reactions of nuclear recoil particles in gaseous and liquid phases very few clear conclusions can be drawn. This is because of the wide variety of effects that are being studied all at once, and also because research workers have been motivated by different goals, from the practical need for simple direct labelling reactions ranging to the desire to understand more about the fundamental nature of an atom–molecule collision that might lead to a chemical reaction. Some general observations may however be made. For univalent species 'hot' reactions are possible. And in such reactions energy and velocity are both important. Thus the chemistry of 'hot' tritium is unique because it can combine a high velocity with sufficient energy to surmount chemical activation energy barriers to reaction. The heavier univalent halogen atoms with comparable energy will move more slowly enabling the struck molecule to accommodate the excess energy during the collision, without frequent bond rupture. Much lower efficiencies are therefore observed for hot halogen reactions than for hot tritium reactions.

The chemical nature of the recoil species presents great complications when it is polyvalent and the role of excess energy is not clearly defined.

Practical complications arise from the possibility of nuclei being produced in isomeric states with nanosecond or longer half-lives. Distinguishing nuclei activated or in some cases re-activated by such decay processes from these activated by the primary recoil event is extremely difficult, and can give rise to a great deal of confusion in the interpretation of results. Another related practical difficulty is radiation damage associated with the nuclear events that generate the recoil particles. In some cases, as with the $^{14}N(n,p)^{14}C$ reaction, it is possible to calculate that the dose received by the sample will be inadmissably large but it has also been shown that reaction products can often be especially sensitive to very small radiation doses so that this factor can never easily be discounted. Radiation damage to a sample and products can also be caused by the irradiation facility itself, e.g. γ-rays and fast neutrons associated with thermal neutrons in a reactor; such external radiation should always be kept to a minimum.

Now that the general features of recoil atom chemistry are clear and the general problems associated with recoil atom reactions are understood it is possible to hope that this technique will at least be used carefully and profitably to produce pure specifically labelled molecules and also to investigate the physics and chemistry of atom molecule collisions.

While this review was being written the death of Professor Richard Wolfgang whilst sailing was announced. This is a tragic loss to the whole field of recoil chemistry. He had that rare combination of gifts, so necessary to a great scientist, of inspiration and perception, in conjunction with experimental skill and diligence. Much of the basic progress that has been made in the past decade has been directly due to the work, ideas and inspiration of Richard Wolfgang. Those of us who knew him personally mourn the loss of a good teacher and friend. May he rest in peace.

References

1. Szilard, L. and Chalmers, T. A. (1934). *Nature (London)*, **134**, 462
2. *'Chemical Effects of Nuclear Transformations—Prague'*, (1961)—2 vols., *'Chemical Effects of Nuclear Transformations—Vienna'*, (1965)—2 vols. (Vienna: International Atomic Energy Agency)
3. Campbell, I. G. (1963). *Advan. Inorg. Chem. Radiochem.*, **5**, 135
4. Schmidt-Bleek, F. and Rowland, F. S. (1964). *Angew. Chem. Int. Ed. Engl.*, **3**, 769
5. Wolf, A. P. (1964). *Advan. Phys. Org. Chem.*, **2**, 202
6. Filatov, E. S. (1965). *Uspekhi Khim.*, **34**, 1607: (*Russ. Chem. Reviews*, **34**, 680)
7. Wolfgang, R. (1965). *Progr. Reaction Kinetics*, **3**, 97
8. Wolfgang, R. (1965). *Ann. Rev. Phys. Chem.*, **16**, 15
9. Linder, L. (1969). *Chem. Weekbl.*, **65**, 21
10. Wolfgang, R. (1969). *Accounts Chem. Res.*, **2**, 248
11. Wolfgang, R. (1970). *Accounts Chem. Res.*, **3**, 48
12. El-Sayed, M. A., Estrup, P. and Wolfgang, R. (1958). *J. Phys. Chem.*, **62**, 1356
13. Lee, J. K., Musgrave, B. and Rowland, F. S. (1960). *J. Amer. Chem. Soc.*, **82**, 3545
14. Urch, D. S. and Welch, M. J. (1965). *Chem. Commun.*, 126
15. Malcolme-Lawes, D. J., Urch, D. S. and Welch, M. J. (1966). *Radiochim. Acta*, **6**, 184
16. Garland, J. K. and Schroeder, J. W. (1968). *J. Phys. Chem.*, **72**, 2277
17. Rosenberg, A. H. and Wolfgang, R. (1964). *J. Chem. Phys.*, **41**, 2159
18. Lee, E. K. C., Miller, G. and Rowland, F. S. (1965). *J. Amer. Chem. Soc.*, **87**, 190
19. Tachikawa, E. and Rowland, F. S. (1968). *J. Amer. Chem. Soc.*, **90**, 4767
20. Wolfgang, R. and Rowland, F. S. (1968). *Anal. Chem.*, **30**, 903
21. Lee, J. K., Lee, E. K. C., Musgrave, B., Tang, Y. N., Root, J. W. and Rowland, F. S. (1962). *Anal. Chem.*, **34**, 741
22. Stöcklin, G., Cacace, F. and Wolf, A. P. (1963). *Z. Anal. Chem.*, **194**, 406
23. Welch, M. J., Withnell, R. and Wolf, A. P. (1969). *Chem. Instrum.*, **2**, 177
24. Malcolme-Lawes, D. J. and Urch, D. S. (1969). *J. Chromatog.*, **44**, 609
25. Lee, E. K. C. and Rowland, F. S. (1964). *Anal. Chem.*, **36**, 2181
26. Root, J. W., Lee, E. K. C. and Rowland, F. S. (1964). *Science*, **143**, 676
27. Mahan, K. I., Weeks, R. W., Fee, D. and Garland, J. K. (1968). *Anal. Lett.*, **1**, 933
28. Wolfgang, R. and Mackay, C. F. (1958). *Nucleonics*, **16**, 69
29. Welch, M. J., Withnell, R. and Wolf, A. P. (1967). *Anal. Chem.*, **39**, 275
30. Schmidt-Bleek, F. and Rowland, F. S. (1964). *Anal. Chem.*, **36**, 1696
31. Malcolme-Lawes, D. J. and Urch, D. S. (1970). *Radiochim. Acta.*, **13**, 114
32. Urch, D. S. and Welch, M. J. (1968). *Trans. Faraday Soc.*, **64**, 1547
33. Urch, D. S. and Welch, M. J. (1965). *Chemical Effects of Nuclear Transformations—Vienna*, **1**, 71. (Vienna: IAEA)
34. Malcolme-Lawes, D. J. (1971). *Radiochem. Radioanal. Lett.*, **6**, 31
35. Baker, R. T. K., Silbert, M. and Wolfgang, R. (1969). *J. Chem. Phys.*, **52**, 1120
36. Estrup, P. J. (1959). *Ph.D. Thesis*, 23, (Yale University, New Haven, Conn., U.S.A.)
37. Argensinger, W. J. (1963). *J. Phys. Chem.*, **67**, 976
38. Root, J. W. and Rowland, F. S. (1968). *Radiochim. Acta.*, **10**, 104
39. Massey, H. S. W. and Burhop, E. H. S. (1952). *Electronic and Ionic Impact Phenomena.* (Oxford: University Press)
40. Seewald, D. and Wolfgang, R. (1967). *J. Chem. Phys.*, **47**, 143
41. Rowland, F. S., Lee, J. K., Musgrave, B. and White, R. M. (1961). *Chemical Effects of Nuclear Transformations—Prague*, **2**, 67. (Vienna: IAEA)
42. Doucen, R. Le. and Guidini, J. (1969). *Rev. Phys. Appl.*, **4**, 405

43. Shukla, R. V., Jain, S. K., Gupta, S. K. and Srivastava, A. N. (1970). *J. Chem. Phys.*, **52**, 2744
44. Edwards, J. L. and Thomas, E. W. (1970). *Phys. Rev. A.*, **2**, 2346
45. Levy, H. (1970). *Phys. Rev. A.*, **1**, 750
46. Hughes, R. H. *et. al.* (1970). *Phys. Rev. A. (Gen. Phys.)*, **1**, 21
47. Polyakova, G. N., Gusev, V. A., Erko, V. F., Fogel, Y. M. and Zatz, A. V. (1970). *Zh. Eksp. Teor. Fiz.*, **58**, 1186
48. Vitlina, R. Z. and Chaplik, A. V. (1970). *Zh. Eksp. Teor. Fiz.*, **58**, 1798
49. Hyatt, D. and Lacmann, K. (1968). *Z. Naturforsch.*, **23a**, 2080
50. Fleischmann, H. H. and Young, R. A. (1969). *Phys. Rev.*, **178**, 254
51. Libby, W. F. (1947). *J. Amer. Chem. Soc.*, **69**, 2523
52. Friedman, L. and Libby, W. F. (1949). *J. Chem. Phys.*, **17**, 647
53. Miller, J. M., Gryder, J. W. and Dodson, R. W. (1950). *J. Chem. Phys.*, **18**, 579
54. Capron, P. C. and Oshima, T. (1952). *J. Chem. Phys.*, **20**, 1403
55. Estrup, P. J. and Wolfgang, R. L. (1960). *J. Amer. Chem. Soc.*, **82**, 2665
56. Placzek, G. (1946). *Phys. Rev.*, **69**, 423
57. Jurgeleit, H. C. and Wolfgang, R. (1963). *J. Amer. Chem. Soc.*, **85**, 1057
58. Estrup, P. J. (1964). *J. Chem. Phys.*, **41**, 167
59. Hsiung, G. H. and Gordus, A. A. (1964). *J. Amer. Chem. Soc.*, **86**, 2782
60. Felder, R. M. and Kostin, M. D. (1965). *J. Chem. Phys.*, **43**, 3082
61. Amdur, I., Longmire, M. S. and Mason, E. A. (1961). *J. Chem. Phys.*, **35**, 895
62. Amdur, I., Longmire, M. S. and Mason, E. A. (1969). *J. Chem. Phys.*, **51**, 968
63. Amdur, I. and Mason, E. A. (1956). *J. Chem. Phys.*, **25**, 630
64. Felder, R. M. (1967). *J. Chem. Phys.*, **46**, 3185
65. Baer, M. and Amiel, S. (1970). *J. Chem. Phys.*, **53**, 407
66. Thomsen, P. V. (1968). *J. Chem. Phys.*, **49**, 756
67. Muschlitz, E. E. and Simons, J. H. (1952). *J. Phys. Chem.*, **56**, 837
68. Kinsey, B. B. and Bartholomew, G. A. (1954). *Phys. Rev.*, **93**, 1260
69. Campbell, I. G. (1957). *Nucleonika*, **2**, 605
70. O'Connor, D. A. (1958). *Acta Physica Polonica*, **17**, 273
71. Schweinler, H. C. (1961). *Chemical Effects of Nuclear Transformations—Prague*, **1**, 63. (Vienna: IAEA)
72. Snell, A. H., Pleasonton, F. and Carlson, T. A. (1961). *Chemical Effects of Nuclear Transformations—Prague*, **1**, 147. (Vienna: IAEA)
73. Cross, J. and Wolfgang, R. (1961). *J. Chem. Phys.*, **35**, 2002
74. Suplinkas, R. J. (1968). *J. Chem. Phys.*, **49**, 5046
75. Karplus, M., Porter, R. N. and Sharma, R. D. (1964). *J. Chem. Phys.*, **40**, 2033
76. Karplus, M., Porter, R. N. and Sharma, R. D. (1965). *J. Chem. Phys.*, **43**, 3259
77. Karplus, M., Porter, R. N. and Sharma, R. D. (1966). *J. Chem. Phys.*, **43**, 3871
78. Porter, R. N. and Karplus, M. (1964). *J. Chem. Phys.*, **40**, 1105
79. Baer, M. and Amiel, S. (1969). *J. Amer. Chem. Soc.*, **91**, 6547
80. Porter, R. N. and Kunt, S. (1970). *J. Chem. Phys.*, **52**, 3240
81. Kuntz, P. J., Nemeth, E. M., Polyani, J. C. and Wong, W. H. (1970). *J. Chem. Phys.*, **52**, 4654
82. Bunker, D. L. and Pattengill, M. D. (1970). *J. Chem. Phys.*, **53**, 3041
83. Bunker, D. L. and Pattengill, M. D. (1969). *Chem. Phys. Lett.*, **4**, 315
84. Magee, J. L. and Hamill, W. H. (1959). *J. Chem. Phys.*, **31**, 1380
85. Estrup, P. J. (cf ref. 36), (1959), *Ph.D. Thesis*, 57, (Yale University, New Haven, Conn., U.S.A.)
86. Wolfgang, R. L. (1963). *J. Chem. Phys.*, **39**, 2983
87. Milman, M. (1964). *Radiochim. Acta.*, **2**, 180
88. Todd, J. F. J., Colebourne, N. and Wolfgang, R. L. (1967). *J. Phys. Chem.*, **71**, 2875
89. Johnston, A. J. (1966). *Ph.D. Thesis*, 29, (University of London)
90. Malcolme-Lawes, D. J. (1969). *Ph.D. Thesis*, 150, (University of London)
91. Urch, D. S., Welch, M. J. and Arthy, R. J. (1970). *Trans. Faraday Soc.*, **66**, 1642
92. Urch, D. S. (1970). *Radiochim. Acta*, **14**, 10
93. Rowland, F. S. and Coulter, P. (1964). *Radiochim. Acta*, **2**, 163
94. Brodskii, A. M. and Temkin, A. Ya. (1967). *Khim. Vys. Energ.*, **1**, 4
95. Brodskii, A. M. and Temkin, A. Ya. (1970). *Khim. Vys. Energ.*, **4**, 499
96. Leroy, R. L. (1969). *J. Phys. Chem.*, **73**, 4338

97. Menzinger, M. and Wolfgang, R. (1969). *Angew. Chem.*, **81,** 446
98. El-Sayed, M. F. A. and Wolfgang, R. L. (1957). *J. Amer. Chem. Soc.*, **79,** 3286
99. Estrup, P. J. and Wolfgang, R. (1960). *J. Amer. Chem. Soc.*, **82,** 2661
100. Henchman, M., Urch, D. S. and Wolfgang, R. L. (1960). *Can. J. Chem.*, **38,** 1722
101. Henchman, M., Urch, D. S. and Wolfgang, R. L. (1961). *Chemical Effects of Nuclear Transformations—Prague,* **2,** 83. (Vienna: IAEA)
102. Urch, D. S. and Wolfgang, R. L. (1961). *J. Amer. Chem. Soc.*, **83,** 2982
103. Urch, D. S. and Wolfgang, R. L. (1961). *Chemical Effects of Nuclear Transformations—Prague,* **2,** 99. (Vienna: IAEA)
104. Urch, D. S. and Wolfgang, R. L. (1959). *J. Amer. Chem. Soc.*, **81,** 2025
105. Garland, J. K. and Rowland, F. S. (1965). *Radiochim. Acta,* **4,** 115
106. Garland, J. K. and Rowland, F. S. (1966). *J. Phys. Chem.*, **70,** 735
107. Avdonina, E. H. (1970). *Khim. Vys. Energ.*, **4,** 83
108. Winter, G. K. (1971). *Ph. D. Thesis,* 258, (University of London)
109. Ache, H. J., Herr, W. and Thierann, A. (1961). *Chemical Effects of Nuclear Transformations—Prague,* **2,** 111. (Vienna: IAEA)
110. Cetini, G., Gambino, O., Castiglioni, M. and Volpe, D. (1967). *J. Chem. Phys.*, **46,** 89
111. Witkin, J. and Wolfgang, R. L. (1968). *J. Phys. Chem.*, **72,** 2631
112. Tominaga, T., Hosaka, A. and Rowland, F. S. (1969). *J. Phys. Chem.*, **73,** 465
113. Daniel, S. H. and Tang, Y. N. (1969). *J. Phys. Chem.*, **73,** 4378
114. Odum, R. and Wolfgang, R. L. (1963). *J. Amer. Chem. Soc.*, **85,** 1050
115. Tang, Y. N. and Rowland, F. S. (1965). *J. Amer. Chem. Soc.*, **87,** 3304
116. Tang, Y. N. and Rowland, F. S. (1967). *J. Amer. Chem. Soc.*, **89,** 6420
117. Hoff, W. J. Jr. and Rowland, F. S. (1957). *J. Amer. Chem. Soc.*, **79,** 4867
118. White, R. M. and Rowland, F. S. (1960). *J. Amer. Chem. Soc.*, **82,** 4713
119. Rowland, F. S., Turton, C. N. and Wolfgang, R. L. (1956). *J. Amer. Chem. Soc.*, **78,** 2354
120. Keller, H. and Rowland, F. S. (1958). *J. Phys. Chem.*, **62,** 1373
121. Kay, J. G., Malsan, R. P. and Rowland, F. S. (1959). *J. Amer. Chem. Soc.*, **81,** 5050
122. Henchman, M. and Wolfgang, R. L. (1961). *J. Amer. Chem. Soc.*, **83,** 2991
123. Tang, Y. N., Ting, C. T. and Rowland, F. S. (1970). *J. Phys. Chem.*, **74,** 675
124. Urch, D. S. and Wolfgang, R. L. (1961). *J. Amer. Chem. Soc.*, **83,** 2997
125. Odum, R. and Wolfgang, R. L. (1961). *J. Amer. Chem. Soc.*, **83,** 4668
126. Tachikawa, E., Tang, Y. N. and Rowland, F. S. (1968). *J. Amer. Chem. Soc.*, **90,** 3584
127. Tachikawa, E. and Rowland, F. S. (1968). *J. Amer. Chem. Soc.*, **90,** 4767
128. Urch, D. S. and Welch, M. J. (1966). *Trans. Faraday Soc.*, **62,** 388
129. Johnston, A. J., Georgiou, L., Malcolme-Lawes, D. J. and Urch, D. S. (1966). *Abstract RO39 Amer. Chem. Soc. Meeting, New York,* 152
130. Urch, D. S. and Welch, M. J. (1965). *Trans. Faraday Soc.*, **61,** 1411
131. Baker, R. T. K. and Wolfgang, R. L. (1969). *Trans. Faraday Soc.*, **65,** 1842
132. Malcolme-Lawes, D. J. (1969). *Ph.D. Thesis,* 76, (University of London)
133. Hawke, J. and Wolfgang, R. L. (1970). *Radiochim. Acta,* **14,** 116
134. Breckenridge, W., Root, J. W. and Rowland, F. S. (1963). *J. Chem. Phys.*, **39,** 2374
135. Root, J. W., Breckenridge, W. and Rowland, F. S. (1965). *J. Chem. Phys.*, **43,** 3694
136. Tachikawa, E. and Rowland, F. S. (1969). *J. Amer. Chem. Soc.*, **91,** 559
137. Wolfgang, R. L. (1962). *J. Amer. Chem. Soc.*, **84,** 4586
138. Root, J. W. and Rowland, F. S. (1964). *J. Phys. Chem.*, **68,** 1226
139. Tominaga, T. and Rowland, F. S. (1968). *J. Phys. Chem.*, **72,** 1399
140. Baker, R. T. K. and Wolfgang, R. L. (1968). *J. Amer. Chem. Soc.*, **90,** 4473
141. Arthy, R. (1969). *M. Phil. Thesis, The Reactions of Recoil Tritium Atoms with Isobutane,* 97, (University of London)
142. Urch, D. S. and Welch, M. J. (1966). *Radiochim. Acta,* **5,** 202
143. Urch, D. S. and Welch, M. J. (1968). *Trans. Faraday Soc.*, **64,** 1547
144. Johnston, A. J. (1966). *Ph.D. Thesis,* 55, (University of London)
145. Westhead, C. and Urch, D. S. (1971). *Chem. Commun.*, 403
146. Urch, D. S. and Welch, M. J. (1965). *Chemical Effects of Nuclear Transformations—Vienna,* **2,** 71. (Vienna: IAEA)
147. Urch, D. S., Welch, M. J. and Johnston, A. J. (1967). *J. Labelled Compounds,* **3,** 403
148. Seewald, D., Gersh, M. and Wolfgang, R. L. (1966). *J. Chem. Phys.*, **45,** 3870
149. Root, J. W. and Rowland, F. S. (1970). *J. Phys. Chem.*, **74,** 451
150. Root, J. W. and Rowland, F. S. (1967). *J. Chem. Phys.*, **46,** 4299

151. Lee, E. K. C. and Rowland, F. S. (1963). *J. Amer. Chem. Soc.*, **85,** 2907
152. Williams, R. R. and Ogg, R. A. (1947). *J. Chem. Phys.*, **15,** 691
153. Schwartz, H. A., Williams, R. R. and Hamill, W. H. (1953). *J. Amer. Chem. Soc.*, **74,** 6007
154. Carter, R. L., Hamill, W. H. and Williams, R. R. (1955). *J. Amer. Chem. Soc.*, **77,** 6457
155. Martin, R. M. and Willard, J. E. (1964). *J. Chem. Phys.*, **40,** 2999
156. Martin, R. M. and Willard, J. E. (1964). *J. Chem. Phys.*, **40,** 3007
157. Dzantiev, B. G., Eliseeva, N. N., Tantsyrev, G. D. and Shvedchikov, A. P. (1970). *Khim. Vys. Energ.*, **4,** 522
158. Chou, C. C. and Rowland, F. S. (1966). *J. Amer. Chem. Soc.*, **88,** 2612
159. Chou, C. C. and Rowland, F. S. (1969). *J. Chem. Phys.*, **50,** 5133
160. Chou, C. C. and Rowland, F. S. (1969). *J. Chem. Phys.*, **50,** 2763
161. Chou, C. C. and Rowland, F. S. (1967). *J. Chem. Phys.*, **46,** 812
162. Chou, C. C., Smail, T. and Rowland, F. S. (1969). *J. Amer. Chem. Soc.*, **91,** 3104
163. Gann, R. G. and Dubrin, J. (1967). *J. Chem. Phys.*, **47,** 1867
164. Compton, E. L. (1969). *J. Phys. Chem.*, **73,** 1158
165. Dzantiev, B. G., Pershin, A. D., Shimchuk, O. S., Shishkov, A. V. and Yasina, L. L. (1970). *Khim. Vys. Energ.*, **4,** 545
166. Dzantiev, D. G., Lynbimova, A. K. and Shishkov, A. V. (1969). *Khim. Vys. Energ.*, **3,** 478
167. Wooley, G. R. and Cvetanovic, R. J. (1969). *J. Chem. Phys.*, **50,** 4697
168. Chou, C. C. and Rowland, F. S. (1970). *Abstract Amer. Chem. Soc. Meeting*, 59
169. Gann, R. G. and Dubrin, J. (1969). *J. Chem. Phys.*, **50,** 535
170. Gann, R. G., Ollison, W. M. and Dubrin, J. (1970). *J. Amer. Chem. Soc.*, **92,** 450
171. Menzinger, M. and Wolfgang, R. L. (1967). *J. Amer. Chem. Soc.*, **89,** 5992
172. Menzinger, M. and Wolfgang, R. L. (1969). *J. Chem. Phys.*, **50,** 2991
173. Firsova, L. P. and Khar'ono, A. B. (1969). *Radiokhimiya*, **11,** 581
174. Firsova, L. P. and Do-Kim-Tyung (1969). *Radiokhimiya*, **11,** 587
175. Johnston, A. J., Malcolme-Lawes, D., Urch, D. S. and Welch, M. J. (1966). *Chem. Commun.*, 187
176. Winter, G. K. (1971). *Ph.D. Thesis*, 249. (University of London)
177. Baker, R. T. K. and Wolfgang, R. L. (1969). *J. Phys. Chem.*, **73,** 3478
178. Tang, Y. N. and Rowland, F. S. (1968). *J. Phys. Chem.*, **72,** 707
179. Tang, Y. N., Lee, E. K. C. and Rowland, F. S. (1964). *J. Amer. Chem. Soc.*, **86,** 1280
180. Cicolline, R. and Stöcklin, G. (1968). *Radiochim. Acta*, **9,** 105
181. Tang, Y. N. and Rowland, F. S. (1965). *J. Phys. Chem.*, **69,** 4297
182. Lee, E. K. C. and Rowland, F. S. (1963). *J. Amer. Chem. Soc.*, **85,** 897
183. Ting, C. T. and Rowland, F. S. (1970). *J. Phys. Chem.*, **74,** 445
184. Lee, E. K. C., Tang, Y. N. and Rowland, F. S. (1964). *J. Amer. Chem. Soc.*, **86,** 5038
185. Lee, E. K. C., Root, J. W. and Rowland, F. S. (1965). *Chemical Effects of Nuclear Transformations—Vienna*, **1,** 55. (Vienna: IAEA)
186. Avdonina, E. N., Urch, D. S. and Winter, G. K. (1969). *Radiochim. Acta*, **12,** 215
187. Tang, Y. N. and Rowland, F. S. (1968). *J. Amer. Chem. Soc.*, **90,** 570
188. Ting, C. T. and Rowland, F. S. (1968). *J. Phys. Chem.*, **72,** 763
189. Ting, C. T. and Rowland, F. S. (1970). *J. Phys. Chem.*, **74,** 4080
190. Lee, J. K., Musgrave, B. and Rowland, F. S. (1960). *J. Amer. Chem. Soc.*, **82,** 3545
191. Lee, E. K. C. and Rowland, F. S. (1962). *J. Amer. Chem. Soc.*, **84,** 3085
192. Lee, E. K. C. and Rowland, F. S. (1962). *J. Chem. Phys.*, **36,** 554
193. Mahan, K. I. and Garland, J. K. (1969). *J. Phys. Chem.*, **73,** 1247
194. Lee, E. K. C. and Rowland, F. S. (1970). *J. Phys. Chem.*, **74,** 439
195. Kushner, R. and Rowland, F. S. (1969). *J. Amer. Chem. Soc.*, **91,** 1539
196. Lee, J. K., Musgrave, B. and Rowland, F. S. (1960). *J. Chem. Phys.*, **32,** 1266
197. Root, J. W. and Rowland, F. S. (1963). *J. Amer. Chem. Soc.*, **85,** 1021
198. Smail, T. and Rowland, F. S. (1970). *J. Phys. Chem.*, **74,** 1859
199. Lee, E. K. C., Miller, G. and Rowland, F. S. (1965). *J. Amer. Chem. Soc.*, **87,** 190
200. Smail, T. and Rowland, F. S. (1970). *J. Phys. Chem.*, **74,** 456
201. Smail, T. and Rowland, F. S. (1968). *J. Phys. Chem.*, **72,** 1845
202. Root, J. W. and Rowland, F. S. (1963). *J. Chem. Phys.*, **38,** 2030
203. Hsiung, C. H. and Gordus, A. A. (1963). *J. Chem. Phys.*, **39,** 2770
204. Wolfgang, R. (1970). *J. Phys. Chem.*, **74,** 4601

205. Rowland, F. S. (1970). *J. Phys. Chem.*, **74**, 4603
206. Nesmeyanov, An. A., Dzantiev, B. G., Pozdeev, V. V. and Simonov, E. F. (1962). *Radiokhimiya*, **4**, 100
207. Nesmeyanov, An. A., Pozdeev, V. V. and Dzantiev, B. G. (1962). *Radiokhimiya*, **4**, 356
208. Avdonina, E. N., Baranovskii, I. B., Hao-Ming, O. and Nesmeyanov, An. A. (1965). *Radiokhimiya*, **7**, 217
209. Schroeder, J. W., Monroe, N. M. and Garland, J. K. (1969). *J. Phys. Chem.*, **73**, 1252
210. Avdonina, E. N. (1962). *Radiokhimiya*, **4**, 542
211. Avdonina, E. N. (1963). *Radiokhimiya*, **5**, 475
212. Sokolowska, A. (1965). *Chemical Effects of Nuclear Transformations—Vienna*, **1**, 255. (Vienna: IAEA)
213. Sokolowska, A., Haskin, L. A. and Rowland, F. S. (1962). *J. Amer. Chem. Soc.*, **84**, 2469
214. Sokolowska, A. (1968). *Radiochim. Acta*, **10**, 44
215. Avdonina, E. N. and Nesmeyanov, A. N. (1968). *Radiokhimiya*, **10**, 568
216. Avdonina, E. N. and Minkovskii, E. B. (1968). *Radiokhimiya*, **10**, 52
217. Avdonina, E. N., Elzakhir, A. and Nesmeyanov, An. A. (1970). *V. Mosk. Uspek. Khim*, **11**, 42
218. Pozdeev, V. V., Nesmeyanov, An. A. and Dzantiev, B. G. (1962). *Radiokhimiya*, **4**, 351
219. Odell, A., Rosenberg, A., Fink, R. and Wolfgang, R. (1964). *J. Chem. Phys.*, **40**, 3730
220. Root, J. W. and Rowland, F. S. (1962). *J. Amer. Chem. Soc.*, **84**, 3027
221. Root, J. W. (1969). *J. Phys. Chem.*, **73**, 3174
222. Rowland, F. S., Lee, E. K. C. and Tang, Y. N. (1969). *J. Phys. Chem.*, **73**, 4024
223. Chou, C. C. and Rowland, F. S. (1971). *J. Phys. Chem.*, **75**, 1283
224. Tang, Y. N., Lee, E. K. C., Tachikawa, E. and Rowland, F. S. (1971). *J. Phys. Chem.*, **75**, 1290
225. Palino, G. F. and Rowland, F. S. (1971). *J. Phys. Chem.*, **75**, 1299
226. Anbar, M. and Neta, P. (1962). *J. Amer. Chem. Soc.*, **84**, 2673
227. Colebourne, N., Todd, J. F. J. and Wolfgang, R. L. (1965). *Chemical Effects of Nuclear Transformations—Vienna*, **1**, 149. (Vienna: IAEA)
228. Dulmen, A. A. van and Aten, A. H. W. (1971). *Radiochim. Acta*, **15**, 34
229. Dulmen, A. A. van (1968). *Ph.D. Thesis*, Amsterdam University
230. Colebourne, N. and Wolfgang, R. L. (1963). *J. Chem. Phys.*, **38**, 2782
231. Tang, Y. N. and Rowland, F. S. (1967). *J. Phys. Chem.*, **71**, 4576
232. Tang, Y. N., Smail, T. and Rowland, F. S. (1969). *J. Amer. Chem. Soc.*, **91**, 2130
233. McKnight, C. F. and Root, J. W. (1969). *J. Phys. Chem.*, **73**, 4430
234. Smail, T., Miller, G. E. and Rowland, F. S. (1970). *J. Phys. Chem.*, **74**, 3464
235. McKnight, C. F. and Root, J. W. (1969). *Appl. Spectr.*, **23**, 639
236. McKnight, C. F., Parks, N. J. and Root, J. W. (1970). *J. Phys. Chem.*, **74**, 217
237. Spicer, L., Todd, J. F. J. and Wolfgang, R. L. (1968). *J. Amer. Chem. Soc.*, **90**, 2425
238. Aten, A. H. W., Koch, B. and Kommandur, I. (1955). *J. Amer. Chem. Soc.*, **77**, 5498
239. Brinkman, G. A. and Pauwels, E. K. J. (1970). *Progress Report, Institut voor Kernphysisch onderzoek, Amsterdam (IKO)*, **3**, section IV.6.2.b
240. Ledrut, J. H. T. (1970). *Radiochim. Acta*, **14**, 55
241. Elias, H. and Ndiokwere, C. L. (1970). *Radiochim. Acta*, **14**, 160
242. Richards, A. E. and Wolfgang, R. (1970). *J. Amer. Chem. Soc.*, **92**, 3480
243. Wai, C. M. and Rowland, F. S. (1968). *J. Amer. Chem. Soc.*, **90**, 3638
244. Spicer, L. and Wolfgang, R. (1969). *J. Chem. Phys.*, **50**, 3466
245. Willard, J. E. (1961). *Chemical Effects of Nuclear Transformations—Prague*, **1**, 215. (Vienna: IAEA)
246. Gordon, A. A. and Willard, J. E. (1957). *J. Amer. Chem. Soc.*, **79**, 4609
247. Chien, J. C. W. and Willard, J. E. (1953). *J. Amer. Chem. Soc.*, **75**, 6160
248. Evans, J. B., Quinlan, J. E., Sauer, M. C. and Willard, J. E. (1958). *J. Phys. Chem.*, **62**, 1351
249. Miller, J. M. and Dodson, R. W. (1950). *J. Chem. Phys.*, **18**, 865
250. Kontis, S. S. and Urch, D. S. (1971). *Radiochim. Acta*, **15**, 21
251. Spicer, L. and Wolfgang, R. L. (1968). *J. Amer. Chem. Soc.*, **90**, 2426
252. Black, A. and Morgan, A. (1970). *Int. J. Appl. Rad. Isotopes*, **21**, 5
253. Wai, C. M. and Rowland, F. S. (1967). *J. Phys. Chem.*, **71**, 2752
254. Wai, C. M. and Rowland, F. S. (1970). *J. Phys. Chem.*, **74**, 434
255. Rowland, F. S., Wai, C. M., Ting, C. T. and Miller, G. (1965). *Chemical Effects of Nuclear Transformations—Vienna*, **1**, 333. (Vienna: IAEA)

256. Wai, C. M., Ting, C. T. and Rowland, F. S. (1964). *J. Amer. Chem. Soc.*, **86**, 2525
257. Hammill, W. H. and Williams, R. R. (1954). *J. Chem. Phys.*, **22**, 53
258. Aten, A. H. W. and Raaphorst, J. G. van (1961). *Chemical Effects of Nuclear Transformations—Prague*, **1**, 203. (Vienna: IAEA)
259. Dulmen, A. A. van and Aten, A. H. W. (1971). *Radiochim. Acta*, **14**, 26
260. Brinkman, G. A. and Pauwels, E. K. J. (1970). *Progress Report. Institut voor Kernphysisch Onderzoek, (IKO), Amsterdam*, **3**, section IV, p.7
261. Vasáros, L. (1965). *Chemical Effects of Nuclear Transformations—Vienna*, **1**, 301. (Vienna: IAEA)
262. Wai, C. M. and Rowland, F. S. (1969). *J. Amer. Chem. Soc.*, **91**, 1053
263. Stöcklin, G. and Tornau, W. (1968). *Radiochim. Acta*, **9**, 86
264. Stöcklin, G. and Tornau, W. (1968). *Radiochim. Acta.*, **9**, 95
265. Berei, K. and Stöcklin, G. (1971). *Radiochim. Acta*, **15**, 39
266. Anders, O. U. (1965). *Phys. Rev.*, **138B**, 1
267. Emery, J. F. (1965). *J. Inorg. Nucl. Chem.*, **27**, 903
268. Carlson, T. A. and White, R. M. (1965). *Chemical Effects of Nuclear Transformations—Vienna*, **1**, 23. (Vienna: IAEA)
269. Wexler, S. (1961). *Chemical Effects of Nuclear Transformations—Prague*, **1**, 115. (Vienna: IAEA)
270. Kazanjian, A. R. and Libby, W. F. (1965). *J. Chem. Phys.*, **42**, 2778
271. Silbert, M. D. and Tomlinson, R. H. (1966). *Radiochim. Acta*, **5**, 217
272. Merrigan, J. A., Nicholas, J. B. and Rack, E. P. (1966). *Radiochim. Acta*, **6**, 94
273. Spicer, L. D. and Gordus, A. A. (1965). *Chemical Effects of Nuclear Transformations—Vienna*, **1**, 185. (Vienna: IAEA)
274. Rack, E. P. and Gordus, A. A. (1961). *J. Phys. Chem.*, **65**, 944
275. Tachikawa, E. and Okamoto, J. (1970). *Radiochim. Acta* , **13**, 159
276. Okamoto, J. and Tachikawa, E. (1969). *Bull. Chem. Soc. Jap.*, **42**, 2404
277. Tachikawa, E., (1970). *Bull. Chem. Soc. Jap.*, **43**, 63
278. Nicholas, J. and Rack, E. P. (1968). *J. Chem. Phys.*, **48**, 4085
279. Tachikawa, E. and Kahara, T. (1970). *Bull. Chem. Soc. Jap.*, **43**, 1293
280. Okamoto, J. and Tachikawa, E. (1969). *Bull. Chem. Soc. Jap.*, **42**, 1504
281. Aditya, S. and Willard, J. E. (1957). *J. Amer. Chem. Soc.*, **79**, 3367
282. Milman, M. (1963). *J. Phys. Chem.*, **67**, 537
283. Stamouli, M. I. and Katsanos, N. A. (1965). *Z. Phys. Chem., N.F.*, **47**, 306
284. Gavoret, G. and Wanoff, N. (1953). *J. Chim. Phys.*, **50**, 183, 434, 524
285. Nesmeyanov, An. A., Guseva, L. I., Tikhonova, L. I. and Zaborskii, A. K. (1955) *Dokl. Acad. Nauk.*, **103**, 1041
286. May, S., Roux, M., Buu-Hoi, Ng. Ph. and Daudel, R. (1949). *Compt. Rend.*, **228**, 1865
287. Stamouli, M. I. and Katsanos, N. A. (1968). *Radiochim. Acta*, **9**, 13
288. Stamouli, M. I. and Katsanos, N. A. (1967). *Radiochim. Acta*, **7**, 177
289. Iyer, R. M. and Willard, J. E. (1967). *Radiochim. Acta*, **7**, 175
290. Apers, D. J., Capron, P. C. and Gilly, L. (1957). *J. Chim. Phys.*, **54**, 314
291. Apers, D. J. and Capron, P. C. (1956). *J. Inorg. Nucl. Chem.*, **2**, 219
292. Chien, J. C. W. and Willard, J. E. (1954). *J. Amer. Chem. Soc.*, **76**, 4735
293. Chien, J. C. W. and Willard, J. E. (1957). *J. Amer. Chem. Soc.*, **79**, 4872
294. Roy, J. C., Williams, R. R. and Hammill, W. H. (1954). *J. Amer. Chem. Soc.*, **76**, 3274
295. Cole, A. J., Mia, M. D., Miller, G. E. and Shaw, P. F. D. (1966). *Radiochim. Acta*, **6**, 140
296. Harris, W. E. (1961). *Chemical Effects of Nuclear Transformations—Prague*, **1**, 229. (Vienna: IAEA)
297. Harris, W. E. (1961). *Can. J. Chem.*, **59**, 121
298. Milman, M. and Shaw, P. F. D. (1957). *J. Chem. Soc.*, 1303
299. Milman, M., Shaw, P. F. D. and Simpson, I. P. (1957). *J. Chem. Soc.*, 1310
300. Milman, M. and Shaw, P. F. D. (1957). *J. Chem. Soc.*, 1317
301. Milman, M. and Shaw, P. F. D. (1957). *J. Chem. Soc.*, 1325
302. Mia, M. D. and Shaw, P. F. D. (1966). *Radiochim. Acta*, **6**, 172
303. Cole, A. J., Mia, M. D., Miller, G. E. and Shaw, P. F. D. (1968). *Radiochim. Acta*, **9**, 194
304. Miller, G. E., Patterson, B. C. and Shaw, P. F. D. (1966). *Radiochim. Acta*, **6**, 178
305. Milman, M. (1967). *J. Chim. Phys.*, **64**, 658

306. Abedinzadeh, Z., Radicella, R., Tanka, K. and Milman, M. (1969). *Radiochim. Acta,* **12,** 4
307. Gilroy, T. E., Miller, G. and Shaw, P. F. D. (1964). *J. Amer. Chem. Soc.,* **86,** 5033
308. Merrigan, J. A., Ellgren, W. K. and Rack, E. P. (1966). *J. Chem. Phys.,* **44,** 174
309. Geissler, P. R. and Willard, J. E. (1963). *J. Phys. Chem.,* **67,** 1675
310. Chien, J. C. W. and Willard, J. E. (1955). *J. Amer. Chem. Soc.,* **77,** 3441
311. Fouseca, A. J. R. de, Fuller, K., Lathane, A. and Shaw, P. F. D. (1969). *Radiochem. Radioanal. Lett.,* **2,** 69
312. Iyer, R. M. and Willard, J. E. (1965). *J. Amer. Chem. Soc.,* **87,** 2494
313. Katsanos, N. A. and Stamouli, M. I. (1967). *Radiochim. Acta,* **7,** 126
314. Hornig, J. F., Levey, G. and Willard, J. E. (1952). *J. Chem. Phys.,* **20,** 1556
315. Levey, G. and Willard, J. E. (1956). *J. Chem. Phys.,* **25,** 904
316. Rack, E. P. and Gordus, A. A. (1961). *J. Chem. Phys.,* **34,** 1855
317. Rack, E. P. and Gordus, A. A. (1962). *J. Chem. Phys.,* **36,** 287
318. Rack, E. P. (1961). *Ph.D. Thesis,* 70. (University of Michigan, Ann Arbor, Mich., U.S.A.)
319. Schroth, F. and Adloff, J. P. (1964). *J. Chim. Phys.,* **61,** 1373
320. Welch, M. J. (1970). *J. Amer. Chem. Soc.,* **92,** 408
321. Paiss, U. and Amid, S. (1964). *J. Amer. Chem. Soc.,* **86,** 2332
322. Denschlag, H. O., Henzel, N. and Herrmann, G. (1963). *Radiochim. Acta,* **1,** 172
323. Cross, R. J. and Wolfgang, R. L. (1964). *Radiochim. Acta,* **2,** 122
324. Laurence, G. S. and Stranks, D. R. (1962). *Radioisotopes in the Physical Sciences and Industry,* **3,** 483. (Vienna: IAEA)
325. Gordus, A. A. and Hsiung, C. (1962). *J. Chem. Phys.,* **36,** 954
326. Reid, A. F. (1946). *Phys. Rev.,* **69,** 530
327. Shaw, P. F. D. (1963). *Radiochim. Acta,* **2,** 76
328. Geissler, P. R. and Willard, J. E. (1962). *J. Amer. Chem. Soc.,* **84,** 4627
329. McCauley, C. E. and Schuler, R. H. (1956). *J. Chem. Phys.,* **25,** 1080
330. Vlatkovic, M. and Willard, J. E. (1970). *Radiochim. Acta,* **14,** 19
331. Levey, G. and Willard, J. E. (1952). *J. Amer. Chem. Soc.,* **74,** 6161
332. Brustad, T. and Baarli, J. (1954). *J. Chem. Phys.,* **22,** 1311
333. Stöcklin, G., Schmidt-Bleek, F. and Herr, W. (1961). *Chemical Effects of Nuclear Transformations—Prague,* **1,** 245. (Vienna: IAEA)
334. McCauley, C. E., Hilsdorf, G. J., Geissler, P. R. and Schuler, R. H. (1956). *J. Amer. Chem. Soc.,* **78,** 3246
335. Macrae, J. E. C. and Shaw, P. F. D. (1962). *J. Inorg. Nucl. Chem.,* **24,** 1337
336. Iyer, R. M. and Martin, G. (1961). *Chemical Effects of Nuclear Transformations—Prague,* **1,** 281. (Vienna: IAEA)
337. Willard, J. E. and Chang, H. M. (1965). *Chemical Effects of Nuclear Transformations—Vienna,* **1,** 226. (Vienna: IAEA)
338. Schuler, R. H. (1954). *J. Chem. Phys.,* **22,** 2026
339. Alimarin, I. P. and Svoboda, K. F. (1958). *Atomnaya Energya,* **5,** 73
340. Svoboda, K. F. (1963). *Nature (London),* **198,** 986
341. Macrae, J. E. C. and Shaw, P. F. D. (1962). *J. Inorg. Nucl. Chem.,* **24,** 1327
342. Wheeler, O. H. and Lecumberry, C. (1970). *Radiochim. Acta,* **14,** 119
343. McCauley, C. E. and Schuler, R. H. (1958). *J. Phys. Chem.,* **62,** 1364
344. Ayres, R. L., Lemnitz, E. J., Lambrecht, R. M. and Rack, E. P. (1969). *Radiochim. Acta,* **11,** 1
345. Ormond, D. and Rowland, F. S. (1961). *J. Amer. Chem. Soc.,* **83,** 1006
346. Halpern, A. and Sochacka, R. (1961). *J. Inorg. Nucl. Chem.,* **23,** 7
347. Halpern, A. and Sochacka, R. (1961). *Chemical Effects of Nuclear Transformations—Prague,* **2,** 223. (Vienna: IAEA)
348. Adloff, M. and Adloff, J. P. (1966). *Bull. Soc. Chim. Fr.,* 3304
349. Llabador, Y. and Adloff, J. P. (1967). *Radiochim. Acta,* **8,** 41
350. Llabador, Y. and Adloff, J. P. (1968). *Radiochim. Acta,* **9,** 171
351. Llabador, Y. and Adloff, J. P. (1967). *Radiochim. Acta,* **7,** 20
352. Llabador, Y. and Adloff, J. P. (1966). *Radiochim. Acta,* **6,** 49
353. Halpern, A. (1963). *J. Inorg. Nucl. Chem.,* **25,** 619
354. Kontis, S. S. and Katsanos, N. A. (1969). *Z. Phys. Chem. N.F.,* **63,** 161
355. Shaw, P. F. D. (1951). *J. Chem. Soc.,* 443
356. Peacocke, T. A. H. and Walton, G. N. (1969). *J. Chem. Soc. A.,* 1264

6
Mössbauer Spectroscopy in the Study of the Chemical Effects of Nuclear Reactions in Solids

A. G. MADDOCK
University of Cambridge

6.1 INTRODUCTION

During the last decade two techniques have raised hopes that they might lead to a much deeper insight into the processes following nuclear transformations in solids. One of these, using angular correlation measurements, is considered in another chapter (Chapter 2): the other is the application of Mössbauer spectroscopy, which will be treated in this chapter. Four reviews have appeared on this subject[1–4], but only one[3] has a similar coverage to this review.

The attractive feature of both the above techniques is that they are *in situ* procedures; they enable one to learn something about the effects of the transformation without destroying the crystal lattice in which it has taken place. The earlier established radiochemical procedures always lead to destruction of the lattice, either by solution preceding analysis or, less commonly, by volatilisation for analysis. It has been known for a long time that the more reactive fragments produced by the nuclear reaction frequently react with the solvent so that the identification of the fragments is at best indirect.

The emission spectrum of a Mössbauer source is determined by the environment of the nuclei when they emit the Mössbauer photons. The lifetime of the excited nuclear state emitting the photons is usually quite small, generally between 10^{-9} and 10^{-6} s, so that the excited state must be continuously replenished if a steady emission of Mössbauer photons is desired. Such replenishment can be achieved by feeding the Mössbauer excited state either by radioactive decay or some other nuclear reaction. In the usual absorptiometric application of Mössbauer spectrometry one chooses as a source a substance in which the prior nuclear reaction does not modify the situation of the Mössbauer nucleus and in which the maximum degeneracy of the nuclear states is preserved, so that a single line emission spectrum is obtained. Such a source can be used to establish the Mössbauer absorption spectrum of compounds containing the ground-state species of this Mössbauer transition.

If, however, the preceding nuclear transformation does lead to changes in the source material, these may be investigated *in situ*, by exploring the emission spectrum of the source using a single line absorber chosen on the basis of experiments with a single line source. Thus, some knowledge of the effects of nuclear reactions in solids is essential to the preparation of satisfactory Mössbauer sources. Conversely, a study of the emission from unsatisfactory sources provides information about the effects of the nuclear reactions feeding the excited Mössbauer states.

6.2 MÖSSBAUER SOURCES

Radioactive decay processes are generally chosen for feeding the excited state. The figure (Figure 6.1) shows a number of examples of how different radioactive decay processes can be used.

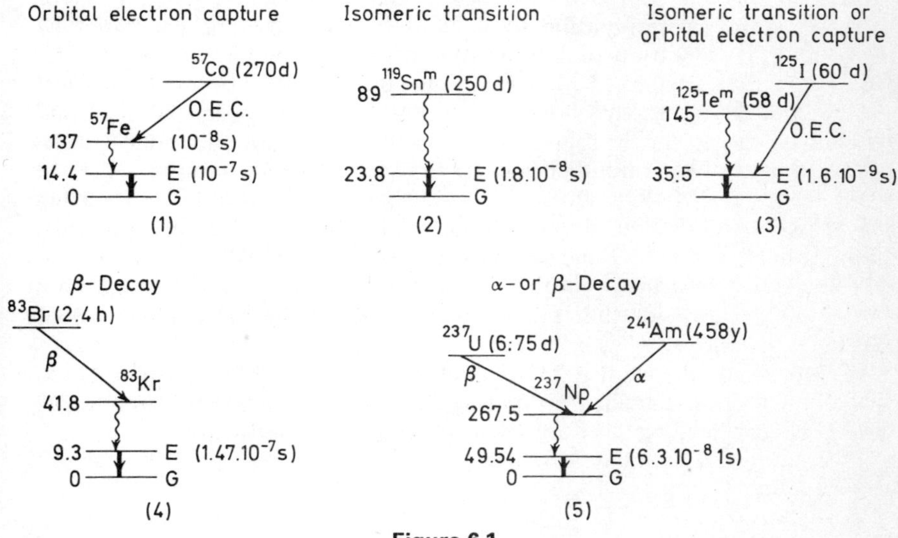

Figure 6.1

The parent isotope is sufficiently long-lived in these cases that it can be converted into a chemically well-defined state, as the element, in an alloy or solid solution, or as a compound, before use as a Mössbauer source. The Mössbauer emission spectrum will be determined by the environment and state of the atom containing the excited Mössbauer nucleus (denoted by E in Figure 6.1) in the period immediately preceding emission of the Mössbauer photon. Since the half-lives of these excited states are very short, the spectrum will reflect the situation of the Mössbauer atom very soon after the preceding radioactive decay process. As the decay schemes show this may be either a single process, an isomeric transition in the case of $^{119}Sn^{m}$, or a combination of such a process with photon emission from a higher excited state than the Mössbauer state. Thus $^{57}Co^{m}$ decays by orbital electron capture (O.E.C.) to an excited state of ^{57}Fe which rapidly emits a photon and gives $^{57}Fe^{m}$, the Mössbauer state. The half-lives for the primary photon emitting excited states are generally very small, $\sim 10^{-12}$ s.

It is also possible to regenerate the excited Mössbauer state by irradiation. A more general representation of the possibilities is given in Figure 6.2. The stable nuclear species, P, can be converted by a nuclear reaction, N_1 to the radioactive species, I, which feeds the excited Mössbauer state, E_1, either

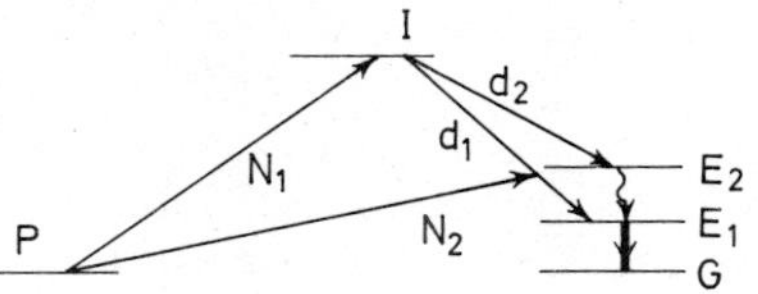

Figure 6.2

directly, d_1, or indirectly, d_2, by radioactive decay. For instance, ^{57}Fe (P) can be converted to ^{57}Co (I) by the (p, n) reaction. Alternatively, by 'in beam' measurements one can avoid the use of the intermediate species, I. For instance, states corresponding to E_1 and E_2 can often be generated by (a) coulombic excitation or bombardment of the ground-state species, G, with heavy ions, such as 25 MeV O^{8+} ions. In this case P becomes identical with G, and N_2 represents heavy ion excitation. (b) They can be produced by neutron capture or the related (d, p) reaction. These use the stable isotope of the element of mass number one less than the Mössbauer isotope as P. In these processes it is often necessary to reduce the background by measuring the Mössbauer emission in coincidence with some preceding photon emission. In the case of $^{57}Fe^{m}$ one can measure the 14.4 keV Mössbauer photons in coincidence with the preceding 122 keV radiation, using not too good a resolution to allow for the appreciable life-time of the Mössbauer excited state.

In either case the excited Mössbauer state is generated continuously in the source and no chemical processing can intervene between formation of the Mössbauer feeding state and measurement of the spectrum.

6.3 PREDICTED EFFECTS

Since many of the earlier investigators using this technique completely ignored the previous radiochemical and other work on this subject, it is perhaps appropriate to outline the effects predicted in different materials and following different kinds of nuclear reactions. The nuclear reactions are conveniently classified into those leading to (a) a negligible or small recoil of the affected atom (recoil comparable to or less than chemical bond strengths). This group includes most cases of β-decay, isomeric transitions and orbital electron capture. (b) A substantial mechanical recoil (recoil usually much larger than bond energies). A detailed account of the primary processes following the different nuclear reactions has been given by Wexler [5].

6.3.1 Negligible or small recoil

The most important mechanism effecting chemical changes following such reactions must be ionisation.

6.3.1.1 Isomeric transition and orbital electron capture

If an isomeric transition is not internally converted, the emission of the photon leads to a small recoil given in electron volts by 536 E_γ^2/M, where E_γ is the photon energy in MeV and M the mass number of the atom affected. Since E_γ for these processes is not usually more than 100 keV, then if $M = 100$ the recoil energy is less than 0.1 eV. This will lead to modest local heating but should produce no chemical changes.

If, however, the radiation is internally converted, the situation will resemble that of orbital electron capture; in both cases an atom is produced with a low-lying orbital electron vacancy and this will give rise to an Auger cascade. The orbital electron capture process also gives rise to a modest recoil, because of conservation of momentum with the neutrino emitted. In the case of the formation of $^{57}Fe^{m}$ from ^{57}Co, the recoil amounts to about 3.5 eV. It is given approximately by 536 $E_v{}^2/M$, where E_v is the neutrino energy.

The Auger cascade will lead to a rapid build-up of positive charge and emission of low-energy electrons by the affected atom. The number and energy of the electrons depends on the atomic number of the species concerned. At iron the most probable charge is about 4–5 units and by bromine it has risen to about 7. The maximum energy of the electrons ($E_K - 2E_L$) ranges from about 11 keV at krypton to about 24 keV at xenon. The charging process is rapid and should be complete in about 10^{-15} s.

These facts are well established, the uncertainty begins when one considers the process taking place in a solid. If the affected atom was initially combined in a molecule and the system is a gas, or indeed, possibly, even in a liquid, it has been suggested that the positive charge will quickly migrate to the periphery of the molecule and the repulsive forces arising between the different similarly-charged parts of the molecule will lead to a 'coulombic explosion'[6]. In a solid, and possibly even in a liquid, it is quite likely that electrons can return and neutralise this charge before the explosion can occur. It should be noted that the ranges of the Auger cascade in the solid will be very small so that even the original electrons have not moved very far from the affected atom.

None the less the charge neutralisation process itself will be very energetic and must lead to considerable excitation, and very possibly bond rupture, of the affected molecule. The fate of the Auger electrons will depend on the nature of the solid, the incidence of holes and electron traps in the vicinity of the affected molecule, and the possibility of producing local radiolytic effects.

There is a further possibility that the Auger cascade will be less disruptive in solids than in liquids or gases. It is quite reasonable to expect that in some solids it will proceed until the last electrons affected are no longer to be considered as associated with individual atoms or molecules of the solid. The excitation of these electrons, or the neutralisation of the holes produced, will probably be less disruptive than the events taking place with individual molecules.

All these ionisation, excitation and neutralisation processes should generally be complete within 10^{-9} s of the radioactive change so that by the time the Mössbauer photons are emitted the charge state and environment of the atoms concerned should be well defined. In the case of $^{57}Co \longrightarrow {}^{57}Fe^{m}$ there is a further delay introduced by the appreciable lifetime of the 122 keV emitting state ($\sim 10^{-8}$ s). This radiation is very little internally converted and unlikely to lead to further chemical changes. No other processes changing the charge or environment with half-lives comparable to that of the Mössbauer excited state need be taking place. This is equivalent to stating that the rapid annealing processes that determine the initial retention and yields of other products following an Auger cascade are complete in 10^{-9} s, and that the other annealing processes generally have half-lives greater than, say, 10^{-6} s. This may well be true for many systems but one might expect at least a few exceptions.

Unless the local radiolysis produced by the Auger electrons yields reducing species, one should always expect that these two radioactive decay processes may lead to oxidation of the affected atom, but oxidation implies trapping out at least one of the Auger electrons.

6.3.1.2 β-Decay

The immediate effect of negatron emission is to increase, and of positron emission to decrease, the positive charge of the affected atom by one unit. This is followed in about 10% of the events by the loss of another orbital electron due to the process of shaking[5] and in a further, slightly smaller, proportion of events by still greater charging. The processes of electron and neutrino emission both lead to mechanical recoil, but for elements of medium or high atomic number this recoil will not often exceed chemical bond energies, say 3 or 4 eV. β-Decay should therefore be expected to lead to oxidation but not much more drastic changes in the charge state and environment of the effected atom.

6.3.2 Processes with substantial recoil

The other processes used, α-decay, coulomb excitation, radiative thermal neutron capture, etc. all lead to substantial mechanical recoil. The processes involving heavy charged particles often produce a recoil of several thousand electron volts. Even the recoil following thermal neutron capture, which arises from conservation of momentum with the photons of the γ cascade, can amount to several hundred electron volts, although it may range down to values not greatly exceeding bond energies.

Existing analyses of this problem[7, 8c] suggest that the final environment of the affected atom is not very sensitive to its initial recoil energy. The original molecule will, of course, be ruptured and the recoiling atom is projected through the lattice producing displacements and fragmenting molecules until it has only one or two hundred electron volts of kinetic energy. The remaining energy is then deposited in a small region of the lattice at the end of the recoil track. Displacements, replacements and fragmentation take place in this region, and the effective temperature of the zone rises above the melting point of the solid. The loss of the recoil kinetic energy and the creation of the molten zone takes place in about 10^{-12} s. Calculations suggest that the quasi-molten zone persists for a little more than 10^{-11} s. During this period, and perhaps for a little while afterwards, chemical reactions can take place between the recoil atom and its surroundings. Recrystallisation then occurs and by 10^{-9} s the environment of the recoil atom has been determined and further change can only take place by the slower annealing reactions. The greatest uncertainty in this picture concerns the size of the molten zone; the original estimates suggested a content of about 1000 atoms, but more recent investigation indicates that it may be considerably smaller[8].

It must always be remembered that the processes leading to mechanical recoil may be accompanied by events leading to ionisation. For instance one or more of the photons emitted following radiative thermal neutron capture may be internally converted. The resultant Auger cascade will then take place after the recoil and its immediate consequences are complete, because the lifetime of the converting state may well exceed 10^{-9} s[9].

6.3.3 Time scale

Since the Mössbauer spectrum is determined by the state and environment of the atoms about $10^{-8}-10^{-6}$ s after formation, it should reveal whether there are annealing processes affecting the atom with half-lives in this range. It should also serve to distinguish that contribution to the retention that is due to a rapid replacement or reformation reactions in the hot zone.

6.4 MÖSSBAUER DATA

A Mössbauer spectrum yields six kinds of parameters, the chemical isomer shift, the quadrupole splitting, the magnetic splitting, the line widths, the magnitude of the Mössbauer effect and the area under the line or intensity of the line. Measurements at different temperatures will add temperature coefficients for each parameter.

The chemical isomer shift, δ, is determined by the valence state, or effective charge, of the Mössbauer atom and, to a smaller extent, by the number and character of the ligands. The quadrupole splitting, Δ, is determined by the asymmetry of the electric field at the Mössbauer nucleus; this in turn depends in part on the valence and spin state of the atom and in part on the more distant charged species in the lattice. It is, therefore, also affected by the oxidation state, the number, character and symmetry of the ligands and other components of the lattice. A permanent magnetic field at the Mössbauer nucleus will produce a magnetically split spectrum, from which the magnitude of the field at the nucleus and the angle it makes with the principal axis of the electric field gradient may be calculated. Equally important from the point of view of these studies is the line width, Γ, and magnitude of the Mössbauer effect, f; though these quantities are less frequently measured or discussed. Line widths greater than those found for the source material when functioning as an absorber can arise because of (a) unresolved quadrupole splitting or differences in chemical isomer shift due to the Mössbauer atom becoming stabilised in two or more rather similar environments. One might expect differences in the number of neighbouring vacancies to give rise to such effects. (b) The existence of processes changing the charge state or the environment of the Mössbauer atom with a probability per unit time comparable with that for the Mössbauer emission. Amongst such processes would be rapid diffusional jumps. (c) With paramagnetic species, the Mössbauer atom may become stabilised in a site at which the relaxation time becomes comparable to the mean life of the excited Mössbauer state. (d) With thick, or aged sources, which give substantial line broadening, this can be detected by measurements at different source thickness and it will be assumed that this possibility has been excluded by suitable experiments.

The Mössbauer fraction, *f*, a parameter rather less easily measured than those already described[10], is determined by the mean-square amplitude of vibration of the Mössbauer atom. Finally, the ratio of intensities of a pair of quadrupole split lines is significant in that when different from one it may be evidence of an anisotropy of the recoil-free fraction for the Mössbauer atom in its new site, the Goldanskii–Karyagin effect[11].

6.4.1 Correlation with radiochemical data

It would be very interesting to discover a system where the interpretation of the Mössbauer emission spectrum could be directly compared with the results of a radiochemical analysis and where there are no other changes that might invalidate the comparison. Unfortunately, there are serious difficulties in making such a comparison.

It might appear possible if the ground state of the Mössbauer nucleus, although long-lived, was radioactive; as, for instance, in the case of ^{129}I. One might suppose that a radiochemical analysis of the source ought to correspond to the indications of the Mössbauer spectrum. This is not so: all Mössbauer transitions are substantially internally converted so that most of the ground-state atoms are produced by an internally converted decay process which, as shown above, may lead to chemical changes. It is only the minority of ground-state atoms that must be in the same chemical state as the parent excited state and it is these that have been formed in Mössbauer emission events.

An alternative approach is to seek a system where the Mössbauer excited state is fed by a radioactive process that is unlikely to lead to chemical change. This would include not too energetic photon emission processes (<0.5 MeV), provided they are not substantially internally converted. However, if the photon processes proceed slowly enough to permit an initial radiochemical analysis it will most probably be internally converted. Thus although both ^{119}Sn^{m} and ^{125}Te^{m} can be produced by radiative thermal neutron capture, and one can explore their chemical state by radiochemical analysis, further chemical changes are possible and indeed probable in the decay process feeding the excited Mössbauer state.

6.5 THE EFFECTS FOLLOWING ISOMERIC TRANSITION AND ORBITAL ELECTRON CAPTURE

6.5.1 Metals

In metals, where there is a free supply of electrons, these processes have a negligible affect, so that if the ^{57}Co occupies a site of cubic symmetry the source will give a single-line emission of width not much greater than the theoretical value, as calculated from the half-life of the excited Mössbauer state.

6.5.2 Oxides

The first Mössbauer emission spectrum showing evidence of the chemical effects of a nuclear transformation was obtained with ^{57}CoO [12]. Nearly a decade of further work, reported in numerous papers, cannot be said to have resolved all the problems associated even with the CoO system.

6.5.2.1 Effects in cobalt oxides

The first paper on ^{57}CoO reported a two-line spectrum at or above 243 K. It was suggested[12] that these lines were due to emission by ferrous and ferric ions in normal lattice sites. No evidence of higher or lower charge states has been found in this, or subsequent papers. A typical spectrum is shown in Figure 6.3, taken from a more recent paper.

The Fe^{2+} line appears in the position found in the absorption spectrum of cobalt oxide doped with ferrous iron. Ferric doped samples have not been prepared.

At lower temperatures magnetically split spectra are obtained. Since the width of the ferric line was much greater than that of the ferrous line, the

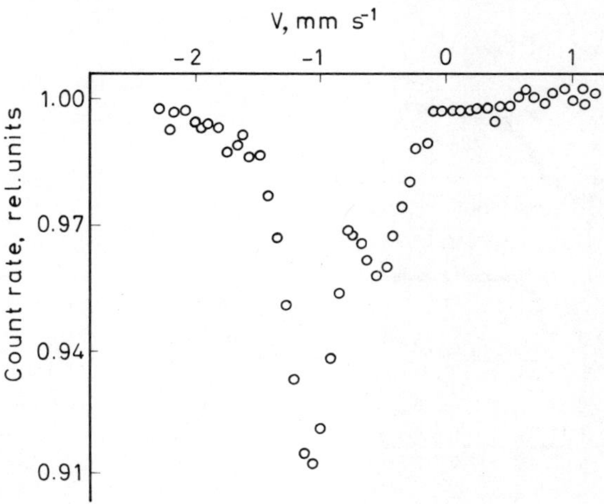

Figure 6.3 A typical spectrum. (From Murin *et al.*[29])

latter being normal, Wertheimer suggested that the $^{57}Fe^{m\,3+}$ might be unstable, with a probability for electron capture comparable with the probability of photon emission. Alternatively, it was possible that there was an unresolved quadrupole splitting or multiplicity of rather similar sites for the $^{57}Fe^{m\,3+}$. Similar conclusions were reached by other investigators[13,14] and it was noted that the ratio of intensities of the ferric to ferrous lines, approximately a measure of the $^{57}Fe^{m\,3+}/^{57}Fe^{m\,2+}$ ratio, increased with increasing temperature[14].

However, some investigators only found a single Fe^{2+} line emission from ^{57}CoO [15, 16]. The reduction of the Néel temperature by pressure was investigated using such a sample[16], but some irreproducibility of the material was observed and some preparations gave two-line spectra.

The possible instability of the $^{57}Fe^{m\,3+}$ state on the Mössbauer time-scale has been conclusively disproved by experiments in which the spectra were

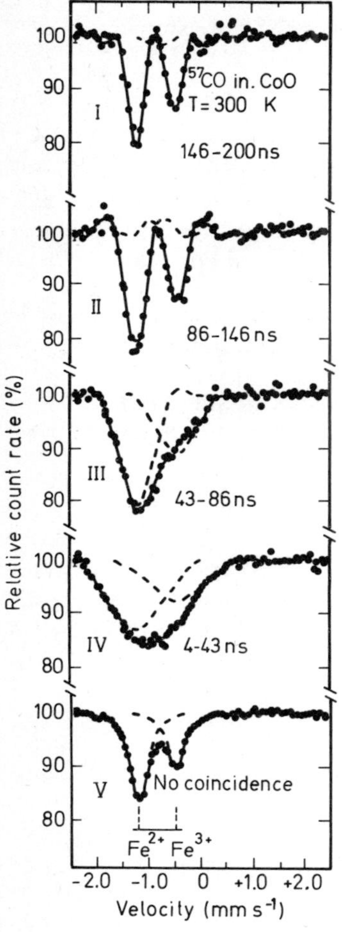

Figure 6.4 Spectra for different ages. (From Triftshauser and Craig[18], by courtesy of The American Institute of Physics)

recorded for different ages of the $^{57}Fe^{m}$ state. The 122 keV emission was used to indicate the formation of the $^{57}Fe^{m}$ species and served to trigger a delay system so that the 14.4 keV radiation would only be recorded in a certain time interval, so that the spectrum recorded would relate to $^{57}Fe^{m}$ atoms that survived for certain periods before photon emission. It was found that the Fe^{3+}/Fe^{2+} ratio was independent of the age of the $^{57}Fe^{m}$ at the time of photon emission[17, 18]. Spectra for different ages are shown in Figure 6.4.

Experiments were made both on CoO and ^{57}Co doped NiO. (The line broadening effects arise from the 'time filtering' effects of the delayed coincidence measurements; it is the constancy of ratio of line intensities that concerns us here.)

In the same paper[18] attention is drawn to the question of local charge compensation for the $^{57}Fe^{m\,3+}$ by vacancies in the lattice and it is suggested that the non-stoichiometry and the defect character and concentrations in the lattice might determine the Fe^{3+}/Fe^{2+} ratio. Wertheim had indeed already drawn attention to a role for cation vacancies. Pursuing this interpretation the reversible effect of temperature on the Fe^{3+}/Fe^{2+} ratio can be attributed to the higher concentration and greater mobility of the cation vacancies at the higher temperature increasing the likelihood of stabilisation as $^{57}Fe^{m\,3+}$. The lower third ionisation potential of the iron compared with the cobalt (30.7 eV and 33.5 eV respectively) may well stabilise a Fe^{3+} ion and a vacancy-trapped electron where a Co^{3+} ion would not survive. The greater line width of the Fe^{3+} line might then be due to the differing configuration of the Fe^{3+} and its associated vacancy giving a unresolved superpositioning of lines.

This approach, relating the appearance of the Fe^{3+} to the non-stoichiometry and defect structure of the source substance, was at first rejected by Ok and Mullen[19–21]. They obtained two forms of cobalt oxide. Both had a composition very close to CoO and possessed a face-centred cubic lattice, but one gave only an Fe^{2+} emission line whereas the other, which had half the density of the first preparation, gave only an Fe^{3+} line. The heavier form had all the characteristics of normal cobalt(II) oxide and the authors regarded their sample as possessing a cubic lattice with a low concentration of defects. The lighter form they supposed had half the anion and cobalt sites vacant with, possibly some short-range ordering of the system. They did not account for the considerable instability of the system in respect of lattice energy of this light form. Samples of cobalt(II) oxide giving two-line spectra were regarded as mixtures, possibly solutions, of these two forms. The light form readily absorbed oxygen and could be converted to heavy form by heating *in vacuo* above 300 °C. The transformation was slow.

The Mössbauer fraction, and hence the Debye temperature, was considerably less for the light than the heavy oxide and the transition to a magnetically split spectrum, on lowering the temperature of the oxide, took place over a wider range of temperatures than for the heavy oxide, which changed at about the accepted Néel temperature for CoO.

This interpretation, in terms of two forms of cobalt(II) oxide was soon challenged. It was pointed out that the behaviour of the light oxide suggested the superparamagnetic behaviour associated with samples of very small particle size. The lower Debye temperature of the light sample could also be explained in terms of very small particles. The preparation of the light material, precipitation of cobalt carbonate and thermal decomposition *in vacuo* at not above 300 °C, could be expected to lead to very small particles[22]. Ok *et al.* have subsequently modified their position[23], although they have excluded the superparamagnetic explanation by showing that for applied fields ($\leqslant 50$ kOe) the effects of the external and internal fields are additive at the Mössbauer nucleus. They now suggest that their second form of cobalt oxide has a sponge-like structure with a high concentration of vacancy defects.

Trousdale and Craig showed that the temperature dependence of the Fe^{3+}/Fe^{2+} ratio showed no appreciable hysteresis effects, even on rather rapid thermal cycling[24]. This seems hardly consistent with Ok and Mullen's original interpretation of the effect, which they supposed was due to interconversion and interaction between the two forms of the oxide. But it does not exclude the stabilisation of $^{57}Fe^{m\ 3+}$ by a balancing cation vacancy mechanism.

The behaviour of the other oxides of cobalt is also interesting in this connection. Like CoO the results reported for Co_3O_4, which has a spinel structure, were not always the same. With some samples a three-line spectrum[25,26] was obtained at 296 K. This was interpreted as arising from tetrahedrally coordinated $^{57}Fe^{m\ 3+}$, giving a single line, and a quadrupole-split pair from the octahedrally coordinated $^{57}Fe^{m\ 3+}$. However, in some samples, lines attributed to $^{57}Fe^{m\ 2+}$ were observed[26]. At low temperatures superparamagnetic effects were found because of the small particle size of the samples[26].

On the other hand, a study of ^{57}Co doped oxides ranging from composition CoO to Co_2O_3 always gave both $^{57}Fe^{m\ 3+}$ and $^{57}Fe^{m\ 2+}$ lines[27,28] and the proportions of the Fe^{3+} increased with increasing proportion of parent Co^{3+} [29]. These investigations, in keeping with a number made on the simpler salts of cobalt (Section 6.5.3), indicated that the form of the $^{57}Fe^m$ was substantially determined by the parent cobalt, Co^{2+} tended to give Fe^{2+} and Co^{3+} gave Fe^{3+}.

Indeed none of the above results are really entirely unexpected. In a simple polar solid containing monatomic cations and anions, such as an oxide, orbital electron capture has been shown (Section 6.3.1.1) to lead to multiple ionisation of the $^{57}Fe^m$ daughter together with modest local thermal excitation. In the close-packed lattice, implied by the polarity of the solid, the electrons lost by the $^{57}Fe^m$, which have low energies, do not move far from the affected atom and are unlikely to escape from the affected crystallite. There is no reason why the $^{57}Fe^m$ should be displaced from the site occupied by the parent cobalt. Thus in a perfect lattice, free of all imperfections, both impurities and defects, charge balance will require that the $^{57}Co^{2+}$ decay will yield $^{57}Fe^{m\ 2+}$ by the time the electronic processes following the orbital electron capture event have completely ceased, probably in less than 10^{-11} s. Similarly $^{57}Co^{3+}$ will yield $^{57}Fe^{m\ 3+}$, and so on. In real crystals, and especially in non-stoichiometric systems, there will be some probability that one, or more, of the Auger electrons are trapped and that some proportion of $^{57}Fe^m$ may survive in a higher oxidation state. But it is quite unlikely that such states will include any oxidation states that are not found in other non-stoichiometric compounds of iron.

The probability that an oxidised species will survive will depend not only on the exact composition of the polar solid, if it is markedly non-stoichiometric, but also on the concentration and nature of the point defects in the crystals and on the nature and concentration of impurities, especially aliovalent species. Further it may be influenced by the number and character of the dislocations in the crystals especially if the material is poorly crystallised and the crystallites are small. In such cases the surface characteristics and morphology of the crystal may be important. Attention to this latter point has been drawn by Schroeer and Niningen[30].

Which of these factors will predominate in a given system can also be predicted with some confidence. Markedly non-stoichiometric systems of substantial intrinsic disorder, like transition metal oxides, will generally provide intrinsic possibilities of stabilisation of $^{57}Fe^{m\,3+}$. But the proportion surviving will depend on defect density in the sample. In the more perfect halides the extent of oxidation should be more sensitive to impurities. A theoretical study of the stability of the $^{57}Fe^{m\,3+}$–vacancy complex has been made[31].

It is much more difficult to see how such a process should ever lead to reduction in such solids.

The most recent work on the CoO system is in substantial agreement with the above analysis[32]. It finds differences in the Fe^{3+}/Fe^{2+} ratio with the defect concentration in the oxide for material of precisely defined stoichiometry. It shows that the removal of defects during recrystallisation reduces

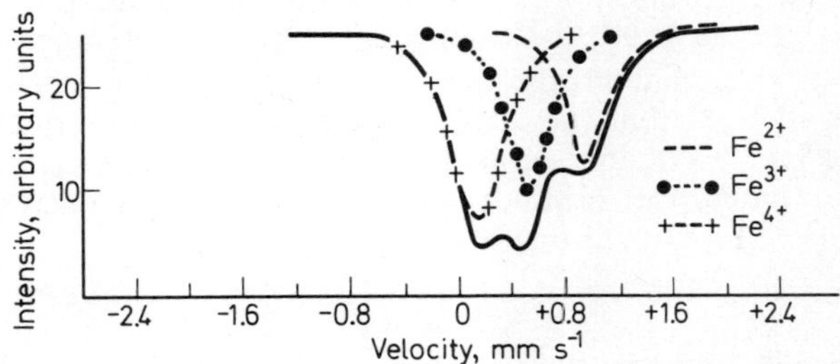

Figure 6.5 (From Jagannathan and Mathur[33], by courtesy of Pergamon Press)

this ratio, but that in equilibrium system of the same composition it will increase with temperature. Finally the ratio for samples of composition $Co_{1-x}O$, is a function of x.

It should be noted that Ok and Mullen's experimental observations are in no way questioned. Their light oxide has been duplicated by other workers[32] who find it to consist of aggregates of small particles with a high concentration of point defects. Such a material should have a high probability of stabilising $^{57}Fe^{m\,3+}$. Conversely the most perfect CoO should yield largely $^{57}Fe^{m\,2+}$.

Two significant investigations have been carried out using cobalt spinels. The first dealt with the normal spinel $CoCr_2O_4$, where the divalent cobalt occupies the tetrahedral sites[33]. This mixed oxide is known to be a p-type conductor and can probably accommodate cation vacancies. The spectra obtained, one of which is shown in Figure 6.5 have been interpreted in terms of tetrahedrally coordinated $^{57}Fe^{m\,2+}$, $^{57}Fe^{m\,3+}$ and $^{57}Fe^{m\,4+}$, the two last being stabilised by cation vacancies. The line widths, particularly for the Fe^{3+} and Fe^{4+} lines are large, which is compatible with the varying configurations of associated stabilising vacancies.

The appearance of $^{57}Fe^{m\,4+}$ is perhaps surprising. The authors did not have facilities for a detailed computer analysis of their data and the possibility that the Fe^{4+} line is part of a quadrupole-split pair should, perhaps, be explored more exhaustively. However, *a priori* there seems no reason to exclude the survival of $^{57}Fe^{m\,4+}$ in favourable situations, other compounds of this oxidation state are known and the higher charge will be compensated to some extent by more covalent bonds to the oxygen. At higher temperatures, above 200 °C, the Fe^{4+} disappeared and the Fe^{3+} line showed a corresponding increase. It was suggested that the di-vacancies necessary to stabilise the Fe^{4+} begin to dissociate. Unfortunately, the emission spectrum was not re-measured on cooling the sample to room temperature.

Cruset and Friedt have studied the ferrimagnetic spinel $CoFe_2O_4$. In this largely inverse spinel 96% of the divalent cobalt occupies octahedral sites and the remaining 4% tetrahedral sites. The emission spectrum of the ^{57}Co labelled spinel consisted of two six-line magnetically split spectra[34]. The magnetic fields at the $^{57}Fe^{m}$ nuclei were 489 and 509 kOe at 295 K and the chemical isomer shifts, as well as their temperature coefficients, showed that the spectra arose from tetrahedral and octahedral $^{57}Fe^{m\,3+}$.

Measurements have also been reported on some sulphides. ^{57}CoS gave a rather asymmetric quadrupole-split pair. Change of temperature had no effect on the ratio of intensities of the two lines[35]. The chemical isomer shifts suggested Fe^{3+}, but the quadrupole splitting was more than 1 mm/s, indicative of Fe^{2+}. Possibly rapid Fe^{2+}–Fe^{3+} exchange can occur in this system. The spinel-type sulphide $CoCr_2S_4$, gave lines corresponding to Fe^{2+} and Fe^{3+} at tetrahedral sites[33].

6.5.2.2 *Cobalt doped oxides*

The interpretation of the behaviour of the ^{57}Co doped systems is still more difficult because the initial site and charge state of the cobalt is often in doubt None the less, there are numerous data on such systems.

The closely related transition metal oxides are perhaps the least obscure. ^{57}Co doped NiO and MnO exhibit similar properties to ^{57}CoO. Thus earlier investigators reported the doped oxides gave entirely $^{57}Fe^{m\,2+}$ [36], or mixtures of $^{57}Fe^{m\,2+}$ and $^{57}Fe^{m\,3+}$ [14,17,36]. In the latter case a similar temperature dependence of the Fe^{3+}/Fe^{2+} ratio was observed[37]. More recently, it has been shown that in these systems, as in ^{57}CoO, the defect structure and degree of non-stoichiometry has a profound influence on the proportion of $^{57}Fe^{m\,3+}$ stabilised. In 'pure' nickel oxide in the absence of air very little Fe^{3+} is formed[38]. As yet, the study of these systems has added little to the picture developed from the cobalt oxide work. It has not been pursued so thoroughly as the more recent CoO work[32].

The results on ^{57}Co doped magnesium and calcium oxides are very interesting, but their interpretation is equivocal and not easily accommodated by the above model for the process. Two-[39,40,42] or three-[41] line spectra have been obtained. The relative intensities, but not the positions of the lines, change with the temperature of the sample. In each case one line appears at a very negative velocity, about −1.5 to −1.6 mm/s, with respect to soft iron. It has been suggested that this line must be due to $^{57}Fe^{m+}$, the other line or lines being due to $^{57}Fe^{m\,2+}$, or $^{57}Fe^{m\,2+}$ and $^{57}Fe^{m\,3+}$. The appearance of such a low oxidation state in materials like magnesium or calcium oxide is very difficult to understand.

Perhaps even more remarkable results have been reported for ^{57}Co doped cuprous oxide and aluminium oxide[42]. The former gave a four-line spectrum at 298 K with chemical isomer shifts of −2.01, −0.77, +0.14 and +1.14 mm/s with respect to soft iron. Since the positions of these four lines were not markedly dependent on the temperature it was argued that a quadrupole-split Fe^{2+} emission was not involved and the lines were assigned to Fe^{+}, Fe^{2+}, Fe^{3+} and Fe^{4+} respectively. The aluminium oxide gave lines at −1.10, −0.33 and +0.60, assigned to Fe^{2+}, Fe^{3+} and Fe^{4+} respectively. Delayed coincidence spectra showed that all these species were stable, on the Mössbauer time-scale, in the lattice. There was some evidence of line narrowing (beyond the usual 'time filtering' effects) at greater $^{57}Fe^{m}$ ages. This could indicate that the coordination environment of the $^{57}Fe^{m}$ became better defined on this kind of time-scale. These results are so curious that further work on this kind of system seems very desirable. There are also some data for $^{57}Co/CeO_2$ and other rare earth oxides[43] and for $^{57}Co/ThO_2$ [44]. The line widths in these systems are especially large. In both the cobalt oxides and the cobalt doped oxides it is curious that the $^{57}Fe^{m\,2+}$ usually gives a single-line spectrum; indeed, the ferrous line is usually narrower than any accompanying $^{57}Fe^{m\,3+}$ line. This seems to imply that the coordination octahedron is sufficiently rigid and exactly octahedral to suppress appreciable Jahn–Teller splitting and consequent quadrupole splitting in the spectrum.

The perovskites are well-defined solids perhaps better suited to investigation. In ^{57}Co doped $BaTiO_3$ the lattice is tetragonal at room temperature but becomes cubic above 107 °C. The Mössbauer emission spectrum changes from a quadrupole split $^{57}Fe^{m\,3+}$ emission below the transition to a single, broader line above, suggesting the ^{57}Co replaces the Ti, with vacancy compensation, and decays to yield $^{57}Fe^{m\,3+}$ in the same lattice site[45]. Later work suggested that more than one site might be involved[46]. $SrTiO_3$ has also been investigated[47]. Both Fe^{2+} and Fe^{3+} emissions are found, the proportion depending on the temperature of the sample. The spectrum changes at the transition temperature, which for this perovskite lies below room temperature (110 K). The spectrum could be changed by treatments affecting the stoichiometry of the sample. The authors propose that one of the emissions is due to a low-spin Fe^{3+} species, but this seems rather improbable. The behaviour of these compounds is in general agreement with what might be expected from the cobalt oxide studies.

6.5.2.3 Other oxide systems

Isomeric transition should resemble orbital electron capture and lead to little effect in oxides of highest oxidation state of an element. Certainly with $^{119}Sn^{m}$ the internally converted 65 keV photon emission feeding the Mössbauer excited state produces little effect in SnO_2 [48] or $BaSnO_3$ [49]. Similarly,

β TeO_3 has been proposed a suitable source material for $^{125}Te^m$ Mössbauer spectroscopy, implying that the isomeric transition produces little or no effect[50, 51]. Tellurium dioxide has also been investigated on a number of occasions[52, 53, 54]. There appears to be no evidence of Te^{VI} production[54], but the natural line width with $^{125}Te^m$ spectra is rather large so that the sensitivity for Te^{VI} production would be poor. The reports do not state whether computations were carried out to set limits to the possible production of Te^{VI}. A similar situation[55] pertains with Sn^{4+} production by $^{119}Sn^mO$ decay. Orbital electron capture in ^{153}Gd in an oxide matrix gives two lines[56] one of which may be due to Eu^{4+}.

The only remaining study on an oxide is noteworthy because it relates to the very narrow lines found for $^{67}Zn^m$ decay[57]. The experiments were carried out with a ^{66}ZnO source, which was deuteron-bombarded to produce ^{67}Ga by the (d, n) reaction. The irradiated sample was carefully annealed before use, the annealing treatment in fact determining whether a spectrum could be obtained. Such a source was examined in conjunction with a ^{67}ZnO absorber and a quadrupole-split spectrum obtained. No evidence of effects due to the orbital electron capture in the ^{67}Ga was obtained. Supposing that the annealing treatment allows the $^{67}Ga^{3+}$ to find a cationic lattice site and an associated compensating cation vacancy, the nascent $^{67}Zn^m$ will need to recapture all the electrons lost in the Auger cascade and acquire another to reach $^{67}Zn^{m\,2+}$. One might, perhaps, have expected that this might be a sufficient local disturbance to be reflected in the Mössbauer spectrum.

6.5.3 Polar compounds with monatomic ions

The principal source matrices in this category are the halides. Early work soon established that hydrated halides showed effects more closely related to other compounds with some covalently bonded structures (Section 6.5.4).

It might be expected that the halides would resemble the oxides, but that stabilisation of higher oxidation states might be more difficult, partly because of the generally lower intrinsic density of defects in the halides and partly because in the iodides, for example, simple charge transfer would take place between anion and cation with Fe^{3+}. Indeed the first results showed that most cobaltous halides[58, 59] yielded only a $^{57}Fe^{m\,2+}$ spectrum, but that the fluoride[60] gave both $^{57}Fe^{m\,2+}$ and $^{57}Fe^{m\,3+}$ emissions. These, and subsequent results for cobalt chloride[61] and bromide[62], as well as for cobaltous fluoride[63, 64] and $KCoF_3$ [65, 66] and cobaltic fluoride[67] and K_3CoF_6 [66], led to the very reasonable conclusion[68] that a cobalt halide matrix yields ^{57}Fe in the oxidation state of the parent cobalt and sometimes in the next higher oxidation state. Thus the best established evidence for $^{57}Fe^{m\,4+}$ was obtained in K_3CoF_6, but CoF_3 yields only $^{57}Fe^{m\,3+}$ [66, 67].

Data have also been reported on several ^{57}Co doped fluorides. ZnF_2 and NaF gave both $^{57}Fe^{m\,2+}$ and $^{57}Fe^{m\,3+}$. The proportion of the latter species was small and it was most easily seen in the magnetically split spectrum that it gave at low temperatures. The spectrum from the doped sodium fluoride was very complex and seemed to comprise several quadrupole-split doublets due to $^{57}Fe^{m\,2+}$. These were attributed to different configurations of a

Fe^{2+}-compensating cation vacancy combination[69]. Similar experiments on $^{57}Co/CaF_2$ also gave a mixed ferrous and ferric emission and the spectrum was observed to be changed by heating the source[70].

A much more extended study has included ^{57}Co doped ZnF_2, FeF_2, MnF_2, CoF_2, NiF_2 and MgF_2. FeF_2 and MnF_2 gave no measurable proportion of $^{57}Fe^{m\,3+}$ (<1%). ZnF_2 gave only a little, but the remaining fluorides gave appreciable proportions. Inspection of the results showed that the fraction of ^{57}Co decays stabilised as $^{57}Fe^{m\,3+}$ increased as the ionic radius of the host lattice cation decreased[71]. This implies that the fraction of Fe^{3+} increases with the lattice energy of the host lattice. Indeed Pollak[13] had pointed out that restricted space would favour Fe^{3+}, because of its smaller radius.

There are two other very significant observations on the halide system. In contrast to the oxide systems the Fe^{3+}/Fe^{2+} ratio decreases with increasing temperature[63–66, 71]. The change is reversible with no apparent hysteresis[65]. Finally, Cavanagh[60] has found that the proportion of $^{57}Fe^{m\,3+}$ formed in a CoF_2 matrix is much reduced if the host also contains inactive ^{57}Fe.

Friedt *et al.*[62, 66] have tried to give a more quantitative expression to the effect of the lattice energy of the host. The experimental data shown in

Table 6.1

Compound	CoF_2	β-$Co(OH)_2$	$CoCl_2$	$CoBr_2$
% $^{57}Fe^{m\,3+}$	44	22	0	0
Lattice energy (range of published values)	739–690	724–656	649–608	633–609

Table 6.1 suggest that Fe^{3+} stabilisation requires a lattice energy greater than about 650 kcal mol^{-1}. It will be noted that cobalt hydroxide behaves like a halide in this respect. Friedt *et al.* suggest that the Fe^{3+} state will be stabilised if $\Delta U > U_i - E_{exc.}$, where ΔU is the difference in lattice energy on replacing an Fe^{2+} ion in the matrix by an Fe^{3+} ion, U_i is the energy necessary to remove one electron from an Fe^{2+} ion *in situ* in the host lattice and $E_{exc.}$ is the energy of excitation available from the orbital electron capture event, assumed for this model to be constant. Since $U_i = I_3 - \Delta U$, where I_3 is the third ionisation potential of iron, presumably the above expression could be transformed to $2\Delta U > I_3 - E_{exc.}$.

This model, although pointing in the right direction, can be criticised in various respects. Most fundamentally, one may question whether equilibrium energetics are really relevant to this problem? The system at the time of Mössbauer emission may still be in a metastable state, although further change might be slow on the Mössbauer time-scale. It might perhaps be compared with the situation of a sample of γ-irradiated sodium chloride at room temperature. The energy $E_{exc.}$ is not very clearly defined, presumably the Auger cascade nearly always leads to at least momentary formation of Fe^{3+}, the question is how rapidly does the last recombination event take place? One would suppose that two factors are involved. It is reasonable that the instability of the Fe^{3+} ion, represented by U_i and equal to $I_3 - \Delta U$, is one of them; but the neutralisation event is also dependent on the supply of the electron, so that the depth of the trap that captured the last of the Auger electrons must have some importance.

The difference in the temperature dependence of the oxide and halide systems seems important. In the former systems it has been suggested that increased temperature increases both the defect density and mobility in the matrix and thus facilitates charge compensation for the Fe^{3+}. In the halide systems similar effects lead to increased charge neutralisation. Different matrices will have different requirements in respect of charge compensation. In some it will have to be very close to the Fe^{3+} site, in others it can be much more distant. In most papers attention has been focused on compensation by cation vacancies, this ignores the fate of the last Auger electron and more attention should perhaps be given to compensation by a trapped electron. It may be argued that competitive processes must be involved. Although reproducibility in these data leaves much to be desired, an observation in itself pointing to a role of defects in the matrices, the proportions of Fe^{3+} do not change sharply from 0 to 100% in different hosts. The line widths in these systems are invariably wide but if we can assume that the $^{57}Fe^{m\ 3+}$ is undergoing no other changes on the Mössbauer time-scale, for there is no direct experimental data in these systems, the effect of temperature can only reflect the differing competition between two very fast processes leading to Fe^{2+} and Fe^{3+}, that are essentially complete in 10^{-9} s.

Not surprisingly there are no detectable effects in $^{119}Sn^mI_4$ [55]. A detailed undergoing no other changes on the Mössbauer time-scale, for there is no the most interesting of the current objectives in this field. One might speculate that experiments ought to be made with crystalline material in which more is known about the nature and concentration of defects, including materials doped to contain deliberate electron and hole traps.

6.5.3.1 *Rubidium halides*

The emission spectrum of $^{83}Kr^m$ formed by orbital electron capture in the rubidium halides from the orbital electron capturing parent ^{83}Rb has been investigated. A first report[72,73,74] that quadrupole-split spectra were obtained has been disproved[75], the earlier results being due to splitting in the krypton clathrate used as absorber. Thus, as expected, there are no chemical after-effects in this system.

6.5.3.2 *Doped alkali and silver halides*

The behaviour of doped aliovalent host lattices has an additional complication in its interpretation: in such matrices the situation and charge state of the parent species is often either speculative or unknown. Indeed the objective of these investigators has been rather the solution of this problem than the elucidation of the effects of the nuclear transformation. These studies have been the subject of a recent review[76].

(a) $^{119}Sn^m$ *doped systems*—Parent $^{119}Sn^m$ has been introduced into lithium, sodium, potassium and silver chlorides and silver iodide[78, 79] by fusing the salt *in vacuo* in the presence of tin containing $^{119}Sn^{m}$. Silver chloride has been doped with stable ^{119}Sn in the same way[74] so that absorptiometric measurements on the state of the tin could be made.

In the silver halides both the emission and absorption spectra[77] gave only a single, rather broad line, corresponding to Sn^{4+}.

In the alkali halides, optical [76] and other data suggest tin can enter the lattice as both Sn^{2+} and Sn^{4+}. The emission spectrum obtained[78, 79] consisted of an Sn^{4+} emission together with a poorly resolved, quadrupole-split doublet due to Sn^{2+}. On preparing or annealing the doped crystals in air the Sn^{2+} lines disappear. The same effect can be produced by irradiation of the vacuum doped salts.

The authors[77] suggest that the tin is present in the silver halides in interstitial sites associated with three cation vacancies which lead to an unresolved splitting of the $^{119}Sn^{m\,4+}$ emission. In the alkali halides it would seem likely that the tin will occupy cationic lattice sites as Sn^{2+} or it might form $SnCl_3^-$ and occupy an anionic site. The isomeric transition could then lead to the formation of Sn^{4+}. For the moment, the interpretation of the results must remain speculative.

Murin and Seregin in their review draw attention to yet another of the difficulties attending these studies. Unless macrocrystalline material has been prepared there is always some possibility that the results will be determined by forms of the dopant species present on surfaces, at dislocations etc. and hardly present in the interior of the lattice.

(b) ^{57}Co *doped systems*—The data on the ^{57}Co doped alkali and silver halides further emphasises all the difficulties referred to above.

The earlier work of de Coster and Amelinckx[80] was reviewed by Mullen[81], who showed that unless the concentration of cobalt introduced in doping the halide was kept below 10^{-3} mol % the cobalt was either not homogeneously incorporated, or possibly it entered as $CoCl_4^{2-}$, and a quadrupole-split emission spectrum corresponding to the absorption spectrum found for iron in a cobalt chloride lattice was obtained.

Several other papers have appeared reporting results using alkali halides[82–84] and silver halides[82, 83, 85–90] as host lattices. The lines in the emission spectra obtained are always broad, the spectra are temperature dependent, but not always reversibly so, and in some samples of ^{57}Co doped single-crystals of silver chloride the spectrum changes with the age of the source. At low concentrations of cobalt two kinds of spectra are found. In one, obtained both with NaCl and AgCl, two lines of unequal intensity are emitted. The separation of the lines hardly changes with temperature until above about 400 K when it gradually falls to zero. The other kind of spectrum only shows quadrupole splitting below room temperature, but increases rapidly with falling temperature. These observations have been interpreted in a number of ways. It has been suggested[87] that the first spectrum corresponds to $^{57}Fe^{m\,2+}$ and $^{57}Fe^{m\,3+}$ emissions. However, application of a magnetic field to the source broadens one line much more than the other which might indicate a quadrupole-split pair[90]. Simultaneous doping of a silver chloride crystal with both ^{57}Co and divalent cadmium considerably reduces the asymmetry of the first kind of spectrum[88].

It is still far from certain whether Fe^{2+}, Fe^{3+} or both are produced and what kind of site, or sites, are involved. Even $^{57}Fe^{m\,+}$ has been suggested[81] and although this might seem improbable one can envisage a reasonable mechanism for its stabilisation. The short range shower of Auger electrons can produce local radiolysis of the lattice (*see* Section 6.5.4) surrounding the nascent $^{57}Fe^{m}$, if this produces a species like Cl_2^-, or indeed any of the V type defects, another electron will be released for neutralisation of the nascent iron ion, so that $^{57}Fe^{m\,+}$ could be produced from $^{57}Co^{2+}$. It is only necessary that the energy of activation for migration of the hole at the V defect to the iron atom be not too low for the $^{57}Fe^{m\,+}$ to survive till Mössbauer emission.

A Fe^{2+} or Fe^{3+} plus cation vacancy seems a very likely product and it has been suggested that the change from a quadrupole-split spectrum to a line at elevated temperatures might result from the increasing mobility of the vacancy at the higher temperature[83, 85, 88, 89].

The combination of the uncertainty in the location and charge on the parent cobalt and the incomplete understanding of the effects of the nuclear transformation make this a very difficult area of study.

The most recent results of Hennig and Kim Yung[91] still further complicate the issue. Some of their spectra are shown in Figure 6.6.

They find that their spectra are not only modified by heating *in vacuo* but even more profoundly by heating in a moist inert atmosphere or in air. It seems most probable that impurity defects associating oxide or hydroxide ions with the ^{57}Co defect are formed and that these lead to quite different spectra. The definitive identification of the cobalt centres must await more extensive studies and these results are most significant in underlining the difficulties that beset such studies.

6.5.4 Compounds with covalent bonds

In solids in which there is some covalent bonded structure, especially where the bonds to the Mössbauer atom have appreciable covalent character, the model involving a coulombic explosion combined with local radiolysis of the surrounding lattices, suggested by Willard[6b] to account for the chemical effects following isomeric transition in organic liquids, generally fits the pattern of the data. However, the Mössbauer results have led to some important new ideas regarding such a process in aromatic or extensively π-bonded solid complexes and there are several observations that are still difficult to understand.

6.5.4.1 ^{57}Co compounds or doped compounds

It was expected from *a priori* considerations that if the Mössbauer atom was covalently bound the overlap of its orbitals with that of the ligands would facilitate the return of the Auger electrons and favour stabilisation in the same oxidation state as the parent atom. By contrast, experimental data on hydrated salts of cobalt showed that some $^{57}Fe^{m\,3+}$ emission was nearly always found in such systems[27, 59, 63, 68, 92–94]. As in most of these systems, the line widths in

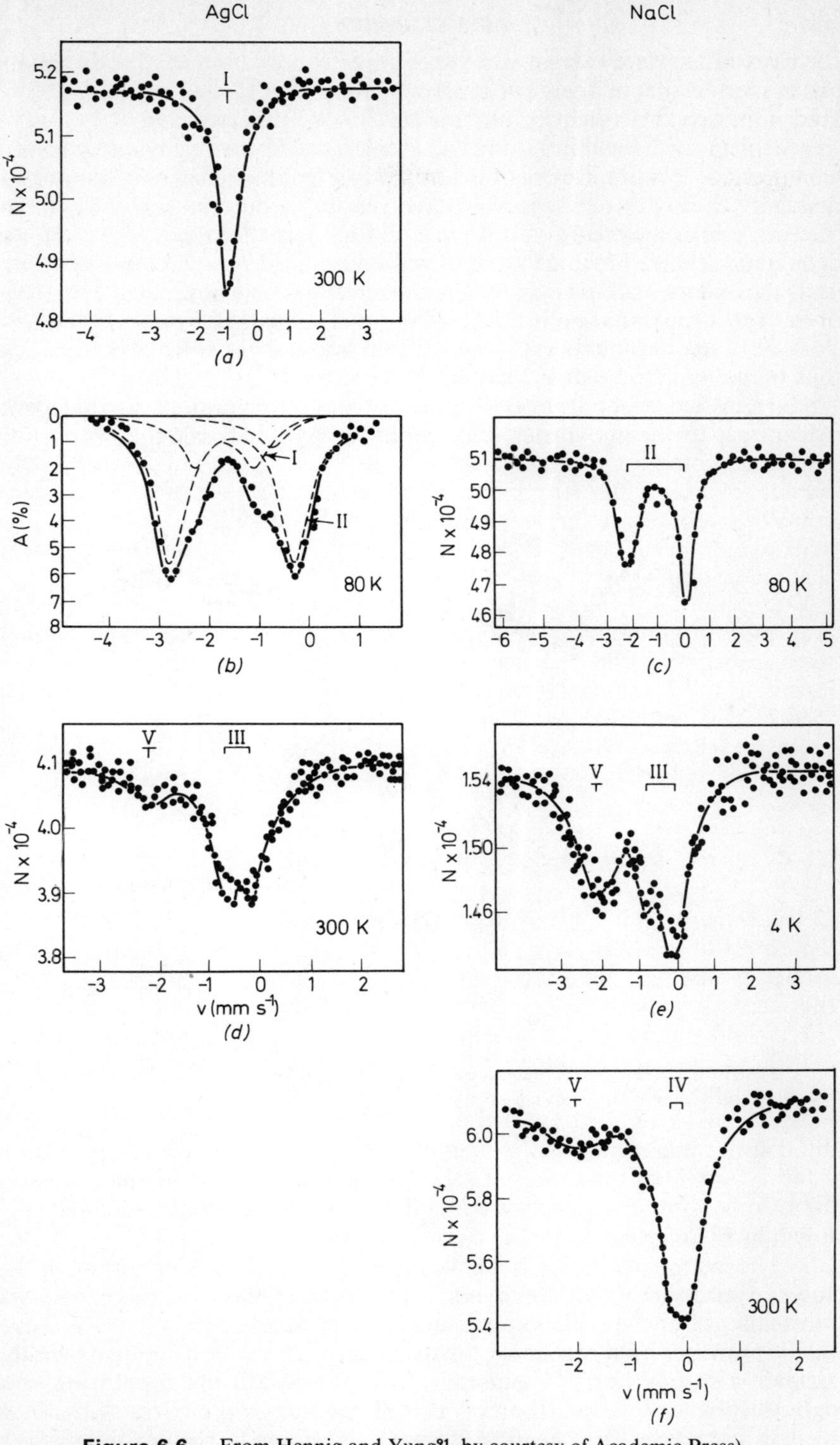

Figure 6.6 From Hennig and Yung[91], by courtesy of Academic Press)

the emission spectra were large, but at least in the case of hydrated ferrous ammonium sulphate[18] this cannot be due to a short lifetime of the $^{57}Fe^{m\,3+}$ state with respect to changes other than photon emission.

The proportion of $^{57}Fe^{m\,3+}$ found increases with the degree of hydration when several hydrates can be prepared[59] and a halide such as cobalt chloride, which gives no $^{57}Fe^{m\,3+}$ as an anhydrous host, does so when hydrated. These observations led Wertheim and Buchanan[68] to propose that the behaviour of these hydrated systems was determined by a sequence of events such as Willard had suggested. The Auger electrons cause local radiolysis of the lattice surrounding the affected atom. The hydroxyl radical produced from the water in the process provides deep traps for the Auger electrons and thus higher oxidation states may be stabilised.

The emission spectra are both pressure and temperature[63, 93] dependent. In contrast to the anhydrous fluoride, in the hydrated salt the proportion

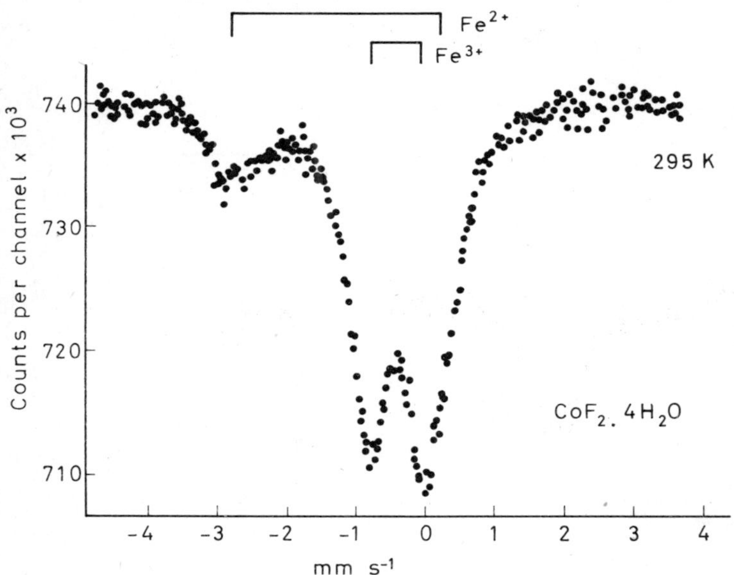

Figure 6.7 Mössbauer spectra of $^{57}CoF_2{\cdot}4H_2O$. (From Friedt and Adloff[63], by courtesy of Gauthier-Villars)

of $^{57}Fe^{m\,3+}$ increases with temperature[63]. Possibly charge compensation of the $^{57}Fe^{m\,3+}$ by migration of OH^- becomes quicker at higher temperatures. Some hysteresis effects are found[93]. The spectrum[63] of $CoF_2{\cdot}4H_2O$ is shown in Figure 6.7.

Early experiments on ^{57}Co labelled complexes showed that although the emission spectrum sometimes included lines corresponding to the immediate iron analogue of the parent compound several other species of $^{57}Fe^m$ were usually present[95, 96]. In particular, the spectrum of a cobaltic complex generally included a strong $^{57}Fe^{m\,2+}$ emission. It is known that the application of a high pressure to a ferric absorber changes the electron density at the iron nucleus and moves the chemical isomer shift closer to the ferrous value[97].

Hazony and Herber noted that the effective radius of the iron in ferric acetylacetonate was greater than that of the cobalt in cobaltic acetylacetonate and suggested that the apparent $^{57}Fe^{m\,2+}$ emission from the labelled complex might be due to the compression of $^{57}Fe^{m\,3+}$ by the lattice of cobalt complex[98, 99]. However, a ferrous emission is equally prominent in the spectrum obtained with some ^{57}Co doped ferric complexes[100].

The Willard model implies that the emission spectrum for a ^{57}Co labelled source might resemble the absorption spectrum of the partially radiolysed analogous iron complex[68]. This resemblance has been remarked on by a number of workers in cobalt acetyl acetonate[101], in potassium ferrioxalate[100, 102] and in hydrated salts[103]. One might expect ferri- or cobalti-oxalates to give large yields of reduced $^{57}Fe^{m}$, since, because of their internal instability, they are 'trigger' systems that can be initiated thermally, photolytically or radiolytically. The data for labelled hydrated and anhydrous cobalt oxalate are, at present, conflicting. Recent work[104, 105] reports both $^{57}Fe^{m\,2+}$ and $^{57}Fe^{m\,3+}$ from each compound, while an earlier paper[106] found only $^{57}Fe^{m\,2+}$. The model could accept Fe^{3+} in the hydrated salt, but it is difficult to account for its formation from the anhydrous salt.

A comparison has been made of the proportion of ferrous and ferric iron formed following orbital electron capture in ^{57}Co doped hydrated ferrous oxalate, hydrated ferric oxalate (ferrioxalic acid?) and hydrated potassium ferrioxalate, on the one hand, and by radiative thermal neutron capture in the

Table 6.2

	^{59}Fe *by* (n,γ)	$^{57}Fe^{m}$ *from* O.E.C.
Fe^{3+} in $FeC_2O_42H_2O$	9%	10%
Fe^{2+} in $Fe_2(C_2O_4)_3xH_2O$	3%	80%
Fe^{2+} in $K_3Fe(C_2O_4)_33H_2O$	60%	69%

iron in the same compounds[107]. The results are shown in Table 6.2. As might be expected orbital electron capture gave the largest amount of Fe^{2+} in the ferric complexes.

The cobaltous ammines behave like hydrated salts[108], the radiolysis of the ammonia presumably producing the electron-trapping radical NH_2. Thus $Co(NH_3)_6Cl_2$ gives a ferric emission in addition to ferrous doublets. As might be expected, if the covalent structure is resistant to radiolytic change, the $^{57}Fe^{m}$ will generally assume the valence of the parent cobalt, as happens in $^{57}CoCO_3$ [109]. But cobalt sulphate gives some $^{57}Fe^{m\,3+}$ [59, 93].

Perhaps the most unexpected and significant results have followed the work of Amar Nath and others on ^{57}Co labelled complexes, especially those with extensively π-bonded structures. In $Co(dipy)_3(ClO_4)_3$, for example, a surprisingly high proportion of the $^{57}Fe^{m}$ seems to appear as the corresponding iron complex[110, 111], but with $Co(1.10\ phenan)_3(ClO_4)_32H_2O$ nearly half the $^{57}Fe^{m}$ gives a high spin ferrous emission[112]. The ethylenediaminetetracetate complex, the indenyl complex and the bis-salicylaldehyde triethylenetetramine complexes all fragment substantially and yield $^{57}Fe^{m\,2+}$ in addition to other species[115]. But most interestingly and rather unexpectedly, the labelled phthalocyanine, vitamin B_{12} and the cobalamins and the tris-dipyridyl complex seem to escape fragmentation and the ^{57}Co decay yields a large proportion of the $^{57}Fe^{m}$ in form of the analogue of the parent compound[113–115].

It is well known that complexes such as those in the latter group are very radiation resistant, so that the Auger electrons produce little effect and may be available for neutralisation, but it is surprising that the coulombic explosion seems somehow to be avoided. Presumably electrons are able to flow back and effect neutralisation before any coulombic disruption, nor apparently does the neutralisation process dissociate the molecule. There is indeed some evidence that a similar survival process can occur to some extent in solution[116]. It was observed that yield of the $^{57}Fe^{m}$ in the complexed form was much reduced when these complexes were dispersed in ethylene oxide and the frozen system examined[117].

All these systems give broad-line spectra which are difficult to analyse and interpret and at present it is perhaps premature to conclude that the spectrum seen proves that the corresponding $^{57}Fe^{m}$ complex is formed, but it does seem safe to conclude that far less disruption and production of high-spin $^{57}Fe^{m\,2+}$ and $^{57}Fe^{m\,3+}$ occurs in these complexes than in the simpler complexes. It is also desirable that measurements of the Mössbauer factions should be made in these systems to find out whether other states of $^{57}Fe^{m}$ are produced which do not appear in the emission spectrum.

Another very important result has been obtained during these experiments. By examining the emission spectrum of cobalt(III) tris-dipyridyl perchlorate doped with $^{57}Co^{2+}$ it has been shown that an exchange can occur in the solid state at 25 °C. After a few days storage an emission spectrum attributed to the $^{57}Fe^{m\,III}$ tris-dipyridyl perchlorate appeared[115, 117]. Such an exchange may well occur more readily and quickly in the excited zone following orbital electron capture in the cobalt.

In ^{57}Co labelled cobalt(II) bis-pyridine chloride the iron appears in the ferrous state[118]. Two complexes with this empirical formula are known. The violet octahedrally coordinated polymeric form is found to give a larger Mössbauer fraction than the blue tetrahedrally coordinated monomeric form. The other pyridine complexes of cobalt chloride also give simple ferrous emissions[119].

Soluble Prussian blue, $KFe^{3+}Fe^{II}(CN)_6H_2O$ has been doped separately with ^{57}Co in the uncomplexed cationic sites and in the complex[120]. The former yields principally an $^{57}Fe^{m\,3+}$ emission with a little $^{57}Fe^{m\,2+}$. The latter gave lines corresponding to a mixture of $^{57}Fe^{m\,II}(CN)_6$ and either $^{57}Fe^{m\,3+}$ or more likely a cyanide deficient $^{57}Fe^{m\,II}(CN)_n^{(n-2)-}$. These results suggest that orbital electron capture seldom displaces an atom from its lattice site.

6.5.4.2 *Other elements*

Results involving elements other than iron are fragmentary. Labelled potassium stannioxalate yields some quadrupole split emission due to $^{119}Sn^{m\,2+}$ in addition to a stannic line[211]. The emission spectrum of labelled stannous oxalate gives broader lines than the absorption spectrum[122] possibly indicating some effects due to radiolysis of the surrounding oxalate. $^{119}Sn^{m}SO_4$ gives some tetravalent tin emission[123] and $^{119}Sn^{m}(C_6H_5)_4$ some stannous emission[55].

The larger part of the remaining data refer to ^{125}Te. Orbital electron capture in ^{125}I as CuI gives a single line of not much greater than theoretical width and has been used for $^{125}Te^{m}$ sources[53, 124–126]. Labelled iodides, iodates and periodates have been investigated. The iodates gave complex spectra with quadrupole split and single-line components. The rather large natural line widths complicates their analysis. $Na_3H_2IO_6$ and sodium and potassium iodides gave single, but exceptionally broad, line spectra. $NaIO_4$ on the other hand gave a complex spectrum[127, 128]. A wider range of $^{125}Te^{m}$ labelled compounds have also been examined. Telluric acid and sodium tellurate, $Na_2H_4TeO_6$, give a mixture of a quadrupole split and a single-line spectrum. The former is presumably due to Te^{IV} production, which is to be expected in view of the radiochemical results on the effects of isomeric transition in these compounds[50, 53, 129, 130]. Labelled magnesium tellurate, unaccountably, is reported to give a single-line emission[131].

A final example of the reducing effect of an oxalate lattice is the $^{157}Eu^{m\ 2+}$ emission found in ^{151}Gd doped hydrated erbium oxalate[132].

6.5.5 Semiconductor matrices

A number of investigations have been made on ^{57}Co doped semiconductors. A simple doublet with $\delta = 0.51$ mm/s (w.r.t. stainless steel) and $\Delta = 0.45$ mm/s has been found in both n- and p-type InAs[133]. Belozerskii, Nemilov and their collaborators[134–138] have reported the data collected in Table 6.3. All are single line spectra. All values are in mm/s and δ is with respect to stainless steel.

A more recent paper[139] finds two kinds of quadrupole split pairs in the spectra. Both n- and p-type doped compounds of the type $A^{III}B^{V}$ have been measured. In both n- and p-type InAs and InSb a spectrum attributed to

Table 6.3

Compound	InSb	InP	GaAs	ZnS	GaSb	Ge (n-*type*)
δ	0.44	0.36	0.40	0.76	0.46	0.40
Γ	0.51	0.59	0.51	0.71	0.58	1.00

$^{57}Fe^{m\ 3+}$ with $\delta = 0.3$–0.4 and $\Delta = 0.5$ was observed. But n- and p-type InP, GaAs and GaSb gave different spectra. The p-type gave the same unresolved pair as the InAs and InSb but the n-type gave, in addition for InP and GaAs, or instead for GaSb and some samples of InP, another resolved pair with $\delta = 0.5$–0.6 and $\Delta = 1.2$, presumably due to $^{57}Fe^{m\ 2+}$. The ^{57}Co was electro-deposited on the semiconductor and then diffused in by heating *in vacuo*. One of the earlier papers[137] demonstrated the importance of excluding air during the diffusion.

This last result is most interesting and important and further experiments of this kind are desirable.

6.5.6 Conclusions

In concluding this section it must be acknowledged that these results pose almost as many new problems as they solve. The data suggest that by the time Mössbauer emission occurs after either of these two decay processes, the atom has already reached a well-defined and much longer lived state. On the other hand, radiochemical data on similar processes show that further changes are still possible, these have been studied as annealing reactions and tend to form the analogue of the parent compound. One must therefore conclude that in some temperature ranges and for some compounds one should find that the half-life of the Mössbauer atom with respect to these processes becomes comparable with that appropriate to photon emission. The temperature dependence of some of the spectra tends to support this conclusion, although its detailed interpretation seems somewhat uncertain.

Finally, since defects must be important in stabilising the charge state of the Mössbauer atom, more systematic studies of their effect are very desirable.

6.6 EFFECTS OF β-DECAY

Far fewer results are available relating to these processes. Supposing the β-particle leaves the source, charge conservation requires that negative β-decay should increase the oxidation state of the affected atom by one unit. Except in a very small proportion of the events no other highly disruptive events are expected. The results so far obtained are compatible with this picture.

6.6.1 Rare gas systems

6.6.1.1 Producing xenon

A Mössbauer excited state of ^{129}Xe can be produced by β-decay in ^{129}I. The most interesting feature of experiments with ^{129}I labelled compounds[140–143] lies in their use for the matrix synthesis of xenon compounds, including some that cannot, as yet, be made in any other way. Thus the interest lies rather in the failure of nuclear event to rupture chemical bonds! The Mössbauer emission spectrum produced by labelled $K^{129}IO_3$ includes, in addition to a line due to atomic xenon, a quadrupole-split pair that can be identified as arising from XeO_3; in this case the product is a known compound whose absorption spectrum can be measured[141]. Still more interesting are those cases where the product has not been obtained in any other way. Thus the quadrupole-split spectrum observed in $K^{129}ICl_4$ indicates the formation of $XeCl_4$ [140, 142], and that in $K^{129}ICl_2$ suggests $XeCl_2$. With $Na_3H_2{}^{129}IO_6$, a single-line emission was obtained but the rather large Mössbauer fraction could be due to perxenate formation. The most unexpected and most recent results report a quadrupole-split spectrum from $K^{129}IBr_2$ corresponding to the formation of $XeBr_2$. The $XeBr_2$ did not appear to form in impure $KIBr_2$, nor in the caesium salt[144].

6.6.1.2 Producing krypton

The Mössbauer level of ^{83}Kr can be fed, indirectly, by β-decay of ^{83}Br, or even by successive β-decays from ^{83}Se. However, in each case a comparatively long-lived (half-life 1.86 h) isomeric state of krypton, which feeds the Mössbauer state by a heavily internally converted transition, is formed. The decay of this latter species may well dominate the chemical effects in these systems.

No evidence has been found for KrO_3 production in $K^{83}BrO_3$. Rather broad single-line spectra are obtained, as is the case with the alkali bromides[145]. Zinc, lead and tin selenides, labelled with ^{83}Se, give rather sharp single-line spectra and are useful sources for ^{83}Kr Mössbauer spectroscopy[146]. Labelled selenium dioxide and ammonium selenate give rather broader lines, but the comparatively large Mössbauer fraction in the dioxide may indicate some Kr—O bonding[147].

6.6.2 Other systems

Zinc telluride labelled with ^{127}Te, $^{127}Te^m$, ^{129}Te and $^{129}Te^m$ gives a line emission with a width not much greater than the theoretical value; it has been used extensively as a source in iodine Mössbauer spectrometry[148–150]. A labelled sample of telluric acid gives a single, rather broad, line spectrum[151].

A very unexpected result has been reported for a rare-earth system. It was found that ^{151}Sm doped lanthanum chloride yields some $^{151}Eu^{2+}$ emission. A repetition of this experiment seems desirable[152]. A more reasonable result has been obtained for ^{161}Tb doped ceric oxide where some $^{161}Dy^{m\,4+}$ seems to survive[153].

In some systems, however, the result obtained are much closer to those predicted and the daughter products appear to be isoelectronic with the parent, corresponding to an increase of one unit in oxidation state. A number of ^{193}Os labelled osmium compounds have been investigated[154]. The oxidation state of the Mössbauer emitting $^{193}Ir^m$, produced by β-decay, has been determined in the usual way. In OsO_4, $K_2OsO_42H_2O$, and osmocene the principal products indeed seem to be isoelectronic with the parent, that is IrO_4^+, IrO_4^- and $IrCp_2^+$ seem to be produced. But in the complex halides, K_2OsX_6, derivatives of Ir^{IV}, Ir^{III} and possibly Ir^{II} seem to be formed. Some examples of the spectra are shown in Figure 6.8. One can suppose that the primary IrX_6^- is unstable with respect to processes such as $IrX_6^- \rightarrow IrX_4^- + X_2$.

Closely related to these results, but in contrast to the last observations, in $(NH_4)_2$ $TeCl_6$ labelled with either ^{129}Te or $^{129}Te^m$, a single-line spectrum corresponding to ICl_6^- is obtained. The Mössbauer fraction is essentially the same for the two sources (isomer and ground-state labelling). This result

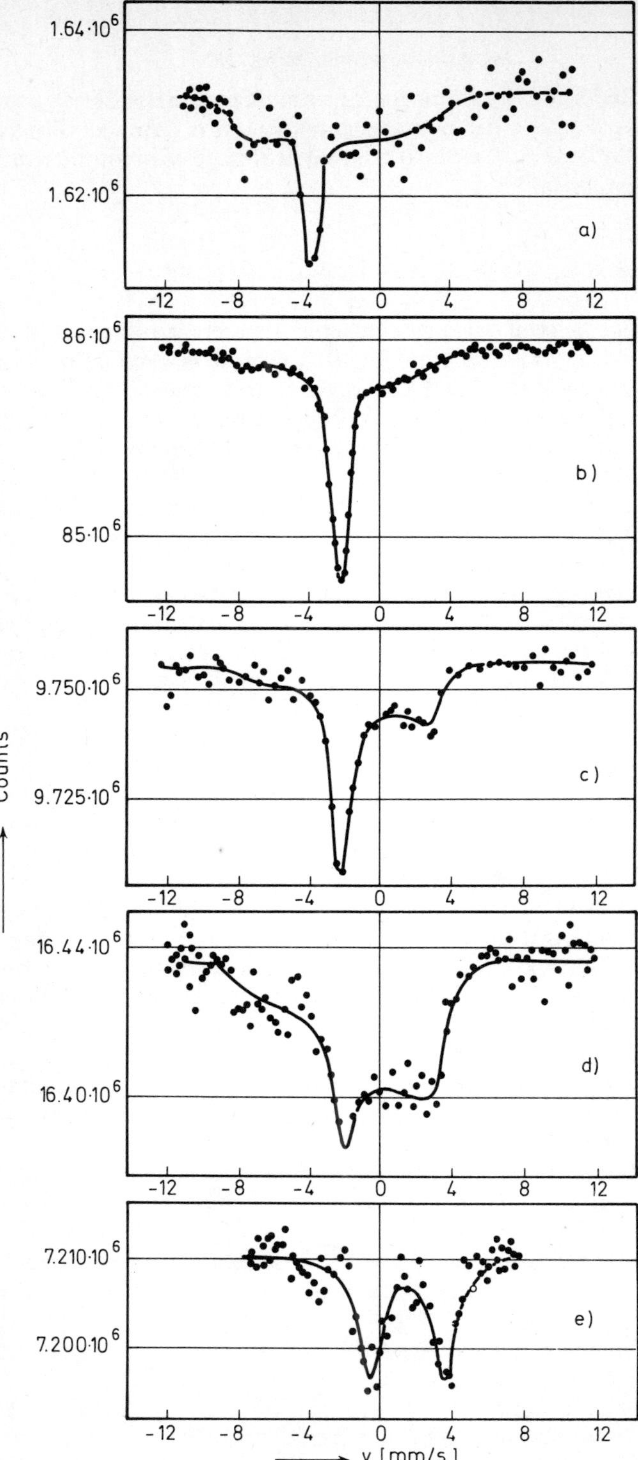

Figure 6.8 Mössbauer spectra obtained with sources of OsO_4(a) and $K_2OsO_4{\cdot}2H_2O$ (b) showing the consequences of β-decay only, of neutron irradiated and annealed (c) or unannealed (d) $K_2OsO_4{\cdot}2H_2O$, and of osmocene (e). (From Rother, Wagner and Zahn[154], by courtesy of Academic Press)

requires that the $TeCl_6^{2-}$ can escape rupture during the isomeric transition; another example where the coulombic explosion is avoided. The result is also noteworthy in that it leads to the matrix synthesis of another otherwise unknown entity, ICl_6^- [155].

6.7 PROCESSES WITH SUBSTANTIAL RECOIL

Much of the work under this heading has had the production of suitable sources for Mössbauer spectrometry as its prime objective, so that its interest has lain in avoiding chemical effects. But investigations have also been made to try to learn something about the radiation damage that occurs at the end of the trail of the recoiling particle. On the whole, comparatively little has been learnt about the last objective.

6.7.1 Metals

One might predict that by the time Mössbauer emission occurs in a metallic matrix the atom concerned will have reached a normal lattice site and charge state. This prediction has been confirmed even for the very large recoil following coulombic excitation as a means of production of the Mössbauer excited state. In coulombic excitation the nucleus under study is hit by an energetic charged particle but compound nucleus formation is avoided. The bombarding ion may be, for example, a 3 MeV α-particle or a 25 MeV oxygen ion.

When metallic iron targets were used, the spectrum obtained was always the same in every detail as if ^{57}Co had been alloyed in the metal. Nor is there appreciable difference if the recoiling iron atom is allowed to embed itself in another metallic matrix by virtue of its recoil energy. The source so formed emits a spectrum like the corresponding ^{57}Co alloy[157–159]. Recoiling $^{73}Ge^m$ can be implanted in this way in chromium and other metals or in germanium itself[160–164]. The line width observed with such sources are not greatly in excess of the theoretical values, although some broadening often occurs. Similar results have been obtained with coulombically excited $^{61}Ni^m$ [164,165]. Naturally these experiments are 'in beam' experiments, the spectrum must be recorded while the irradiation proceeds and appropriate experimental precautions taken to avoid interference effects. Evidence that the site occupied by these implanted atoms may not be in all respects the same as the average lattice site can be derived from the generally lower Mössbauer fractions for these sources[161,162,164].

Bondarevskii, Murin and Seregin argue that these results exclude the production of a tiny molten zone at the end of the track of the recoiling atom[3]. But this conclusion is not entirely secure. The results only require that solidification is complete by the time Mössbauer occurs. This is indeed suggested by calculations.

The Mössbauer excited state of iron can also be produced by the (d,p) reaction in ^{56}Fe in natural iron. The recoil from this reaction is very large (> 100 keV) but the emission from an irradiated metal target is essentially the same as for an analogous ^{57}Co alloy source[166–168].

With these results one would not expect any effect following production of $^{57}Fe^m$ by thermal neutron capture in ^{56}Fe. In general, this must be true[169], but significant effects can be produced in certain ordered iron alloys, where the recoiling atom does not always re-enter the same kind of site as it originally occupied[170,171].

6.7.2 Coulombic excitation in compounds

Even in oxides the spectrum obtained following production of the Mössbauer species by coulombic excitation is not very different from that of the oxide used as an absorber. In the iron oxides the most noticeable effect is a reduction in the Mössbauer fraction. In Fe_2O_3 it is about half the value the oxide displays as an absorber[158]. In Fe_3O_4 there is some disagreement as to whether much difference exists[157,172].

Several observations have been made on other rare earth oxides[173–176]. These oxides seem singularly unsuited to such studies because of the low symmetry of their lattices. The line widths observed are generally rather large. But it is claimed that the emission spectra following coulombic excitation are not essentially different from the spectra the oxides show as absorbers[174].

There are very few data for any other compounds, but the spectra on coulombic excitation of hafnium have been studied in HfB_2, HfC, HfN and HfO_2. The spectra are not at all similar to those shown by these substances as absorbers. Line widths are large and the quadrupole splitting suggests the implanted hafnium becomes associated with some defect, probably a vacancy[177,178].

Since an alien atom violently injected into a lattice seems to reach a well defined site, one might use an alien atom as a Mössbauer probe to study the environment. A beginning to such studies has been made in which ^{57}Co ions have been implanted at 60 keV into a diamond. The emission spectrum was a quadrupole split pair with $\Delta = 1.88$ mm/s and $\delta = -0.32$ with respect to ferrocyanide. The line widths were large. One can expect more studies using such techniques[179].

6.7.3 Effects of the (n,γ) reaction

Much of the data are fragmentary and the largest body relate to the tin oxides.

6.7.3.1 Compounds other than tin

Continuous production of $^{57}Fe^m$ by the neutron irradiation of Fe_2O_3 gives an emission spectrum with a slightly smaller magnetic splitting than when the oxide is used as an absorber[169]. With $FeSO_47H_2O$ there is some $^{57}Fe^{m\,3+}$ emission; radiochemical measurements in the same system, using the ^{59}Fe production, show 3.0% of $^{59}Fe^{3+}$ [180].

The Mössbauer spectroscopy of potassium can only be studied by generating the $^{40}K^m$ by the (n,γ) reaction in ^{39}K. For all the compounds studied, potassium halides, KC_8, $KNiF_3$, $KCoF_3$, KCN, KH, KHF_2 and KN_3 the values of δ and Δ were approximately zero. The Mössbauer fractions varied considerably. The line widths were about 1.3 times the theoretical value[181,182].

Production of the 1.86 h $^{83}Kr^m$ by thermal neutron capture in ^{82}Kr in the β quinol clathrate gave the same spectrum as if the clathrate was synthesised from $^{83}Kr^m$ [183,184]. Much the same was true following neutron irradiation of Gd_2O_3 but the line widths in the emission spectrum were greater[185].

Rather more surprisingly it was found that production of ^{193}Os compounds by direct neutron irradiation yielded the same iridium emission spectra as if the osmium compound was synthesised from ^{193}Os [154].

Some observations on $^{125}Te^m$ produced by the (n,γ) effect from ^{124}Te are conflicting. Stepanov and Aleksandrov[186] found that the chemical isomer shift of the emission from neutron irradiated lead telluride decreases with time with a half-life of 10 d at room temperature. But Ullrich and Vincent find no difference, nor did annealing make any difference. They tried TeO_2 and Te in addition to PbTe. They attributed the previous results to fast neutron bombardment effects[54].

6.7.3.2 *Effects of (n, γ) reaction in tin oxides*

A large number of papers has been published on the effects of the (n,γ) reaction in the tin oxides. The Mössbauer parent $^{119}Sn^m$, can be produced by thermal neutron capture in ^{118}Sn, and separated ^{118}Sn has usually been used in these experiments. There is one logical weakness of these experiments, it has been shown above that the isomeric transition, by which the $^{119}Sn^m$ feeds the Mössbauer excited state, does not produce effects in the tin oxides. However, it is still possible that if the $^{119}Sn^m$ produced by the (n,γ) reaction is in an abnormal site or charge state the isomeric transition could still modify the state of the tin or even, conceivably, return it to its original condition.

Unfortunately, some of the papers are conflicting. Nesmeyanov, Babeshkin and collaborators have carried out a series of studies, exploring the state of the $^{119}Sn^m$ radiochemically and by the Mössbauer emission spectrum[187–194]. Their materials have usually received a very large dose of thermal and fast neutrons as well as ionising radiation during production of the $^{119}Sn^m$. Irradiation of 2 months in the reactor were used. They found modest proportions of Sn^{4+} and Sn^0 in the SnO and of Sn^{2+} and Sn^0 in SnO_2 [191]. They obtained radiochemical evidence for the same species[190] but the agreement between the two methods of estimation was not very good. They also found both radiochemical and Mössbauer evidence for annealing processes[194]. In SnO_2 there were two stages, the first between 150 °C and 350 °C and the second from 400 to 600 °C; SnO could only be studied in the lower temperature region because it disproportionates above 400 °C.

Yoshida and Herber have also studied these two oxides and they attribute the Sn^{4+} production in SnO to a radiolytic process[195]. But Bondarevskii and Seregin found no evidence of macroscopic production of Sn^{4+} in their irradiated SnO [196], although they found that the proportion of $^{119}Sn^{m\,4+}$ increased with the length of irradiation[197].

Measurements of the Mössbauer fraction have proved useful in these studies. Yoshida and Herber find that the Mössbauer fraction increases on annealing[195] and Murin, Bondarevskii and Seregin note that the Mössbauer fraction for the $^{119}Sn^{m\,2+}$ produced in the irradiation of SnO is less than if SnO is used as an absorber[198]. They also find that the $^{119}Sn^{m\,4+}$ spectrum

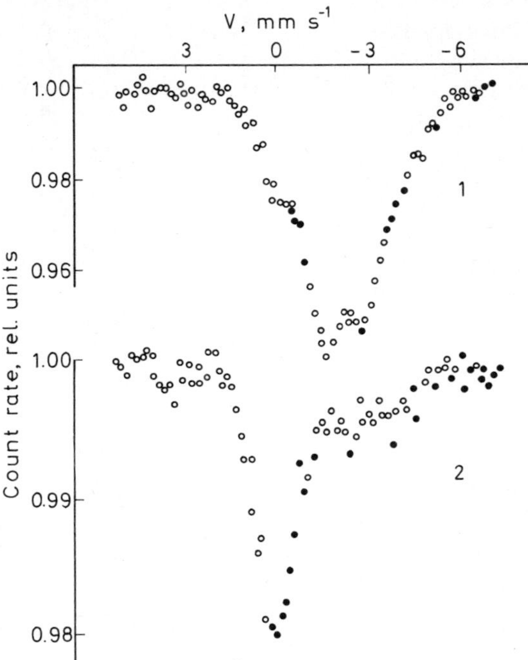

Figure 6.9 NGR spectra of ^{118}SnO, irradiated in a reactor, at 80 K (1) and a hydrochloric acid solution of the same specimen at 80 K (2). Absorber calcium stannate. (From Murin *et al.*[198])

in the irradiated SnO is greatly intensified if the oxide is dissolved in hydrochloric acid and the frozen solution measured. (Figure 6.9).

This observation seems very significant and confirms the suggestion made that some (n, γ) products might end up in such a variety of situations that they only contribute to background, that is to say their lines become very broad indeed.

Several observations have been made on magnesium stannate[199–201]. Lines attributed to Sn^{2+} are produced and can be removed by annealing. The Mössbauer fraction for the Sn^{2+} species, estimated from its contribution to Sn^{4+} after annealing, is low (0.15 as against 0.5 for the Sn^{4+}). Andersen and Østergaard have made the interesting discovery that the tin species yielding the supposed Sn^{2+} spectrum does not appear in the stannous fraction on radiochemical analysis[200]. They also find evidence for some other tin species.

Further studies of these systems are necessary.

6.7.4 Some effects of α decay

These studies arose from the necessity of obtaining simple sources for feeding the $^{237}Np^m$ Mössbauer excited state. ^{241}Am in ThO_2, AmO_2 or NpO_2 gives a single but very broad line. It is significant that the Mössbauer fraction is substantially less than a similar oxide in which the $^{237}Np^m$ is produced by β-decay of ^{237}U. Clearly the recoiling neptunium atoms from α-decay do not all reach normal lattice sites[202–208].

A theoretical analysis of the fate of the recoiling ^{237}Np atom, from α decay of ^{241}Am, along similar lines to the analyses already referred to, confirms that the atoms should reach resting positions before Mössbauer emission[209].

A closer examination of the oxide systems shows that some tri-, tetra- and penta-valent neptunium may be produced[210], although other investigators only report tetra- and penta-valent forms[208]. In frozen solutions of americium salts only trivalent neptunium seems to form, a rather surprising result[211].

A plutonium oxide source has also been used to feed the Mössbauer excited state of ^{238}U by α-decay. A line width of 1.5 times the natural value is reported[212].

6.8 CONCLUSION

Few, if any, of the results so far reported are seriously inconsistent with other existing ideas of the effects of nuclear reactions in crystalline solids. On the other hand, there is a very large number of observations whose significance is far from clear and a very large number of problems which would repay a much more detailed investigation using this technique. So far much of the work has been exploratory and sometimes a little superficial. What is needed now is a few well chosen investigations in depth.

References

1. Danon, J. (1968). *Notas De Fisica,* **14,** 239
2. Nath, A., Klein, M. P., Kündig, W. and Lichtenstein D. (1970. *Mössbauer Effect Methodology,* Ed. I. J. Gruverman, **5,** 173. (New York: Plenum Press)
3. Bondarevskii, S. I., Murin, A. N. and Seregin, P. P. (1971). *Uspekhi Khim.,* **40,** 95
4. Adloff, J. P., *Mössbauer Spectroscopy Panel,* I.A.E.A. April 1971, to be published
5. Wexler, S. (1965). *Actions Chimiques et Biologiques des Radiations,* Ed. M. Haissinsky, Vol. 7, p. 105 *et seq.* (Paris: Masson)
6. (a) Campbell, I. G. (1963). In *Advances in Inorganic and Radio-chemistry,* Eds. A. G. Sharpe and H. J. Emeléus, Vol. 5. (New York: Academic Press)
   (b) Geissler, P. R. and Willard, J. E. (1963). *J. Phys. Chem.,* **67,** 1675
7. Harbottle, G. and Sutin, N. (1958). *J. Phys. Chem.,* **62,** 1344

8. (a) Müller, H. (1965). *Proc. Conf. Chem. Effects Nuclear Transformations,* I.A.E.A. Vienna 1964, STI/PUB 91, **2,** 359
(b) Müller, H. (1965). *J. Inorg. Nucl. Chem.,* **27,** 1745
(c) Müller, H. (1967). *Angew. Chem. Int. Ed. Engl.,* **6,** 133
9. Thompson, J. L. and Miller, W. W. (1963). *J. Chem. Phys.,* **38,** 2477
10. Housley, R. M., Erickson, N. E. and Dash, J. G. (1964). *Nucl. Inst. Methods,* **27,** 29
11. (a) Goldanskii, V. I., Gorodinskii, G. M., Karyagin, S. V., Korytko, L. A., Krizhanskii, L. M., Makarov, E. F., Suzdalev, I. P. and Khrapov, V. V. (1962). *Dokl. Akad. Nauk. SSSR,* **147,** 127
(b) Goldanskii, V. I., Karyagin, S. V., Makarov, E. F. and Khrapov, V. V., *Materialy Rabochego Soveshchaniya Po Effectu Mössbauera,* Dubna July 1962, p. 189 (Trans. Consultants Bureau; N.Y. 1963)
12. Wertheim, G. K. (1961). *Phys. Rev.,* **124,** 764
13. Pollak, H. (1962). *Phys. Stat. Solidi,* **2,** 720
14. Bhide, Y. G. and Shenoy, G. K. (1966). *Phys. Rev.,* **147,** 306
15. Bearden, A. J., Mattern, P. L. and Hart, J. R. (1964). *Rev. Mod. Phys.,* **36,** 470
16. Coston, C. J., Ingalls, R. and Drickhamer, H. G. (1966). *Phys. Rev.,* **145,** 409
17. Triftshauser, W. and Craig, P. P. (1966). *Phys. Rev. Lett.,* **16,** 1161
18. Triftshauser, W. and Craig, P. P. (1967). *Phys. Rev.,* **162,** 274
19. Mullen, J. G. (1966). *Phys. Rev. Lett.,* **17,** 287
20. Ok, H. N. and Mullen, J. G. (1968). *Phys. Rev. Lett.,* **21,** 823
21. Ok, H. N. and Mullen, J. G. (1968). *Phys. Rev.,* **168,** 550 and 563
22. Schroeer, D. and Triftshauser, W. (1968). *Phys. Rev. Lett.,* **20,** 1242
23. Ok, H. N., Helms, W. R. and Mullen, J. G. (1969). *Phys. Rev.,* **187,** 704
24. Trousdale, W. and Craig, P. P. (1968). *Phys. Rev. Lett.,* **27A,** 552
25. Kündig, W., Kobelt, M., Appel, H., Constabaris, G. and Lindquist, R. H. (1967). *Bull. Amer. Phys. Soc.,* **12,** 1149
26. Kündig, W., Kobelt, M., Appel, H., Constabaris, G. and Lindquist, R. H. (1969). *J. Phys. Chem. Solids,* **30,** 819
27. Murin, A. N., Lur'e, B. G. and Seregin, P. P. (1968). *Soviet Phys. Solid State,* **10,** 1000 trans. (1254 orig.)
28. Efimov, A. A., Shipatov, V. T. and Seregin, P. P. (1970). *Soviet Phys. Solid State,* **11,** 2462 trans. (3032 orig.)
29. Murin, A. N., Lur'e, B. G. and Seregin, P. P. (1968). *Soviet Phys. Solid State,* **10,** 2066 trans. (2620 orig.)
30. Schroeer, D. and Niningen, R. C. (1967). *Phys. Rev. Lett.,* **19,** 632
31. Leon, V. and Negrete, P. (1970). *Solid State Comm.,* **8,** 749
32. Blomquist, J., Grapengiesser, S. and Söderquist, R. (1971). *Phys. Stat. Solidi,* **44A,** 435
33. Jagannathan, R. and Mathur, H. B. (1969). *J. Inorg. Nucl. Chem.,* **31,** 3363
34. Cruset, A. and Friedt, J. M. (1971). *Phys. Stat. Solidi,* **45B,** 189
35. Seregin, P. P., Bondarevskii, S. I. and Efimov, A. A. (1970). *Soviet Phys. Solid State,* **12,** 1462 trans (1841 orig.)
36. Siegwarth, J. D. and Dash, J. G. (1966). *Bull. Amer. Phys. Soc.,* **11,** 474
37. Ando, K. J., Kündig, W., Constabaris, G. and Lindquist, R. H. (1967). *J. Phys. Chem. Solids,* **28,** 2291
38. Siegwarth, J. D. (1967). *Phys. Rev.,* **155,** 285
39. Chappert, J. and Frankel, R. B. (1967). *Bull. Amer. Phys. Soc.,* **12,** 352
40. Chappert, J., Frankel, R. B. and Blum, N. A. (1967). *Phys. Lett.,* **25A,** 149
41. Frankel, R. B. and Blum, N. A. (1967). *Bull. Amer. Phys. Soc.,* **12,** 24
42. Triftshauser, W. and Schroeer, D. (1969). *Phys. Rev.,* **187,** 491
43. Clifford, A. F. (1967). *A.C.S. Advan. Chem. Ser.* No. 68
44. Shecter, H., Dash, J. G., Erickson, G. A. and Ingalls, R. (1970). *Phys. Rev.,* **B2,** 613
45. Bhide, V. G. and Multani, M. S. (1965). *Phys. Rev.,* **139A,** 1983
46. Bhide, V. G. and Multani, M. S. (1966). *Phys. Rev.,* **149,** 289
47. Bhide, V. G. and Bhasin, H. C. (1967). *Phys. Rev.,* **159,** 586
48. Bocquet, J. P., Chu, Y. Y., Kistner, O. C. and Perlman, M. L. (1966). *Phys. Rev. Lett.,* **17,** 809
49. Sand, H. and Herber, R. H. (1968). *J. Inorg. Nucl. Chem.,* **30,** 409
50. Lebedev, V. A., Lebedev, R. A., Babeshkin, A. M. and Nesmeyanov, A. N. (1968). *Vest. Moskov. Univ. Ser. Khim.,* **4,** 16

51. Lebedev, R. A., Babeshkin, A. M., Nesmeyanov, A. N. and Fateeva, N. L. (1970). *Radiochem. Radioanal. Lett.*, **5,** 83
52. Stepanov, E. P., Aleshin, K. P., Manapov, R. A., Samoilov, B. N., Sklyarevskii, V. V. and Stankevich, V. G. (1963). *Phys. Lett.*, **6,** 155
53. Violet, C. E. and Booth, R. (1966). *Phys. Rev.*, **144,** 225
54. Ullrich, J. F. and Vincent, D. H. (1969). *J. Phys. Chem. Solids,* **30,** 1189
55. Herber, R. H. and Stöckler, H. A. (1965). *Proc. Conf. Chem. Effects Nuclear Transformations,* I.A.E.A., Vienna, 1964, STI/PUB 91, **2,** 403
56. Steichele, R., Henning, W., Hüfner, S. and Kienle, P. (1966). I.A.E.A., Vienna, 223
57. de Waard, H. and Perlow, G. J. (1970). *Phys. Rev. Lett.*, **24,** 566
58. Mullen, J. G. (1963). *Phys. Rev.*, **131,** 1410
59. Friedt, J. M. and Adloff, J. P. (1967). *Comp. Rend. Acad. Sci.*, **264C,** 1356
60. Cavanagh, J. F. (1969). *Phys. Stat. Solidi,* **36,** 657
61. Friedt, J. M. and Danon, J. (1970). *Radiochem. Radioanal. Lett.*, **3,** 147
62. Cruset, A. and Friedt, J. M. (1971). *Phys. Stat. Solidi,* **44B,** 633
63. Friedt, J. M. and Adloff, J. P. (1969). *Comp. Rend. Acad. Sci.*, **268C,** 1342
64. Friedt, J. M. (1970). *J. Inorg. Nucl. Chem.*, **32,** 431
65. Jagannathan, R., Thaker, R. and Mathur, H. B. (1969). *Indian J. Chem.*, **7,** 353
66. Friedt, J. M. (1970). *J. Inorg. Nucl. Chem.*, **32,** 2123
67. Friedt, J. M. and Adloff, J. P. (1969). *Inorg. Nucl. Chem. Lett.*, **5,** 163
68. Wertheim, G. K. and Buchanan, D. N. E. (1969). *Chem. Phys. Lett.*, **3,** 87
69. Wertheim, G. K. and Guggenheim, H. J. (1965). *J. Chem. Phys.*, **42,** 3873
70. McDonoughs, M. S. and Chen, J. H. (1967). *Bull. Amer. Phys. Soc.*, **12,** 352
71. Wertheim, G. K., Guggenheim, H. J. and Buchanan, D. N. E. (1969). *J. Chem. Phys.*, **51,** 1931
72. Rudy, S. and Selig, H. (1966). *Phys. Rev.*, **147,** 348
73. Krasnoperov, V. M., Lur'e, B. G., Murin, A. N., Cherezov, N. K. and Yutlandov, I. A. (1967). *Atom Energiya,* **23,** 60
74. Krasnoperov, V. M., Murin, A. N., Cherezov, N. K. and Iutlandov, I. A. (1969). *Dokl. Akad. Nauk.* SSSR, **186,** 296
75. Gütlich, P., Odar, S. and Walcher, D. (1970). *Z. Naturforsch.*, **25b,** 1183
76. Murin, A. N. and Seregin, P. P. (1970). *Phys. Stat. Solidi,* **A2,** 663
77. Meisel, W., Hennig, K. and Schnorr, H. (1969). *Phys. Stat. Solidi,* **34,** 577
78. Efimov, A. A., Bondarevskii, S. I., Seregin, P. P. and Shipatov, V. T. (1970). *Fiz. Tverd. Tela,* **12,** 949 (trans 743)
79. Efimov, A. A., Seregin, P. P., Shipatov, V. T. and Bondarevskii, S. I. (1970). *Fiz. Tverd. Tela,* **12,** 1244 (trans. 968)
80. de Coster, M. and Amelinckx, S. (1962). *Phys. Lett.*, **1,** 245
81. Mullen, J. G. (1963). *Phys. Rev.*, **131,** 1415
82. Murin, A. N., Lur'e, B. G. and Seregin, P. P. (1967). *Fiz. Tverd. Tela,* **9,** 1424 (trans 1110)
83. Hennig, K., Yung, Kim and Skorchev, B. S. (1968). *Phys. Stat. Solidi,* **27,** 161K
84. Cappelletti, R., Fieschi, R. and Lamborizio, C. (1970). *Radiation Effects,* **4,** 85
85. Hennig, K., Meisel, W. and Schnorr, H. (1966). *Phys. Stat. Solidi,* **15,** 9K, 199
86. Lindley, D. H. and Debrunner, P. G. (1966). *Phys. Rev.*, **146,** 199
87. Murin, A. N., Lur'e, B. G., Seregin, P. P. and Cherezov, N. K. (1966). *Fiz. Tverd. Tela,* **8,** 3291 (trans 2632)
88. Murin, A. N., Lur'e, B. G. and Seregin, P. P. (1967). *Fiz. Tverd Tela,* **9,** 2428 (trans. 1901)
89. Hennig, K. (1968). *Phys. Stat. Solidi,* **27,** 115K
90. Murin, A. N., Lur'e, B. G. and Seregin, P. P. (1968). *Fiz. Tverd Tela,* **10,** 923 (trans. 728)
91. Hennig, K. and Yung, Kim. (1970). *Phys. Stat. Solidi,* **40,** 365
92. Ingalls, R. and De Pasquali, G. (1965). *Phys. Lett.*, **15,** 262
93. Ingalls, R., Coston, C. J., De Pasquali, G., Drickhamer, H. G. and Pinajian, J. J. (1966). *J. Chem. Phys.*, **45,** 1057
94. Mullen, J. G. and Ok, H. N. (1966). *Bull. Amer. Phys. Soc.*, **11,** 267
95. Wertheim, G. K., Kingston, W. R. and Herber, R. H. (1962). *J. Chem. Phys.*, **37,** 687
96. Wertheim, G. K. and Herber, R. H. (1963). *J. Chem. Phys.*, **38,** 2106
97. Champion, A. R., Vaughan, R. W. and Drickhamer, H. G. (1967). *J. Chem. Phys.*, **47,** 2583
98. Hazony, Y. and Herber, R. H. (1969). *J. Inorg. Nucl. Chem.*, **31,** 321
99. Venkateswarlu, K. and Ramshesh, V. (1969). *Proc. Chem. Symp. Chandigarh,* **2,** 129

100. Fenger, J., Siekierska, K. E. and Maddock, A. G. (1970). *J. Chem. Soc. A.*, 1456
101. Friedt, J. M., Baggio-Saitovitch, E. and Danon, J. (1970). *Chem. Phys. Lett.*, **7,** 603
102. Perfilev, Y. D., Kulikov, L. A., Babeshkin, A. M. and Nesmeyanov, A. N. (1969). *Vest. Moskov. Univ. Ser. Khim.*, **24,** 112
103. Gutlich, P., Odar, S., Fitzsimmons, B. W. and Erickson, N. E. (1968). *Radiochim. Acta,* **10,** 147
104. Friedt, J. M. and Asch, L. (1969). *Radiochim. Acta,* **12,** 208
105. Friedt, J. M. (1969). *Thesis,* Strasbourg
106. Sano, H. and Hashimoto, F. (1965). *Bull. Chem. Soc. Japan,* **38,** 1565
107. Siekierska, K. E. and Fenger, J. (1970). *Radiochim. Acta,* **14,** 93
108. Friedt, J. M., Cruset, A., Asch, L. and Adloff, J. P. (1970). *Radiochem. Radioanal. Lett.*, **3,** 81
109. Friedt, J. M. and Adloff, J. P. (1968). *Comp. Rend. Acad. Sci.*, **266,** 1733
110. Nath, Amar, Agarwal, R. D. and Marthur, P. K. (1968). *Inorg. Nucl. Chem. Lett.*, **4,** 161
111. Mathur, P. K. (1969). *Indian J. Chem.*, **7,** 183
112. Jagannathan, R. and Mathur, H. B. (1969). *Inorg. Nucl. Chem. Lett.*, **5,** 89
113. Nath, Amar, Harpold, M., Klein, M. P. and Kündig, W. (1968). *Chem. Phys. Lett.*, **2,** 471
114. Mullen, R. T. (1970). *Mössbauer Effect Methodology,* Ed. I. J. Gruverman, **5,** 95. (New York: Plenum Press)
115. Nath, Amar, Klein, M. P., Kündig, W. and Lichtenstein, D. (1970). *Radiat. Effects,* **2,** 211
116. Lazzarini, E., Private Communication
117. Nath, Amar and Klein, M. P. (1969). *Nature (London),* **224,** 794
118. Sano, H., Aratini, M. and Stöckler, H. A. (1968). *Phys. Lett.*, **26A,** 559
119. Saito, N., Takeda, M. and Tominaga, T. (1971). *Radiochem. Radioanal. Lett.*, **6,** 169
120. Fenger, J., Maddock, A. G. and Siekierska, K. E. (1970). *J. Chem. Soc. A,* 3255
121. Sano, H. and Kanno, M. (1969). *Chem. Commun.*, 601
122. Murin, A. N., Bondarevskii, S. I. and Seregin, P. P. (1969). *Radiokhim.*, **11,** 474
123. Llabador, Y. and Friedt, J. M. (1971). *Chem. Phys. Lett.*, **8,** 592
124. Albanese, G., Lamborizio, C. and Ortalli, L. (1967). *Nuovo Cimento,* **50B,** 65
125. Ullrich, J. F. (1967). *Thesis,* Michigan
126. Pasternak, M. and Bukshpan, S. (1967). *Phys. Rev.*, **163,** 297
127. Jung, P. and Triftshauser, W. (1968). *Phys. Rev.*, **175,** 512
128. Babeshkin, A. M., Lamykin, E. V., Lebedev, V. A. and Nesmeyanov, A. N. (1970). *Vest. Moskov. Univ. Ser. Khim.*, **11,** 117
129. Lebedev, V. A., Babeshkin, A. M., Nesmeyanov, A. N. and Lamykin, E. V. (1969). *Vest. Moskov. Univ. Ser. Khim.*, **24,** 45
130. Lebedev, V. A., Lebedev, R. A., Babeshkin, A. M. and Nesmeyanov, A. N. (1969). *Vest. Moskov. Univ. Ser. Khim.*, **24,** 128
131. Gruchko, Y. S., Lur'e, B. G., Motornii, A. V. and Murin, A. N. (1968). *Radiokhim.*, **10,** 605
132. Glentworth, P., Nichols, A. L., Large, N. R. and Bullock, R. J. (1971). *Chem. Commun.*, 206
133. Bemski, G. and Fernandes, J. C. (1963). *Phys. Lett.*, **6,** 10
134. Belozerskii, G. N., Gusev, I. A., Murin, A. N. and Nemilov, Yu. A. (1965). *Fiz. Tverd. Tela,* **7,** 1254 (trans. 1012)
135. Belozerskii, G. N., Nemilov, Yu. A., Tomilov, S. B. and Shvedchikov, A. V. (1965). *Fiz. Tverd. Tela,* **7,** 3607 (trans. 2908)
136. Belozerskii, G. N., Nemilov, Yu. A. and Tolkachev, S. S. (1966). *Fiz. Tverd. Tela,* **8,** 451 (trans. 358)
137. Belozerskii, G. N., Nemilov, Yu. A., Tomilov, S. B. and Shvedchikov, A. V. (1966). *Fiz. Tverd. Tela,* **8,** 604 (trans. 484)
138. Belozerskii, G. N., Gusev, I. A., Nemilov, Yu. A. and Shvedchikov, A. N. (1966). *Fiz. Tverd. Tela,* **8,** 2112 (trans.1680)
139. Boltaks, B. I., Efimov, A. A., Seregin, P. P. and Shipatov, V. T. (1971). *Fiz. Tverd. Tela,* **12,** 2004 (trans. 1592)
140. Perlow, G. J. and Perlow, M. R. (1964). *J. Chem. Phys.*, **41,** 1157
141. Perlow, G. J. and Perlow, M. R. (1964). *Rev. Mod. Phys.*, **36,** 353
142. Perlow, G. J. and Perlow, M. R. (1965). *Prog. Conf. Chem. Effects Nuclear Transformations I.A.E.A.*, Vienna 1964 STI/PUB 91, **2,** 443

143. Perlow, G. J. and Perlow, M. R. (1968). *J. Chem. Phys.*, **48,** 955
144. Perlow, G. J. and Yoshida, H. (1968). *J. Chem. Phys.*, **49,** 1474
145. Pasternak, M. and Sonnino, T. (1967). *Phys. Rev.*, **164,** 384
146. Bukshpan, S., Goldstein, C. and Sonnino, T. (1968). *Phys. Lett.*, **27A,** 372
147. Hazony, Y. and Herber, R. H. (1971). *J. Inorg. Nucl. Chem.*, **33,** 961
148. de S. Barros, F., Ivantchev, N., Jha, S. and Rama Reddy, K. (1964). *Phys. Lett.*, **13,** 142
149. Perlow, G. J. and Ruby, S. L. (1964). *Phys. Lett.*, **13,** 198
150. Rama Reddy, K., de S. Barros, F. and De Benedetti, S. (1966). *Phys. Lett.*, **20,** 297
151. Perlow, G. J. and Perlow, M. R. J. (1966). *J. Chem. Phys.*, **45,** 2193
152. Probst, C. (1966). Diplomarbeit Tech. Hochschule, München
153. Khurgin, B., Ofer, S. and Rakavy, M. (1970). *Phys. Lett.*, **33A,** 219
154. Rother, P., Wagner, F. and Zahn, U. (1969). *Radiochim. Acta,* **11,** 203
155. Jones, C. H. W. and Warren, J. L. (1970). *J. Chem. Phys.*, **53,** 1740
156. Lee, Y. K., Keaton, P. W., Ritter, E. T. and Walker, J. C. (1965). *Phys. Rev. Lett.*, **14,** 957
157. Goldberg, D. A., Lee, Y. K., Ritter, E. T., Stevens, R. R. and Walker, J. C. (1966). *Phys. Lett.*, **20,** 571
158. Ritter, E. T., Keaton, P. W., Lee, Y. K., Stevens, R. R. and Walker, J. C. (1967). *Phys. Rev.*, **154,** 287
159. Sprouse, G. D., Kalvius, G. M. and Hanna, S. S. (1967). *Phys. Rev. Lett.*, **24,** 1041
160. Czjzek, G., Ford, J. L. C., Obenshain, F. E. and Seyboth, D. (1966). *Phys. Lett.*, **19,** 673
161. Czjzek, G., Ford, J. L. C. and Love, J. C. (1967). *Phys. Rev. Lett.*, **18,** 529
162. Czjzek, G., Ford, J. L. C., Love, J. C., Obenshain, F. E. and Wegener, H. H. (1968). *Phys. Rev.*, **174,** 331
163. Zimmermann, B. H., Jena, H., Ischenko, G., Kilian, H. and Seyboth, D. (1968). *Phys. Stat. Solidi,* **27,** 639
164. Seyboth, D. (1969). *Proc. Roy. Soc.*, **311A,** 119
165. Seyboth, D., Obenshain, F. E. and Czjzek, G. (1965). *Phys. Rev. Lett.*, **14,** 954
166. Goldberg, D. A., Keaton, P. W., Lee, Y. K., Madansky, L. and Walker, J. C. (1965). *Phys. Rev. Lett.*, **15,** 418
167. Christiansen, J., Recknagel, E. and Weyer, G. (1966). *Phys. Lett.*, **20,** 46
168. Christiansen, J., Hindennack, P., Morfield, U., Recknagel, E., Riegel, D. and Weyer, G. (1967). *Nucl. Phys.*, **99A,** 345
169. Berger, W. G., Fink, J. and Obenshain, F. E. (1967). *Phys. Lett.*, **25A,** 466
170. Berger, W. G. (1969). *Z. Physik.*, **225,** 139
171. Czjzek, G. and Berger, W. G. (1970). *Phys. Rev.*, **1B,** 957
172. Ritter, E. T., Lee, Y. K., Stevens, R. R. and Walker, J. C. (1965). *Bull. Amer. Phys. Soc.*, **10,** 1111
173. Eck, J. S., Lee, Y. K., Ritter, E. T., Stevens, R. R. and Walker, J. C. (1966). *Phys. Rev. Lett.*, **17,** 120
174. Eck, J. S., Lee, Y. K., Walker, J. C. and Stevens, R. R. (1967). *Phys. Rev.*, **156,** 246
175. Stevens, R. R., Eck, J. S., Ritter, E. T., Lee, Y. K. and Walker, J. C. (1967). *Phys. Rev.*, **158,** 1118
176. Wilenzick, R. M., Hardy, K. A., Hicks, J. A. and Owens, W. R. (1969). *Phys. Lett.*, **30B,** 167
177. Jacobs, C. G. and Hershkowitz, N. (1970). *Phys. Rev.*, **1B,** 839
178. Jacobs, C. G., Hershkowitz, N. and Jeferies, J. B. (1969). *Phys. Lett.*, **29A,** 498
179. de Barros, F., Guggenheim, H. J. and Buchanan, D. N. E. (1970). *J. Chem. Phys.*, **52,** 2865
180. Fenger, J. and Siekierska, K. E. (1968). *Radiochim. Acta,* **10,** 172
181. Hafemeister, D. W. and Shera, E. B. (1965). *Phys. Rev. Lett.*, **14,** 593
182. Tseng, P. K., Ruby, S. L. and Vincent, D. H. (1968). *Phys. Rev.*, **172,** 249
183. Hazony, Y., Hillman, P., Pasternak, M. and Ruby, S. L. (1962). *Phys. Lett.*, **2,** 337
184. Pasternak, M., Simopoulos, A., Bukshpan, S. and Sonnino, T. (1966). *Phys. Lett.*, **22,** 52
185. Fink, J. and Kienle, P. (1965). *Phys. Lett.*, **17,** 326
186. Stepanov, E. P. and Aleksandrov, A. Yu. (1967). *J.E.T.P. Lett.*, **5,** 83
187. Nesmeyanov, A. N., Babeshkin, A. M., Kosev, N. P., Bekker, A. A. and Lebedev, V. A. (1965). *Proc. Conf. Chem. Effects Nuclear Transformations,* I.A.E.A., Vienna 1964, STI/PUB 91, **2,** 419
188. Nesmeyanov, A. N., Babeshkin, A. M., Bekker, A. A. and Fano, V. (1966). *Radiokhim.*, **8,** 261

189. Nesmeyanov, A. N., Babeshkin, A. M., Bekker, A. A. and Fano, V. (1966). *Radiokhim.*, **8,** 264
190. Nesmeyanov, A. N., Babeshkin, A. M., Pokholok, K. V. and Bekker, A. A. (1967). *Radiokhim.*, **9,** 392
191. Babeshkin, A. M., Pokholok, K. V., Bekker, A. A. and Nesmeyanov, A. N. (1967). *Radiokhim.*, **9,** 668
192. Babeshkin, A. M., Bekker, A. A., Nesmeyanov, A. N., Farbrichnii, P. B. and Pokholok, K. V. (1968). *Radiokhim.*, **10,** 752
193. Babeshkin, A. M., Bekker, A. A., Efremov, E. N. and Nesmeyanov, A. N. (1969). *Vestn. Moskov Univ. Ser. Khim.*, No. 5, 40
194. Babeshkin, A. M., Bekker, A. A., Efremov, E. N. and Nesmeyanov, A. N. (1969). *Vestn. Moskov Univ. Ser. Khim.*, No. 6, 41
195. Hiroyuki, Yoshida and Herber, R. H. (1969). *Radiochim. Acta*, **12,** 14
196. Bondarevskii, S. I. and Seregin, P. P. (1964). *Fiz. Tverd. Tela.*, **10,** 3454 (trans. 2736)
197. Murin, A. N., Lur'e, B. G., Bondarevskii, S. I. and Seregin, P. P. (1968). *Fiz. Tverd. Tela*, **10,** 2803 (trans. 2207)
198. Murin, A. N., Bondarevskii, S. I. and Seregin, P. P. (1970). *Fiz. Tverd. Tela*, **12,** 1095 (trans 857)
199. Hannaford, P., Howard, C. J. and Wignall, J. W. G. (1965). *Phys. Lett.*, **19,** 257
200. Andersen, T. and Østergaard, P. (1968). *Trans. Faraday Soc.*, **64,** 3014
201. Hannaford, P. and Wignall, J. W. G. (1969). *Phys. Stat. Solidi*, **35,** 809
202. Stone, J. A. and Pillinger, W. L. (1964). *Phys. Rev. Lett.*, **13,** 200
203. Mullen, J. G. (1965). *Phys. Lett.*, **15,** 15
204. Stone, J. A. and Pillinger, W. L. (1966). *Bull. Amer. Phys. Soc.*, **11,** 809
205. Belozerskii, G. N., Nemilov, Yu. A., Chaikhorskii, A. A. and Shvedchikov, A. V. (1967). *Fiz. Tverd. Tela*, **9,** 1252 (trans. 978)
206. Bryukhanov, V. A., Dvechkin, V. V., Peryshkin, A. I., Rzhekina, E. I. and Shpinel, V. S. (1967). *Fiz. Tverd. Tela*, **9,** 1519 (trans. 1189)
207. Aleksandrov, B. M., Kalyamin, A. V., Krivokhatskii, A. S., Lur'e, B. G., Murin, A. N. and Romanov, Y. F. (1968). *J.E.T.P. Lett.*, **8,** 111
208. Aleksandrov, B. M., Kazyamin, A. V., Krivokhatskii, A. S., Lur'e, B. G., Murin, A. N. and Romanov, Y. F. (1968). *Fiz. Tverd. Tela*, **10,** 1896 (trans. 1494)
209. Kaplan, M. (1966). *J. Inorg. Nucl. Chem.*, **28,** 331
210. Dunlap, B. D., Kalvius, G. M., Ruby, S. L., Brodsky, M. B. and Cohen, D. (1968). *Phys. Rev.*, **171,** 316
211. Gal, J., Hadari, Z., Yanir, E., Bauminger, E. R. and Ofer, S. (1970). *J. Inorg. Nucl. Chem.*, **32,** 2509
212. Ruby, S. L., Kalvius, G. M., Dunlap, B. D., Shenoy, G. K., Cohen, D., Brodsky, M. B. and Lam, D. J. (1969). *Phys. Rev.*, **184,** 374

7
Positronium and Mesonic Atoms

J. H. GREEN
St Catharine's College, Cambridge

7.1 INTRODUCTION

It has been known for many years that electrically charged elementary particles may become part of atoms and molecules. Negatively charged particles may replace electrons and positively charged particles may capture electrons forming entities, which have properties similar to those of hydrogen atoms. These entities have been called 'exotic atoms' and their properties have been reviewed recently[1]. (Except for entities containing positrons.)

In this review, the term 'mesonic atom' will be used even when the particle is not a true meson (for example, a muon, hyperon, or antiproton). Even though these elementary particles may have short lives, for instance a probable lifetime of about 10^{-10}s, this time is long compared with the rates of many atomic transition processes and atomic reaction rates and very useful information about such processes can be obtained by studying their interactions.

Although mesons have been recognised and available for experimental study for a long time, the readier availability of laboratory sources of positrons has led to a much greater body of studies on their interactions with atoms and molecules. Since the experimental proof of the existence of positronium[2] about 500 papers have described its properties and physical and chemical interactions in matter. The physics and chemistry of positron and positronium annihilation processes have been described in three books[3,4,7] and several recent reviews[5–9].

This review will discuss the principal features of the chemical interactions of positronium and mesonic atoms, especially where they are similar to those of a hot hydrogen atom, and will attempt to summarise the experimental techniques available for their study. An artificial separation, for convenience, will be made in the discussion of positronium and the other more exotic atoms. More emphasis, because of the more extensive literature, will be given to positronium although the development of 'meson factories' is likely to add some highly significant data on meson chemistry, particularly on muonium[10].

7.2 PROPERTIES OF POSITRONS AND POSITRONIUM

7.2.1 Positrons

There are two solutions to the relativistic wave equation for the total energy of an electron: $E = \pm(p^2c^2+m^2c^4)^{\frac{1}{2}}$, where p is the momentum of the electron, m is its rest mass and c is the velocity of light. Thus, in the absence of external electromagnetic fields, electron energies may range from $\pm mc^2$ to ± 00. Electrons normally exist in negative energy states and when an electron is excited to a positive energy state a 'hole' is left in the sea of occupied negative energy states, which appears as an electron of positive charge. This is considered, for practical purposes, as a positive electron, the positron, with the same mass as the negative electron, the negatron, but with opposite charge.

Positrons occur in cosmic radiation and from the decay of many radionuclides. They are produced by the well-known conversion of high-energy photons to mass in the process of pair production, during which a photon with energy above 1.02 MeV interacts with a nuclear field and an electron-positron pair is produced. These positrons may have very high kinetic energies while those from nuclear decay generally have an energy below 2 MeV.

The converse of pair production is the production of electromagnetic radiation by the annihilation of a positron-electron pair. In a collision with a positron an electron has a high probability of making a radiative transition to an unoccupied negative energy level and the transition results principally in the production of annihilation radiation of total energy $E = 2mc^2 + E_+ + E_-$, where m is the rest mass of positron or electron, c is the velocity of light and E_+ and E_- are the kinetic energies of the positron and electron.

One annihilation photon only may be produced, if the interaction with closely bound (K) electrons is strong enough[11, 12], but production of two photons is much more likely. Each photon then has 0.51 MeV (mc^2) energy and the two photons are emitted in nearly opposite directions. Three photons are produced in less frequent events, they have a shared energy of 1.02 MeV and their directions lie at 120 degrees apart in a plane.

The ratio of the probabilities of two-photon to three-photon annihilation is governed by distinct selection rules governing free positron-electron collisions. In a singlet ^1S interaction (spins antiparallel) an even number of annihilation photons, principally two, is allowed. In a triplet ^3S interaction (spins parallel) an odd number of photons, principally three, is allowed. The relative number of two- and three-photon events depends on singlet-triplet interactions and the rate of annihilation from each state. (The average lifetime for triplet annihilation is 1.4×10^{-7} s and the average rate is 7.14×10^6 s^{-1}. For singlet annihilation the lifetime is 1.25×10^{-10} s and the rate is 8×10^9 s^{-1}. The ratio of the rates of singlet to triplet annihilation is 1115:1.) The singlet state has total angular momentum $J = 0$, and hence the z-component of angular momentum, $m = 0$. The triplet state has $J = 1$ and $m = 0, \pm 1$. Thus the statistical ratio of singlet to triplet states, 1:3, combined with the ratio of annihilation rates, 1115:1, gives the ratio of the probability of singlet to triplet annihilations resulting from free positron interactions with electrons to be 1115:3 or 372:1.

In practice, this ratio will vary and so will the free annihilation life time of the positrons. This free annihilation lifetime is the weighted average of the singlet and triplet lifetimes obtained from the decay rates:

$$\{3(7.14 \times 10^6) + 8 \times 10^9\}/4 = 2.0 \times 10^9 \text{s}^{-1}$$

The free annihilation lifetime is then 5×10^{-10}s, but it is dependent on the electron density of the medium and lies between 1 and 5×10^{-10} s in condensed phases. For unpolarised electrons and positrons annihilating with non-relativistic velocities, Dirac[13] calculated the cross-section for two-photon production, $\sigma_s = \pi r_0{}^2 c/v$, where r_0 is the classical electron radius (e^2/mc^2), c is the velocity of light and v is the relative velocity of the positron and electron. The rate of free annihilation depends on σ_s and the number of electrons, n, per cm^3 in the medium. The rate of annihilation is then $\lambda_s = \sigma_s nv = \pi r_0^2 cn = 4.5 \times 10^9 \; dZ/A \, \text{s}^{-1}$ where d, Z and A are density, atomic number and mass number of the medium, respectively. Thus free annihilation of positrons is a function of electron density. The greater the electron density the higher the annihilation rate and the shorter the lifetime of the free positron.

However, other interactions between positrons and electrons may occur and a relatively stable quasi-atom may be formed consisting of a positron and an electron; it is known as positronium[14]. This bound system may have a lifetime as long as 1.4×10^{-7} s and in practical cases in a given material positrons will exhibit lifetimes arising from both free annihilation and annihilation from one or more bound states.

7.2.2 Positronium

Several bound states of positrons and electrons have been shown by theoretical calculations to be stable, e.g. $e^-e^+e^-$ or $e^+e^-e^+$ and $e^+e^-e^+e^-$, but these are rare and have much shorter lifetimes than the simplest e^+e^-, positronium (symbol: Ps).

Ps will exist in the ground state as *para*-Ps (singlet, spins anti-parallel) and *ortho*-Ps (triplet, spins parallel). If formed in any excited state, except possibly the 2S, it will radiate optically to the ground state rather than annihilate directly. Attempts to detect this radiation have had limited success so far. The ground-state splitting of singlet and triplet levels has been measured as 8×10^{-4} eV in excellent agreement with the quantum electrodynamic predictions for this system. The ionisation potential (same as binding energy in this two-particle system) is 6.77 eV (half that of the hydrogen atom) and the inter-particle distance is 1.06 Å (twice the Bohr radius for H). Similarly the Lyman α line would be at 2430 Å.

The fine structure of Ps has been extensively studied[15–18] and the measurement of the fine structure constant agreed with the calculated value to five significant figures[19, 20]. The Zeemann splitting of the ground state of Ps in a magnetic field leads to a magnetic quenching of the triplet state. The sub-levels $m = \pm 1$ of the triplet ($J = 1$) are unaffected, but the sub-level $m = 0$ can mix with the singlet $m = 0$ state. Up to one-third of the triplet Ps can thus be mixed with the singlet state in a magnetic field and decay by two photon annihilation with a shorter lifetime.

If Ps is formed and is unaffected by its environment there will be 75% in the triplet state and 25% in the singlet. Annihilation from these states will then give a ratio of two-photon to three-photon annihilations of 0.33. This is very much less than the ratio, 372, for free positron annihilations and a measurement of a ratio below 372 is an indication that Ps is formed. Again, a measurement of a long lifetime, approaching 1.4×10^{-7} s in gases, is a clear indication of the formation of *o*-Ps. (This method gave the first experimental proof of the existence of Ps[2].) Another method, which has been effectively used[21], depends on the fact that the energy spectrum of the annihilation photons will change if Ps is formed; the peak (0.51 MeV) to valley (0.3 MeV) ratio of events in the energy spectrum of annihilation photons will decrease. (A composite spectrum will generally be observed due to annihilation from both free and bound positron states and parallel experiments are needed in which quenchers of triplet Ps, such as nitric oxide, are added to give a measure of the triplet Ps population.)

7.3 SLOWING OF POSITRONS AND FORMATION OF POSITRONIUM

We are fortunate to have convenient positron sources, such as the radionuclides ^{22}Na and ^{64}Cu, but unfortunate in that the positrons emitted are highly energetic with a range of energies and average energies about 0.5 MeV. (Some progress is being made in the development of mono-energetic positron beams with energies of about 10 eV[22].) These positrons must lose energy to a value of the order 10 eV before the reactions of greatest interest can occur. Energy is lost primarily by ionisation and excitation of atoms and molecules, and by excitation of some collective levels in the condensed phase.

Physical details of these processes have been thoroughly discussed elsewhere[7,23]. It is clear, for example, that less than 5% of the positrons will be lost by annihilation before their energy has dropped to a value comparable to the ionisation energy of atoms and molecules. A detailed computation of the probability of positronium formation then requires a knowledge of the cross-sections for many elementary processes and their energy dependence. These processes include ionisation and excitation of molecules by positrons, electron capture by positrons from molecules to form positronium, elastic and inelastic scattering of positronium by molecules, compound formation between Ps and molecules, dissociation of these compounds and dissociation of Ps in collisions with molecules. Knowledge of these matters is slight, even for gases, and a detailed calculation has been carried out for only one case, the formation of Ps in atomic hydrogen using the Monte Carlo method[24].

A qualitative, but very useful, estimate of the probability of Ps formation in gases is provided in the Oré gap concept[25]. As the positron falls below I_A, the first ionisation potential of the constituents of the medium, the formation of Ps may occur by the reaction

$$e^+ + A \longrightarrow Ps + A^+ \tag{7.1}$$

The threshold energy that the positron must have to permit this reaction is $(I_A - 6.8)$ eV, where 6.8 eV is the binding energy of Ps, and the maximum fraction of positrons forming Ps is $[I_A - (I_A - 6.8)]/I_A$. Losses will occur if an excitation level (E) of the molecules falls between I_A and the threshold. The minimum fraction will then be $[E - (I_A - 6.8)]/E$. For argon gas with $I_A = 15.6$ eV and $E_1 = 11.6$ eV, the threshold is 9.0 eV and the Ps formation fraction lies between 0.23 and 0.43. (Note that any Ps formed with energy above I_A would quickly undergo collisional dissociation.) This concept applies fairly well to simple gases, such as argon, but requires much modification for polyatomic gases and condensed phases[3,7,26].

At a later stage we will discuss the modifications for further analysis of the energetics of positronium formation, inhibition of formation and the kinetics of chemical 'quenching'.

7.3.1 Quenching of positronium

Where Ps is formed, the theoretical ratio 1 *p*-Ps:3 *o*-Ps may not be found exactly in practical cases[27,28], and the Ps atoms will have an energy distribution a little different from that of the parent positrons. The energy distribution of the positrons is not 'rectangular' in the Oré gap region and positrons with the higher energies will rapidly lose energy at first and then form Ps with lower energy, so that the lower energies are slightly favoured in the Ps energy distribution.

Quenching is the term used to describe processes which will alter the annihilation rate of Ps in a given medium. The lifetimes of *p*-Ps and *o*-Ps are 0.125 and 140 ns in free space and, if there are no chemical interactions of the type to be discussed later, Ps will annihilate with a lifetime dependent on the electronic configuration of the medium. If there are no unpaired electrons, the annihilation rate of *o*-Ps will be increased by 'pick-off' quenching and the rate for *p*-Ps will be practically unaltered. If, however, the medium contains unpaired electrons, 'conversion' quenching occurs as well as 'pick-off' quenching, and the annihilation rates of both *p*-Ps and *o*-Ps will be altered. On the other hand, if Ps is energetic enough to take part in chemical reactions to give stable compounds, the annihilation lifetime spectrum will be considerably altered[26].

7.3.1.1 Pick-off quenching

If Ps is formed in a dense molecular lattice, its wave function will overlap those of the electrons in the lattice and the annihilation rate of the positron bound to the electron in Ps will be altered. To put it another way, the positron can sample many electrons in the medium as annihilation partners. This has been called 'pick-off' quenching; its dependence on temperature and state is primarily a function of the 'free volume' available in the substance, where the Ps atoms may be accommodated, and it has been thoroughly studied[29]. Pick-off quenching cross-sections of most compounds are of the order of 10^{-21} cm^2 (Table 7.1), 10^5 times less than their geometrical cross-sections.

Table 7.1 Conversion and pick-off quenching cross-sections

Compound or ion	*Quenching rate* (atm^{-1} s^{-1})	*Quenching cross-section* (cm^2)*	*Suggested mechanisms*
O_2 (gas)	2.34×10^7	1.23×10^{-19}	Conversion†
N_2 (gas)	2.1×10^5	1.06×10^{-21}	Pick-off†
Ar (gas)	2.51×10^5	1.32×10^{-21}	Pick-off†,‡
NO (gas)	3.0×10^9‡	1.6×10^{-17}	Conversion‡
H_2O (liq.)	5.66×10^8	2.3×10^{-21}	Pick-off§,¶
H_2SO_4 (liq.)	3.3×10^8	4.0×10^{-21}	Pick-off§
Fe^{3+} (aq.)	—	3.8×10^{-18}	Conversion¶
Fe^{2+} (aq.)	—	4.8×10^{-19}	Conversion¶
Mn^{2+} (aq.)	—	2.5×10^{-19}	Conversion¶
Co_2^+ (aq.)	—	5.3×10^{-19}	Conversion¶

*Calculated by assuming that thermal equilibrium has been reached.
†Celitans G. J., Tao, S. J. and Green, J. H. (1964). *Proc. Phys. Soc.*, **83**, 833.
‡Heymann, F. F., Osmon, P. E., Veit, J. J. and Williams, W. F. (1961). *Proc. Phys. Soc.*, **78**, 1038.
§Ref. 156.
¶Ref. 157.

(This clearly implies that Ps does co-exist with molecular substances and that its wave function penetrates only a little into neighbouring molecules, if no other interactions are involved.)

Since the process of pick-off quenching involves electrons of the molecules in the medium as a whole, the quenching rate should then depend on the molar density, not on the average electron density. The measurable positron lifetime (modified *o*-Ps lifetime) is generally shorter the higher the molar density. Pick-off quenching occurs generally in all media and alters the annihilation rates of *o*-Ps and *p*-Ps separately because the process does not alter the physical configuration of the Ps.

7.3.1.2 Conversion quenching

This mechanism was suggested by Bell and Graham[30] and Ferrell[31] showed later that the use of concepts such as 'spin-flip' and 'electron exchange' were not needed to explain the physical details. Very clear quantum-mechanical calculations showed that conversion of Ps, either *o*-Ps → *p*-Ps or *p*-Ps → *o*-Ps, depends only on its collision with a system containing one or more unpaired electrons. The configuration of Ps is altered after collision and the annihilation rate is changed.

Most molecules and ions, which have been studied, do not contain unpaired electrons and the electron configuration of Ps is not altered by collision with them. Examples in this category are H_2O, D_2O, HNO_3, HCl, H_2SO_4, Na^+, Cl^-, Ag^+, ClO_4^-, $Fe(CN)_6^{4-}$, CH_4, C_6H_6, C_nH_{2n+2}. Some molecules and radicals have unpaired electrons, e.g. NO, NO_2, O_2, diphenylpicrylhydrazyl, as do transition element ions, e.g. Fe^{3+}, Fe^{2+}, Co^{2+}, $Fe(CN)_6^{3-}$.

Conversion affects the annihilation rates of *o*-Ps and *p*-Ps simultaneously because of the ready reversal of configuration and the dependence on the conversion rates of the mean lifetimes (τ) and intensities (I) of positrons annihilating by the different modes is complicated. Earlier calculations by Dixon and Trainor[32] have been extended by Tao[26].

(In addition to the principal quenching processes described, magnetic and electric field quenching has been observed. This is described elsewhere[33, 34].)

Besides these quenching mechanisms some authors[7, 8] include 'chemical quenching'. We prefer here to discuss this subject in Section 7.5 as part of a general discussion of the chemical reactions of Ps.

7.4 EXPERIMENTAL TECHNIQUES FOR STUDIES OF POSITRONS

The principal methods of experimentation will be outlined and details of the nuclear physical equipment can be found in the references cited. From the point of view of the chemist, a great deal depends on the preparation of the positron source and the materials (other than pure metals and alloys) which are to be studied. For this reason a lengthier discussion of the source-material system is warranted. It should be noted that the procedures are non-destructive in character, weak radiation sources are required and the detection of positron-material interactions is conducted at a distance by transducers, which analyse the annihilation characteristics by virtue of the emitted γ-radiation.

7.4.1 Positron sources and materials

It is fortunate for positron lifetime measurements that the nuclide ^{22}Na is available. It has a conveniently long half-life (2.6 years) so that decay corrections are not required during a measurement, positrons are emitted with 542 keV maximum and 120 keV average energy, and the de-excitation of the excited state of the daughter ^{22}Ne occurs within 10^{-12} s by emission of a 1.28 MeV γ-photon. This energetic photon signals the birth of the positron (for all practical purposes) and the end of its life is signalled by annihilation radiation of maximum energy 0.51 MeV, which can be readily differentiated from the 1.28 MeV photon. ^{64}Cu has advantages in the study of angular correlation and the energy spectra of the annihilation radiation, because it only emits one γ-photon (1.34 MeV) for every 20 positrons. Interference from Compton scatter is minimised and, although the 12.9 h half-life is short, the ^{64}Cu is readily obtained and replenished by thermal neutron irradiation of a small piece of copper foil.

A ^{22}Na source for positron lifetime studies is prepared by evaporating a drop of carrier-free ^{22}NaCl solution on aluminium or mica of about 10 μm thickness. This is folded or sealed against leakage before insertion into the system to be studied. In the case of aqueous solutions it is sometimes convenient to dissolve the source in the system. About 5 μCi is sufficient source strength and 10–50% of the positrons are absorbed in the mount. Since no long-lived Ps is formed in these materials, there is no interference with lifetime measurements, but the fraction of positrons decaying in the source mount must be known for a correction when determining the intensity of Ps (percentage of positrons forming Ps) in a sample. This is done simply by depositing ^{22}NaCl between two pieces of a suitable solid, measuring the intensity of the long lifetime component and comparing with the intensity observed using the mounted source. This can be avoided in certain cases by using very thin materials such as mylar, but only when a high intensity of long-lived Ps is formed in the sample. These organic thin films generally provide a Ps lifetime of 1–3 ns with up to 30% intensity and, consequently, if 10% of positrons are annihilated in the mount there would be a 3% error in estimating the intensity of short-lived Ps. The source strength suitable for lifetime measurements is a function of the maximum desirable random coincidence rate between scattered 1.28 MeV and 0.51 MeV photons which constitute the principal background events. In general, a maximum of 5μCi is needed and there is no complication arising from radiation damage to the sample under study (e.g. radical production and trapping which would lead to increased conversion quenching during a measurement, especially in solids[35, 36]).

The thickness of the sample studied must be chosen to allow absorption of all positrons which escape the source mount. The maximum range of the positrons can be calculated as for negatrons, but, because of its charge, the positron may have a range in some materials up to 80% greater than that of the electron. In metals the stopping power for electrons is about 25% greater than for positrons. In liquids such as toluene, benzene and water it is about 75% greater[37]. When using ^{22}Na positrons a thickness of 200 mg/cm^2 is sufficient for condensed materials and in gases such as argon and nitrogen the maximum range is about 150 cm atm[38].

It may seem unnecessary to emphasise the necessity for very careful control in the physical and chemical preparation of systems to be studied with positrons, but a great deal of effort has been misspent because this matter was overlooked. As an example, a change in particle size of solid powders will lead to a change in Ps lifetime, because the Ps formed in the lattice will diffuse to the surface and annihilate with a lifetime depending on the gas adsorbed on the surface or in the gas phase[39]. The presence of paramagnetic impurities, particularly oxygen, in gases and liquids has led to a number of errors; a trace of oxygen is a very effective quencher in gases[40] and liquids[41, 42]. In the study of solid polymers inconsistencies have arisen, because the sample chosen, e.g. Teflon, is a function of industrial preparation, storage and distribution which introduce uncontrolled physical and chemical factors into the measurement.

7.4.2 Positron lifetime measurement

The principle is to use two scintillation detectors, one to respond to the 1.28 MeV γ from $^{22}Na \rightarrow {}^{22}Ne$ giving a start pulse, and the other to respond to the 0.51 MeV (max.) γ from annihilation giving a stop pulse. The time interval between these events is converted by a time-to-pulse amplitude-converter into pulses whose amplitudes are proportional to the time interval between arrival of pulses from the detectors. These pulses can be sorted in a multichannel analyser and displayed graphically in a form shown in

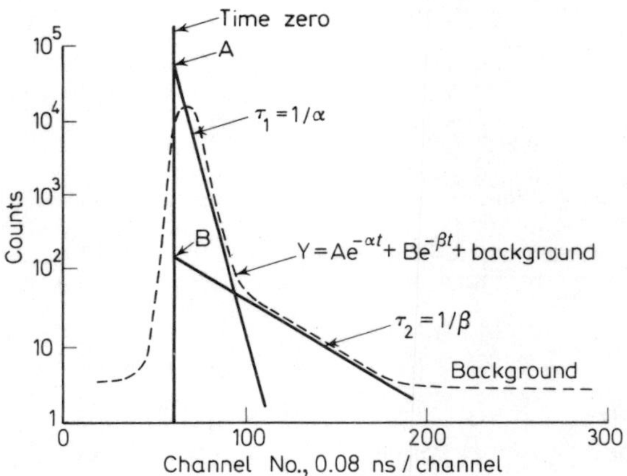

Figure 7.1 Graphical resolution of a two-component positron annihilation lifetime spectrum

Figure 7.1. Typical components and layout[8] are shown in Figure 7.2. Data analysis may be carried out by least-squares fit and more conveniently by computer analysis[8, 43]. Surveys of these methods have been made in a number of extensive articles[3, 7, 8, 44, 45].

The lifetime spectrum in Figure 7.1 consists of two components; the shorter is due to free positron and *p*-Ps annihilation, the longer to *o*-Ps annihilation. Zero time is obtained from the centroid of a 'prompt' peak arising from detection of Compton scatter of simultaneous 1.17 and 1.33 MeV γ-rays from a small ^{60}Co source with all timings spectrometer settings identical with those used for the positron lifetime determination. In practice, the time resolution is about 300 ps as defined by the full width at half-height of the prompt peak. The logarithmic slope of this peak is very important, because it must be small to give a reasonable measure of the short lifetime component. Counts

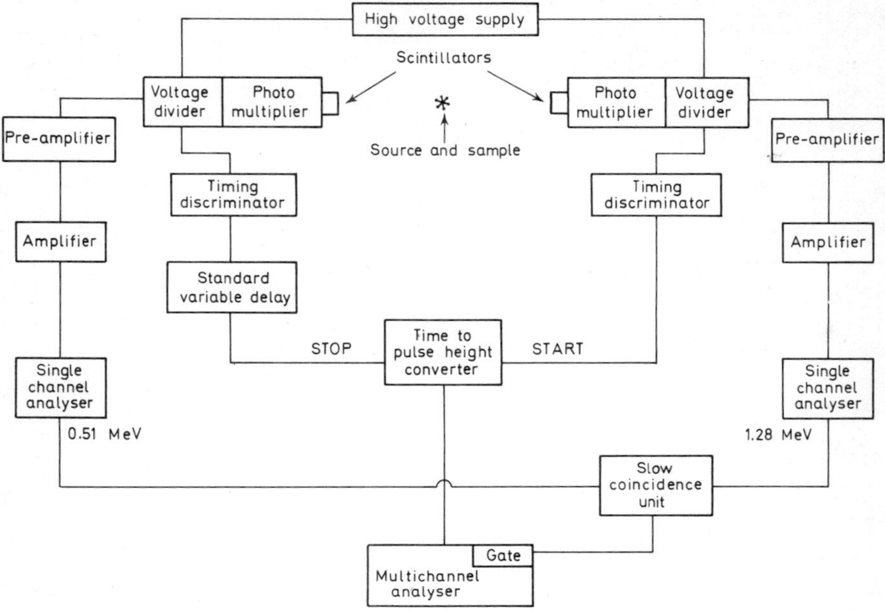

Figure 7.2 Typical layout and components of a fast positron lifetime spectrometer

in any channel on the right-hand side of the spectrum illustrated can be represented by

$$Y = Ae^{-\alpha t} + Be^{-\beta t} + \text{background}$$

where A and B are the zero intercepts of the lifetime components with decay rates α and β respectively and t is the time. The corresponding mean lifetimes are $\tau_1 = 1/\alpha$ and $\tau_2 = 1/\beta$. The intensities (percentages of positrons decaying with a given lifetime) are I_1 and I_2 given by the integrals of $Ae^{-\alpha t}$ and $Be^{-\beta t}$ from zero to infinite time divided by the area under the entire curve, expressed as a percentage.

7.4.3 Two- and three-photon coincidences

The principle here is basically the same as for the determination of lifetime spectra without the requirement of pulse timing channels from the detectors. Indications of the formation of *o*-Ps are found if the ratio of 2γ to 3γ coincidences falls below 372 (as it is, for example in aluminium, where no *o*-Ps is formed). A diagram of a typical spectrometer is given in Figure 7.3. The scintillation detectors are arranged in a plane through the source-sample

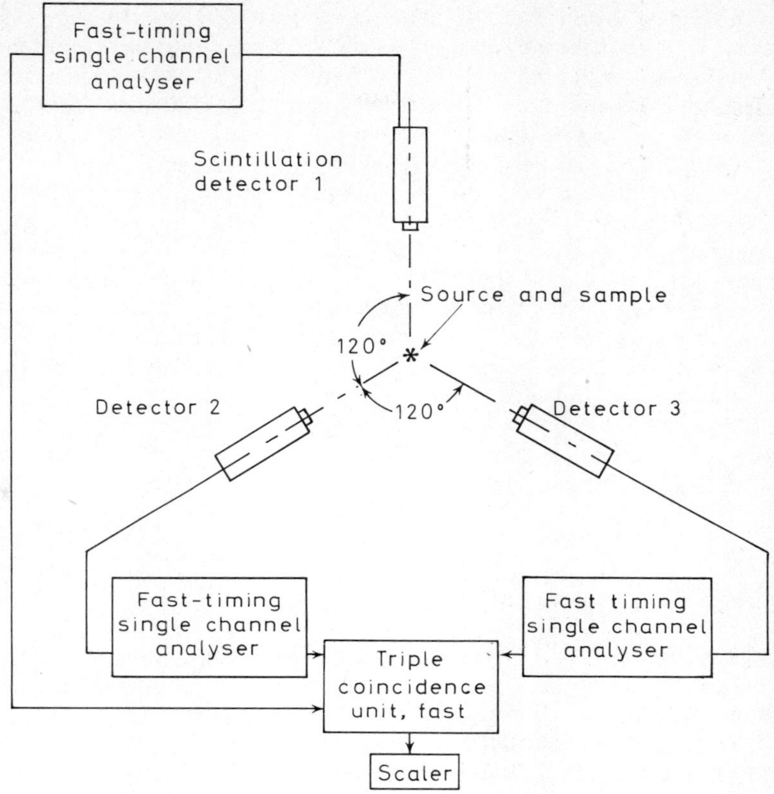

Figure 7.3 Schematic of a two- and three-photon coincidence measurement

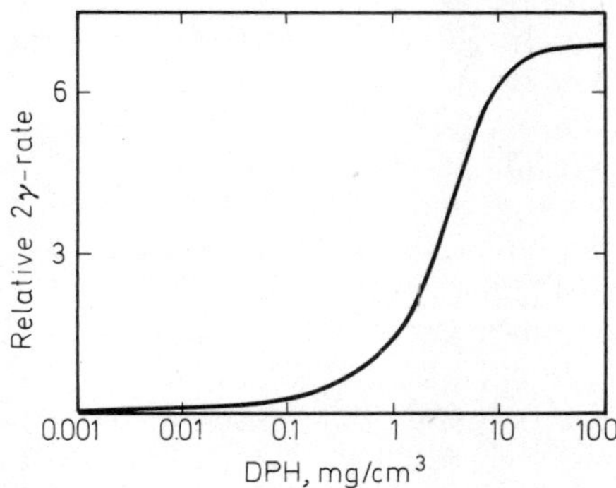

Figure 7.4 Quenching of *o*-Ps in benzene by diphenylpicrylhydrazyl as shown by variation in the relative two-photon coincidence rate

system and well shielded. Two may be used collinear with the sample source to detect 0.51 MeV photons from singlet decay using a double-coincidence supply to the scaler. Three in triple coincidence arranged at 120-degree angles are used for detection of 340 keV photons from triplet decay. All three detectors are heavily shielded and one can be tilted out of the plane for extraneous background measurement. If the single channel analysers are fast in energy selection of each detector pulse, if fast phosphors (e.g. Naton 136) and photomultipliers are used and the coincidence unit is fast, overall resolving times of less than 10 ns are possible. This means that the spurious count rate is kept low and higher source strengths can be used to allow faster accumulation of data, typically a tedious matter in these measurements.

Figure 7.4 shows the effect of an increasing concentration of diphenyl-picrylhydrazyl (DPH) on the relative 2γ rate in benzene solutions. DPH is a stable free radical and an efficient conversion quencher of *o*-Ps. As the concentration increases triplet annihilation decreases and the relative 2γ rate increases very markedly[46]. Similar information has been found on the Ps quenching rate coefficients in gases[47,48] and in insulators and other solid materials[49].

7.4.4 Angular correlations

Coincidence and angular correlation studies have long afforded extensive data in nuclear physics and chemistry (ref. Vargas, this volume). In the case of positrons, annihilation of a singlet positron-electron pair at rest produces two 0.51 MeV photons emitted in opposite directions. If the pair possesses momentum, the angle between the photons will not be exactly 180 degrees and the deviation will give a measure of electron momentum in condensed materials, where the positron is very quickly thermalised before annihilation occurs.

A simple vector diagram shows that the sine of the angle of perpendicular deviation from collinearity, θ, is the ratio of the pair momentum, $\boldsymbol{P}\perp$, to the momentum of an annihilation photon, mc. For small angles, $\sin\theta = \theta$, and hence $\theta = \boldsymbol{P}\perp/mc$, where θ is in radians. Measurement of the angular distribution of annihilation photons, even though the angles are measured in milliradians, thus provides very useful information.

The apparatus is simple, in principle, and consists of a 2γ coincidence spectrometer with slit collimation of the detectors, a scaler-timer-printer and an automatic device to move one detector through angles $\pm\theta$ with respect to the other. Techniques and methods of data correction have been reviewed recently[7,8] and papers by Stewart and Berko should be consulted.

Typical angular correlation curves are shown in Figure 7.5. With horizontal slit geometry the coincidence count rate as a function of the vertical detector displacement, Z, depends only on the Z component of momentum of the centre of mass of the annihilating pair. The count rate at angle θ is proportional to the probability that the momentum vectors, $\boldsymbol{P}$, in momentum space are equal to $\theta.mc$. Assuming that electrons in the sample behave as a Fermi gas, the momentum distribution for zero temperature is given by $N(\boldsymbol{P}) = \boldsymbol{P}^2$ for $P^2 < P_m^2$ and $N(\boldsymbol{P}) = 0$ for $P^2 > P_m^2$, where P_m is the maximum momentum. Since a positron is thermalised before annihilation in a condensed system and taking the annihilation probability to be independent of the electron momentum, the distribution of Z momentum is $N_z(\boldsymbol{P}_z) = (P_m^2 - P_z^2)$ for $\boldsymbol{P}_z^2 < \boldsymbol{P}_m^2$, and 0 for $\boldsymbol{P}_z^2 > \boldsymbol{P}_m^2$. When $\theta > \boldsymbol{P}_m/mc$ the count rate is zero; when $\theta < \boldsymbol{P}_m/mc$, the rate varies as $(\boldsymbol{P}_m^2 - \boldsymbol{P}_z^2)$ and at $\theta = 0$, $\boldsymbol{P}_z = 0$ and the count rate is a maximum. The angular distribution would be expected to be in the form of an inverted parabola where $\boldsymbol{P}_m$ is at the angle where the count rate decreases to zero.

This curve is shown in Figure 7.5 for Fermi electrons in a metal. The count rate is zero at about 4×10^{-3} rad and the Fermi electron energy would be $E = \boldsymbol{P}_m^2/2m = \theta^2(mc)^2/2m = 4.08$ eV. Such curves are commonly found for metals. Bell-shaped curves are found when positrons annihilate with higher momentum electrons, such as core electrons, or electrons so affected by the periodic potential of the lattice that they have an unusually high zero-point motion. (As in the noble metals, Cu, Ag and Au and in ionic crystals and liquids.) The dotted curve in Figure 7.5 represents the 'narrow peak' arising from annihilation of thermalised p-Ps. This peak requires very good angular resolution, at least 0.4 mrad, before its character can be studied as a function of temperature, electric field, mechanism of o-Ps quenching and other parameters. Evidence for positronium-like Bloch states in quartz single-crystals has recently been found by studying the effect of a magnetic field on the system with apparatus of very high resolution[50]. Changes in the angular distribution as a function of partial pressure of O_2 in argon–oxygen mixtures have been studied to determine the quenching effects on o-Ps by oxygen[51]. Goldanskii[7] has stressed that much information on the mechanisms of o-Ps quenching is available by this technique, which cannot be obtained by measurement of the lifetime and intensity decrease of the long lifetime component in positron lifetime spectra. For example

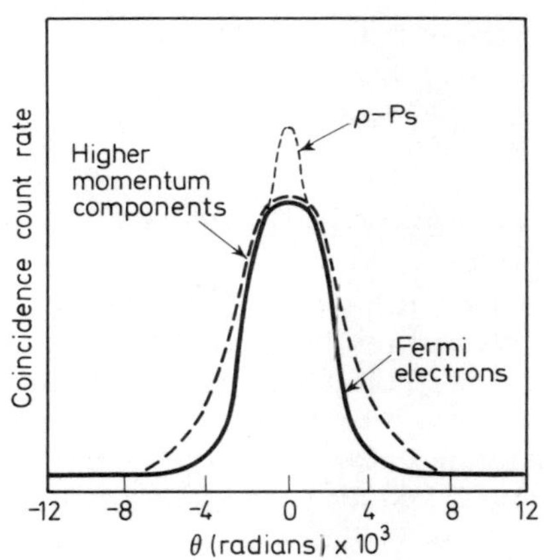

Figure 7.5 Schematic angular correlation curves

conversion quenching will cause a narrow angular distribution while quenching by chemical oxidation will broaden the angular distribution.

7.4.5 Multi-parameter measurements

The techniques described above are now being combined to give much more detailed information on annihilation processes. For example, an angular distribution system may be combined with a lifetime apparatus in such a way as to permit measurement of the lifetime of positrons annihilating with electrons of prescribed energy. Various possible systems have been outlined[8] and Hsu and Wu[52] have obtained positron lifetime spectra at different angles between annihilation photons in a plastic scintillation material. The longer lifetime, τ_2, increased in intensity, I_2, from 20% to 40% as the angle θ was increased from 0 to 5.4 mrad. The indication is that the τ_2 component arises from a bound state, probably *o*-Ps which annihilated by picking off an electron from the environment. The momentum associated with pick-off represents the momentum distribution of available electrons and shows a broad spectrum.

Application of such multi-parameter techniques is spreading but is limited by the need for more extensive and expensive apparatus and by the longer time required to collect sufficient data.

7.5 PHYSICAL CHEMISTRY OF POSITRONS AND POSITRONIUM

Although there are many indications in the extensive literature that the transformations of positrons and positronium can provide information of value to chemists, the study of this branch of nuclear chemistry is only just beginning. The slowing down of positrons in the eV range and the formation of Ps, when interpreted accurately, can give very useful data on the competing processes of ionisation, electronic excitation of molecules and the states of electrons captured by positrons. Once positronium (or mesonic atoms) are formed their reactions may be fast, as discussed some time ago[3, 7], and the kinetic data obtained may be relevant to the corresponding reactions of hydrogen atoms. Alterations in Ps configuration of the type discussed earlier, if studied by lifetime and angular correlation methods, constitute a 'microprobe' of interactions with molecules and ions in the condensed state extending over a range of perhaps 100 Å.

Several schematic diagrams of the interactions of e^+ and Ps in a medium have appeared in the literature and one of these is represented in Figure 7.6 [8]. This is more detailed, as far as chemical reactions are concerned, than the scheme given by Goldanskii[7], but omits the factors causing inhibition of the formation of Ps-slowing of e^+ below the energy region where Ps may be formed, and inhibition due to positron capture and formation of positron-molecule complexes, or compounds. The diagram illustrates the formation and conversion of Ps discussed earlier and introduces the processes of chemical oxidation, reduction and compound formation, which are yet to be treated. The eventual fates of the positrons are indicated, some lifetimes and

$2\gamma/3\gamma$ ratios are given, and the features of the angular correlation of the annihilation radiation should be remembered, as outlined in Section 7.4.4.

7.5.1 Special features of positronium chemistry

In addition to the unique physical properties and interactions, which characterise Ps, there are special problems to be borne in mind when chemical reactions are discussed[7, 8, 26, 53], Ps can be treated as a free radical behaving something like a hydrogen atom, but there are several differences: (a) Ps is usually formed with a kinetic energy well above thermal and may therefore react in the epithermal or 'hot' region, (b) Ps is 919 times lighter than the lightest normal chemical species, H. Before collisional interaction, Ps contributes practically the whole kinetic energy of the system and the energy distribution after reaction depends on the masses of the products. If a Ps

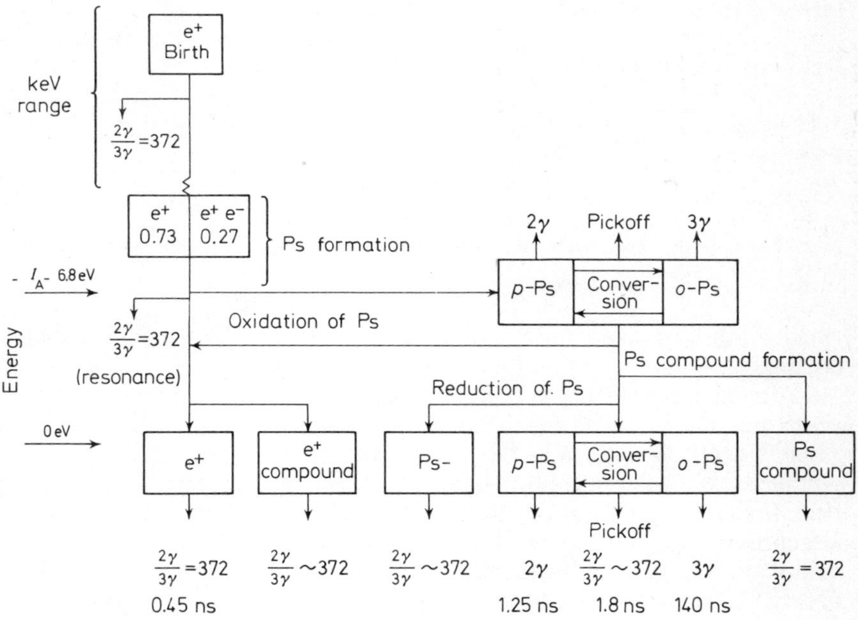

Figure 7.6 Interactions of positrons and positronium in matter

compound is formed, its mass will be high and the total kinetic energy will be shared fairly evenly between the products. If one of the products is a positron, it will have almost all the kinetic energy. From the point of view of the energetics of a reaction only the energy of the Ps has to be considered in discussing the energy requirements for the Ps rection. (c) The bond energies of Ps compounds will be lower than those of corresponding hydrogen compounds, because of the small mass and high mobility of Ps. Ps reactions are therefore energetically more unfavourable than the corresponding H reactions but the activation energies should be fairly low. (d) Chemical equilibrium is seldom reached in a Ps reaction, because of the short lifetime, low concentration and continuous thermalisation. This means that parameters derived from equilibrium measurements, e.g. redox potentials and activity coefficients, cannot be used in Ps chemistry without modification. The chemical kinetics of Ps reactions should not be treated in the classical sense where an approach to equilibrium is assumed. (e) Thermalised Ps will generally be solvated or caged in a liquid, perhaps in a 'bubble'[54], and found in a cavity in solids[29]. However, because its mass is small, the positron part of its wave function may spread over a long distance and exhibit some long-range effects.

7.5.2 Chemical inhibition of positronium formation

As noted earlier, positrons may fail to form Ps, even if the energy region is favourable, for two reasons: slowing down of e^+ below the Oré gap threshold and capture of e^+ by a chemical inhibitor or scavenger. The latter subject has been extensively surveyed by Goldanskii[7] and the former, arising from the work of Tao, Bell and Green[55] will be discussed later in Section 7.5.3.

In summary, chemical inhibition of Ps formation will occur by the basic processes:

(a) $e^+ + AB \longrightarrow ABe^+ \longrightarrow$ annihilation,

which is capture followed by annihilation from the positron complex, and

(b) $e^+ + AB \longrightarrow A + Be^+ \longrightarrow$ annihilation,

which is dissociative capture followed by annihilation from the positron compound e^+B.

Experimental evidence for these processes has been available for many years, but most of it is highly qualitative. Even with the simple diatomic gases, N_2, H_2, D_2 and monatomic rare gases a 100% yield of Ps cannot be obtained, even by application of electric fields to accelerate slowed positrons back to an energy in the Oré gap region. Very early indications of positron attachment to CO_2, CH_4 and Freon were extended by Green and Tao[56], who gave evidence for the formation of the positron complex $e^+CCl_2F_2$ in Freon-12 and particularly by the work of Paul and collaborators on annihilation rates in gases and gas mixtures[57,58]. Collision complexes of positrons with methane, ethane, propane, butane, isobutane and carbon tetrachloride were strongly indicated.

As far as liquids are concerned, the situation is by no means clear. A tremendous amount of work has been done to study positron annihilation in organic liquids and has been thoroughly reviewed[3,4,7,8]. Despite many discrepancies and errors shown[59] to be caused by inadequate removal of the paramagnetic quencher, oxygen, some general conclusions are possible. In many hydrocarbons (saturated and unsaturated) the Ps formation probability is 40–50% (I_2 *c.* 30–40%) and it is roughly what might be expected from the simple Oré model. However Ps formation drops considerably in alcohols and halogen derivatives. This was considered by Hatcher[60] to be explained in terms of the Oré model, if the dissociation energy of the weakest C–halogen bond were taken as the upper limit of the Oré gap. However, later experiments on the halogenated propanes[61] and recent theoretical considerations[26] show that such an explanation is inadequate and the explanation adopted[7] involving self-inhibition by dissociative positron capture, $PhCl + e^+ \longrightarrow Ph + e^+Cl$, may be insufficient. Tao[62] has recently studied reactions between iodine and Ps in organic hydrocarbon and oxyhydrocarbon solvents. He finds that I_2 is *both* a quencher of Ps and an inhibitor of the formation of Ps. The positron reaction is thought to be

$$e^+ + I_2 \longrightarrow e^+I_2$$

and the positronium reaction

$$Ps + I_2 \longrightarrow PsI_2 \text{ (or } PsI + I)$$

quenches the lifetime of *o*-Ps very rapidly. The latter reaction is diffusion controlled and the mean life of the compound is about 0.4 ns. The inhibition coefficients for the first reaction are 20 mol^{-1} in hydrocarbon and 5 mol^{-1} in oxyhydrocarbon solvents. (The influence of charge-transfer complexing between I_2 and the solvent on the behaviour of the positrons is yet to be studied.)

Ps formation decreases in liquids as the dipole character of the medium increases and as localisation of negative charge in the molecules increases. This supports the idea of dissociative positron reaction and has been observed not only in halogen derivatives and alcohols, but also in the isomeric xylenes. In the xylenes the Ps formation probabilities are 32% (*ortho*) 25% (*meta*) and 16% (*para*). Strongest inhibition is found in liquids where the molecules contain two or more halogen atoms – dichlorobenzene, bromoform, CCl_4 – or nitro groups – dinitrobenzene. No Ps at all is formed in these liquids, and the experiments of Ormrod and Hogg[63], which showed extremely strong inhibition of Ps formation in benzene by adding small amounts of CCl_4, are only explicable[7] by the dissociative reactions with positrons.

$$e^+ + CCl_4 \longrightarrow CCl_3 + e^+Cl$$

There are many similar examples where inhibition of Ps formation by positron-molecular interaction is the explanation given but there remains a wide spread of opinion. For example, the decrease in intensity of the long component of positron lifetimes in aqueous solutions containing increasing concentrations of NO_3^- ion has long been explained[64] as due to the formation of $e^+NO_3^-$ by positron attachment. Other interpretations[26] may indeed be more reasonable and will be given later.

7.5.3 Delayed annihilation and Ps formation inhibition

Another form of 'inhibition' of Ps formation is due to the fact that positrons may lose energy by elastic collisions and fall below the energy region where Ps formation is possible. In further slowing down to the thermal energy region they undergo a variety of atomic and molecular collisions, which are of considerable interest in the study of low-energy (5 eV or less) electron and positron collision processes. (In this energy region, some success has been achieved in extracting low-energy mono-energetic e^+ beams from linear accelerators and preliminary total cross-sections for scattering have been presented at the *Sixth Int. Conf. on the Physics of Atomic and Electronic Collisions* (*1968*) but the full details have not been published.)

The possibility of these studies arose from a study[65] of slowing down times of positrons in selected energy ranges in argon gas. For example, the slowing down time in argon from the first excitation level of the gas (11.6 eV) to 0.1 eV is about 100 ns at 10 atm pressure, whereas the Dirac free annihilation lifetime (assuming that the effective number of electrons per atom is 18) is calculated to be only 27.6 ns. This analysis, together with the development of lifetime apparatus of high resolution and long time range, led to experimental results on the fine structure of the lifetime spectra of positrons in

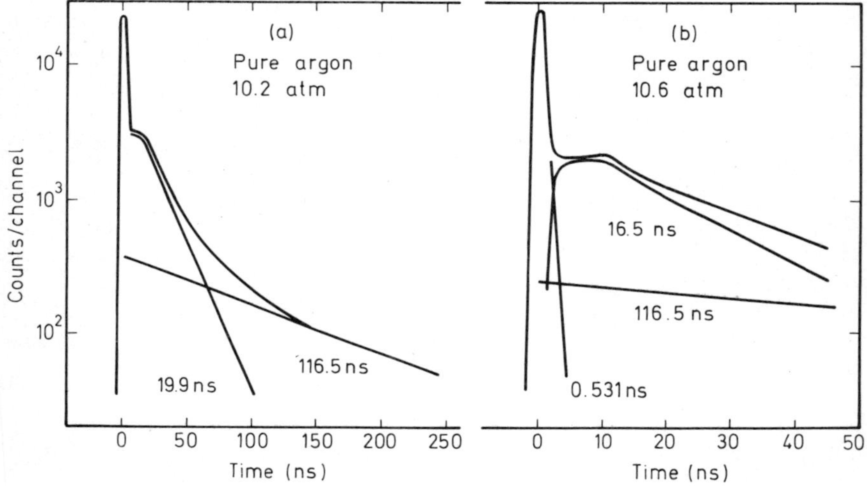

Figure 7.7 Four-component lifetime spectrum of positrons in argon gas

argon, which disclosed for the first time the presence of another lifetime component[66]. This appears as a non-exponential 'shoulder' in the lifetime spectrum and is shown in Figure 7.7. This work of Tao, Bell and Green was almost simultaneously confirmed and developed by Paul[67], also for argon, Falk, Orth and Jones[68, 69], for various inert gases, and by Osmon[70, 71]. All these experiments were carried out at room temperature over a broad range of pressures, but much still remains to be discovered in the kinetics of the interactions involved that might be revealed by a study of the effects of temperature variation.

Figure 7.7 illustrates the component parts of the four-component spectra of annihilation in inert gases (and simple diatomics, such as nitrogen)[72]. First appears the 'prompt peak' due to annihilation of *p*-Ps and all processes of e^+ loss above the Oré gap, including the formation of quasi-bound states, e.g. collision complexes such as e^+Ar. After the prompt peak is a 'shoulder' of time broadness, *b*, characterising the time in which positrons (having failed to form Ps) lose energy below the Ps formation threshold to reach a much lower energy at which annihilation by some attachment process occurs. The length of the shoulder is inversely proportional to the pressure: bp = const. The third component appears after the shoulder (or directly after the prompt peak in cases where no shoulder appears) and reflects exponential positron annihilation with a rate constant, λ_1, proportional to the gas pressure. The high annihilation rates, described in Section 7.5.2, observed in Freon-12 [56], in argon and hydrocarbons and mixtures[57, 58] and in nitrogen and oxygen[40], are observed in this region, and positron–molecule complex formation is believed to be responsible for the high rates. In the region towards the end of the shoulder, at very low energies, the annihilation rate is not constant and is regarded as positron-velocity dependent. The fourth region of the lifetime spectrum (third if there is no shoulder) is due to o-Ps annihilation and the mean lifetime in simple gases, without quenching or impurity effects, will extend to 140 ns.

Without repeating a detailed numerical or graphical analysis of these delayed annihilation phenomena, which may be found elsewhere[43, 67, 73, 74], it may be pointed out that such analysis of the variation of annihilation rate with time reveals most interesting 'resonances' which can readily be associated with e^+ and Ps reactions with the gas medium. Thus Tao[74] has found two distinct resonances in argon gas and two more in argon containing small amounts of Cl_2. The latter only appear as a shoulder on top of a shoulder, where the third component (o-Ps) begins to appear. This component disappears with large concentrations of chlorine or in pure chlorine and the sum of the intensities of the new shoulder and the third component remains constant even though I_3 changes considerably. Hence, the new shoulder must arise from o-Ps and the strong resonance is probably associated with the reaction

$$Ps + Cl_2 \longrightarrow PsCl + Cl$$

which is possible at Ps energies above about 0.5 eV[26]. The four resonances appear at estimated energies of 0.5 and 0.3 eV for Cl_2 and 0.9 and 0.7 eV for A (assuming a constant scattering cross-section for e^+ below the Oré gap threshold for argon). This concept of 'resonance annihilation' is due to Goldanskii and Sayasov[73] and has proved to be most useful.

7.5.4 Selected physical and chemical interactions

Some examples will be chosen from the published literature of the last few years to illustrate progress in positron annihilation studies, particularly those which suggest chemical interactions of Ps and those which shed light on structural and surface properties of solids.

7.5.4.1 Positrons in gases

Recent progress in this area has been slow, possibly because of the near-promise of low-energy mono-energetic positron beams which would enable precise measurement of the all important scattering cross-sections over a range of energies. Theoretical calculations of scattering cross-sections have been intensified since the failure of earlier calculations to match experimental results. Advances were made when the effect of polarisation was included with the adiabatic approximation method[75–77]. Drachman has also calculated the annihilation rate in helium using the same method[78,79]. Some of these results have been reviewed and summarised[80]. Tao and Kelly[81] have made a comparison of experimental results for positron annihilation in He gas at 4.2, 77 and 300 K with results calculated from theoretical values of annihilation and scattering (momentum-transfer) cross-sections. Excellent agreement was found between momentum-transfer cross-sections calculated by the methods of Drachman[77] and Massey[75] (the former fitting better) and the experimental values which range from $0.2\pi a_0^2$ to $0.6\pi a_0^2$ in the range 17.8 to 0.1 eV. Leung and Paul[82] find rather smaller values at the higher energy and much larger (*c.* 3.3 πa_0^2) near 1 eV, as in the experiments of another group[83]. Pressure variations in He gas over the range 20–70 amagat resemble those of Calitans and Green[47] for Ar above 40 amagat, but the annihilation rates at lower pressures are still anomalous. Orth and Jones[84] have calculated the higher value of $33 \pm 5\pi a_0^2$ for the scattering cross-section in argon. In careful experimental Tao[85] has studied the positron annihilation rate in argon gas at pressures from 10 to 65 amagat. He finds that the direct annihilation rate, reciprocal of shoulder width and quenching rate of *o*-Ps are all linearly dependent on the density of Ar in this range. Z_{eff} (the effective number of electrons to be used in determining the annihilation rate from the Dirac equation) is found to be 26.2 up to 40 amagat pressure.

In addition the search for excited states (e.g. 2 $^3S_1$) of Ps still goes on. Urbanovich[86] has made calculations which show that 2^3S_1 Ps can exist in principle in a gas when the pressure is below 10^{-3} Torr and the internal electric field strength of the medium is below 0.1 Vcm^{-1}. Leventhal has made a thorough experimental search for Ps Lyman-α radiation using a 20 Ci source of ^{64}Cu positrons stopping in Ar, Ne and He at pressures from 0.01 to 3.0 amagat. The conclusion is that there is an effective upper limit of 1 positron in 2000 stopped in the gas with resultant emission of Lyman α-radiation[87].

7.5.4.2 *Positrons in solids*

Interesting developments have arisen in recent years in the study of e^+ annihilation in metals, molecular solids, inorganic solids and polymers. Much of the work has depended on the availability of fast timing equipment to determine annihilation rates, in addition to the extensive use of angular correlation techniques.

Angular distributions for more than 30 metals were soon obtained[88,89]. In general, analysis of angular distributions gives information on the band structure and Fermi energies of the electrons and measurements of the short lifetimes has very recently proved most instructive in the study of metal defects and fatigue. Analysis of angular distribution curves gives the momentum space density and the momentum distribution, N($\boldsymbol{P}$), of the electrons with which e^+ annihilates. The experimental curves obtained are illustrated in Figure 7.5 and the value of N($\boldsymbol{P}$) when plotted against θ should show a sharp fall to zero at the Fermi energy. There are generally three categories of behaviour: Ca, Sr, In and Group IA metals show a narrow peak, a sharp fall at the Fermi cut-off and low values of N(***P***) above this point. A second category includes Ba, Hg, Fe and Group IB elements and gives rise to a steep increase in N(***P***) on the low momentum side of the Fermi cut-off, a peak near it and a very slow decline at higher momenta. The third type of momentum distribution is broad and nearly symmetrical with a peak slightly below the Fermi energy position and is found for Be, B, Al, C and Sb.

Annihilation of positrons occurs principally with conduction electrons when the electronic configuration is relatively simple, as in the alkali metals, 'core' electrons in the inner shells are scarcely involved and an almost ideal momentum distribution is found in accordance with theory assuming an ideal Fermi electron gas. Interactions with inner shell electrons are indicated by a spread to a higher momentum tail in the distribution. In metals with a complex electronic structure, e.g. transition metals, inner shell electrons[90] and excluded volume in the metal lattice[91] are important factors. These factors cause an increase in the higher momentum components of the angular distribution curves but there is still no firm agreement on their interpretation. A comparison of results for Cu, for example, shows discrepancies between different experimental methods – point detectors[92], rectangular slits[93, 94], and wide slits[95, 96] – but indicates the possible nature of the 4s and 3d wave functions of metallic copper[97]. Other contributions to higher momentum tails may arise from electron-hole excitation caused by the stopping of positrons in the metal[98], annihilation before thermalisation of the positrons[99] and an abnormally high effective mass[100] of the positron due to association with many electrons.

A great deal of work has been done to study the shape of the Fermi surface in metals by orienting single crystals of the sample at different angles to the angular correlation spectrometer. To get useful results careful experiments with spectrometers employing point slits or very high resolution long slits are required. Much of this research has been reviewed[94] and some examples may be given: Be, Bi, Zn, Ho, Er and Y show anisotropic orientation changes in the angular distribution. Na, Li and Mg give negligible anisotropy and are therefore considered to have a nearly spherical Fermi surface. In Cu there is a large change in angular distribution as the orientation changes and the Fermi surface is not spherical.

As the temperature of the metal varies, positron annihilation angular distributions change, especially at the melting point, primarily because of a decrease in the mean free path of electrons as the temperature increases[101]. Various conditions of metals may be studied by this technique such as plastic deformation in Pt[102], stacking faults, impurity centres, dislocations, effects of magnetic fields[103], the polarisation of electronic bands in ferromagnets[104, 105] and K-space states in superconductors such as Nb_3Sn [106].

The study of positron lifetimes in metals has been complicated by artefacts of sample preparation[107, 108], which introduce oxide layers or adsorbed gases. However there have been some very interesting recent studies of the short lifetimes of bound states of positrons in defect sites in metals. If there is space for it, Ps will be formed and will have a spin-averaged lifetime of 0.5 ns [109]. The physical properties of explosively shocked nickel have been studied[110] in this way, trapping of positrons in dislocations in plastically deformed aluminium single-crystals were revealed by a long lifetime[111] (172 ps in a perfect crystal, 228 ps in a dislocation) and temperature dependence studies[112] have shown trapping of positrons in defects. Technical applications of these advances include the study of early fatigue effects in metals and alloys.

Organic solids have also been studied by the positron lifetime and angular correlation techniques. There are generally two lifetime components; the lifetime and intensity of the longer component are usually less than in the liquid phase. τ_2 is usually less than 2 ns in the solid but may be 1–4 ns in the liquid. τ_1 is about 0.33 ns in the solid and about 0.40 ns in the liquid.

The fate of the positron in these organic solids is particularly dependent on phase transitions and the presence of impurities. Structural changes in the solid are indicated by very marked changes in e^+ annihilation. Thus, in the liquid crystal, cholesteryl myristate, the lifetime of the long component increases markedly at the solid-to-smectic transition, 71 °C, and again at the smectic-to-cholesteric transition, 81 °C[113]. In solid cyclohexane the long lifetime changes at 110 K, probably due to a transition from monoclinic to cubic structure[114]. In very pure, zone-refined anthracene there is no long positron lifetime, but in lower grade and neutron irradiated samples, the trapping centres lead to a long lifetime about 1 ns and 3 % intensity[115]. When the sample is melted, a long lifetime of about 35 % intensity is found[116]. It appears then that the behaviour of positrons in organic substances is very dependent on chemical impurities, radiation damage, crystalline order and phase. Further studies leading to more quantitative correlations are needed. A recent study of Chuang and Tao[117] of temperature and phase dependence of positron lifetimes in CH_3OH and CD_3OD produced interesting results in an attempt at a comparison with H_2O and D_2O [118]. The long lifetime increases and its intensity decreases on melting of all compounds. In the isotropic solid phase below the solid–solid transition (−117 °C for CH_3OH and −110 °C for CD_3OD) there is no difference in τ_2 and I_2 for the isotopic molecules, but above the transition point both τ_2 and I_2 are larger for CH_3OH than for CD_3OD. Molecular rotation and intermolecular interactions clearly affect the formation and annihilation of Ps in these compounds. In the case of choline chloride, there is a transition from orthorhombic to a disordered cubic structure between 73 °C and 80 °C and the observed long lifetime component, τ_2, increases steeply over this temperature range[119].

Studies of inorganic solids, particularly halides and oxides, have long been beset with uncertainties caused by the method of sample preparation—purity, particle size of powders, adsorbed gases and internal physical structure. It is now well-established that Ps is formed in these solids and lifetimes of 0.4–0.7 ns are customary. Any lifetime greater than this would be called a long lifetime component.

In the alkali halides the short lifetime, τ_1, can be separated into short and intermediate lifetimes, τ_1^0 and τ_1'. In each homologous series, τ_1' decreases with increasing molecular density and depends primarily on the nature of the anion[120]. The two-photon angular distribution also depends primarily on the anion[121], but three-photon annihilation rates do not yield any clear correlations with structure[122]. The effect of physical structure is seen in the long lifetime, 0.8 ns found in single-crystals of NaCl and NaF after exposure to ionising radiation. This component appeared in 28% intensity, but this figure was reduced to about 10% after thermal annealing of the crystal[123]. Annihilation rates have been calculated on the assumption that outer-shell electrons in the halide ions can concentrate round the positron as an 'electron plasma', and the values agree well with experiment[124].

There have been some extensive and rewarding recent studies of solid oxides. A very long positron lifetime of about 50 ns is found in SiO_2 powders possibly due to *o*-Ps annihilation in cavities[125] or in the gas on the surface[126]. The intensity of this component varies with the size of the particles and from this variation a diffusion constant for Ps in the lattice has been estimated to be 6×10^{-5} cm^2 s^{-1} [126]. In single-crystals of quartz, positrons annihilate to give angular correlations indicating the presence of Ps[127, 128]. The magnetic field dependence of the angular distribution indicates the presence of positronium-like Bloch states[129]. Positron annihilation in Al_2O_3 depends on crystallinity, gas adsorbed on the surface (e.g. O_2, NO_2, but not H_2, N_2 and Ar) and the particle size[130]. Lifetime and angular correlation measurements with powdered alkaline earth oxides and baked BeO and CaO–BaO show long lifetimes up to 80 ns (8.5%) in the powders but not in the baked oxides. A narrow component in the angular distribution appeared with BeO, MgO, SiO_2 and Al_2O_3 powders but not in Al_2O_3 and SiO_2 single-crystals[131]. A correlation between Ps formation and the band gap (E_g) in some oxides has been made[132]. For example in B_2O_3 (E_g = 9.0 eV) a very intense Ps lifetime of 1.7 ns is found but none in HgO (E_g = 1.0–1.2 eV). Intense Ps is formed in oxides with $E_g > 6.8$ eV and a criterion for Ps formation was derived: for Ps in the 1S state, $E_g \geqslant 6.8$ eV and for the 2S state, $E_g \geqslant 1.7$ eV. It may be that this is the best medium in which to search for the 2S state and some indirect evidence for its formation in HgO, La_2O_3 and Sb_2O_5 has been given[132]. Positron lifetimes in synthetic zeolites 4A and 13X (crystalline aluminosilicates) have been measured[133]. An advantage here is that the structure is well-defined and, after evacuation, the crystals may have up to 25% void space. The long lifetime was 2–4 ns in crystals containing N_2, O_2, He and CH_4 and about 40 ns in evacuated samples. The results can be interpreted in terms of the 'free volume' theory[134]. When Cl_2 was admitted to the samples, both the short lifetime and the long lifetime were quenched and this behaviour suggests compound formation as observed in gas mixtures[74].

Interesting developments have come recently from the study of solid polymers, both organic and inorganic. Biopolymers have yet to be studied. Recent reviews[7, 8, 135] may be consulted for references to the field. In general, annihilation lifetime spectra of positrons in organic polymers can be resolved into three components, if the resolution of the spectrometer is adequate[136, 137]. Angular distribution curves for two-photon annihilation are narrower and more bell-shaped than for Fermi metals and this, together with the long lifetimes, indicates the formation and decay of Ps. A large number of the commoner polymers have been studied and various properties are reflected in the annihilation patterns. An example of lifetime spectra in Teflon samples, untreated, quenched (44% crystallinity) and annealed (76% crystallinity) is given in Figure 7.8 [136]. Three-component resolution of these data gives the results shown in Table 7.2, where it is seen that the three lifetimes do not

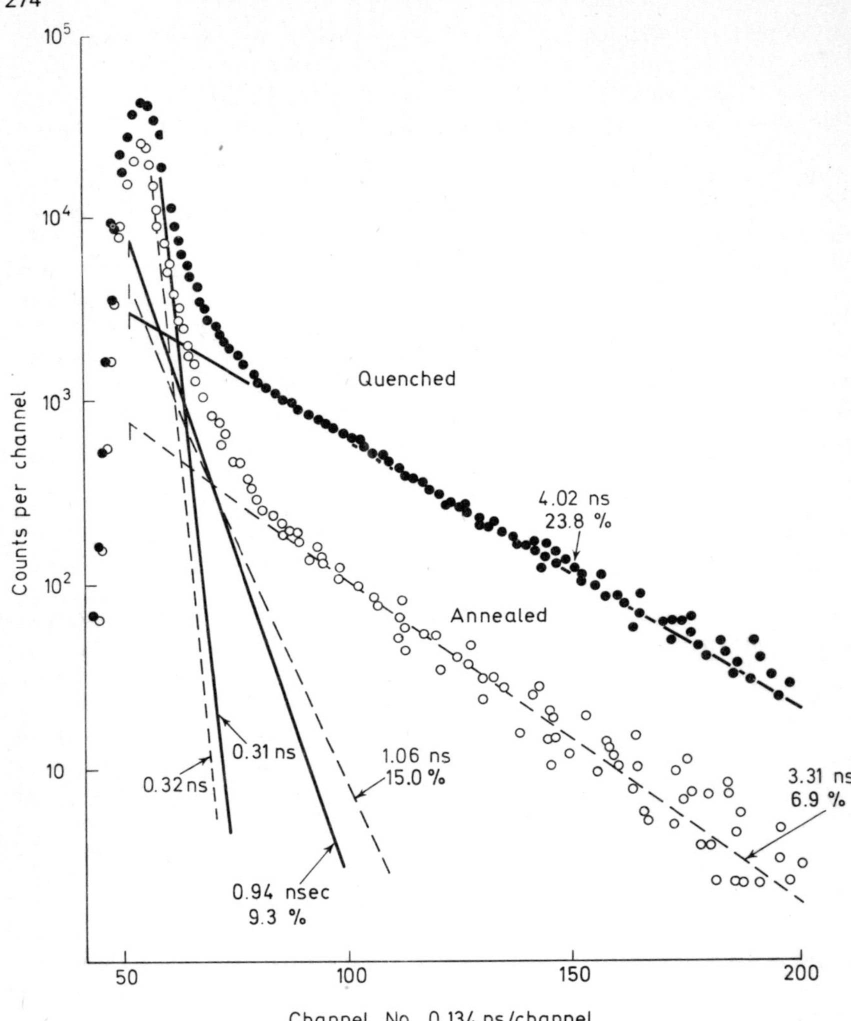

Figure 7.8 Positron annihilation lifetime spectra in Teflon

Table 7.2 Positron annihilation lifetimes and intensities in polymers*

Sample	τ_1, ns	τ_2, ns	I_2, %	τ_3, ns	I_3, %
Teflon					
Quenched, 44 % Crys.	0.32	0.96	9.2	3.95	22.0
Starting material	0.32	0.93	11.4	3.89	16.0
Annealed, 76 % Crys.	0.32	1.03	14.3	3.69	8.0
Polyethylene	0.29	0.70	9.0	2.70	21.5
Polypropylene	0.34	0.84	4.4	2.50	21.4
Paraffin	0.33	0.90	5.5	3.20	19.0
Nylon-6	0.34	0.68	5.5	1.71	15.6

*Ref. 136.

change with crystallinity, but the intensities I_2 and I_3 do. As the crystallinity increases, I_2 increases and I_3 decreases. Thus the long component lifetime (τ_3) arises from annihilation of *o*-Ps in amorphous regions and the second component (τ_2) arises from *o*-Ps interactions in crystalline Teflon. The ratio $I_2/(I_2+I_3)$ is a rough measure of degree of crystallinity and increases in Teflon after neutron- or γ- radiation, while I_3 decreases. Increasing doses of γ- radiation cause an increase in the concentration of free radicals, which eventually quench the long lifetime completely[138]. These changes in crystallinity and their influence on positron annihilation are also seen in the angular distributions, of which an example is shown in Figure 7.9. The sharper peak (more amorphous sample) and the difference between the distributions show that more Ps is formed in the less crystalline sample decaying with a long

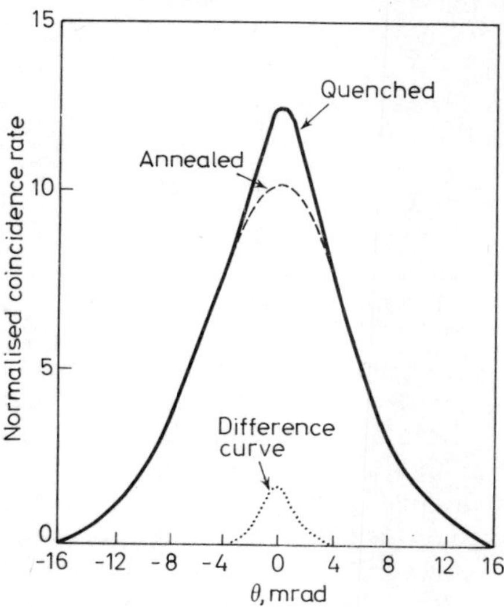

Figure 7.9 Angular distributions for positron annihilation in Teflon

lifetime[139]. Increase in disorder or in free volume as the temperature of the sample is increased is shown by discontinuities in the lifetime-temperature curves and by a general increase in the lifetime[140]. Positron annihilation characteristics in many polymers, including polyvinyl chloride, isotactic and atactic polypropylene, polyacrylonitrile, polystyrene and polyethylene have been compiled[141] and show clearly that pick-off quenching of *o*-Ps depends on the preparation and orientation of the polymer. The lifetime decreases as the molar cohesive density increases in an apparently exponential relation[142]. It seems that the decrease in lifetime is due not only to the effect of microelectric fields caused by permanent dipoles in polar polymers[143], but also to an increase in the total intermolecular forces. The thermal degradation of polyvinyl chloride has recently been carefully studied to attempt to define the mechanism of *o*-Ps quenching[144, 145]. Although capture of *o*-Ps by HCl and Cl to form Ps—Cl in the polymer is an attractive explanation, other mechanisms are possible and further experiments are required. Interesting results were found in studies of *o*-Ps annihilation in polypropylene treated with nitric acid. Selective oxidation of polypropylene with nitric acid has been used to study the morphology of the polymer[146], because the acid removes the amorphous regions quickly and leaves the more crystalline regions behind. As far as the Ps lifetime is concerned, one might expect the microcavities so formed to provide a shelter for *o*-Ps atoms against pick-off quenching by the polymer as a whole. It was in fact found[147] that, after a 3 h treatment with hot 70% HNO_3, the long lifetime increased to as much as 50 ns and it was concluded that this represented annihilation of *o*-Ps which had formed and diffused into the cavities. (Note that the long lifetime found for powdered oxides is believed to be due to formation of *o*-Ps inside the lattice, followed by diffusion *out* of the particle, and annihilation occurs in the space between the particles. The lifetimes observed approach the theoretical for *o*-Ps annihilating in free space, 140 ns.) Taking 50 Å as the rough diameter of the cavity, as determined by x-ray crystallography[146] and using the theoretical approach based on the 'free-volume' theory[126, 134] the diffusion constant for *o*-Ps is found to be 4.5×10^{-4} cm^2 s^{-1} and the potential of the barrier surrounding the *o*-Ps is less than 1 eV. Studies of similar, but more complex, phenomena have been made on the Ps lifetimes in silica gels[148], which have been known and used for a very long time. The physical structure of silica gel can be described as an assembly of closely bonded primary particles with finer particles located within them and larger particles between them[149]. As expected, very long lifetimes were found[150]. In general, the short lifetime component (0.5 ns) is due to annihilation of free positrons and *p*-Ps, the second lifetime (2.0 ns) is attributed to pick-off annihilation of *o*-Ps before diffusion into the pores and the longest lifetime (about 30 ns) is due to annihilation of *o*-Ps in the pores. This study leads to some interesting qualitative and quantitative conclusions, not only about pore sizes but also about quenching of *o*-Ps by adsorbates and free radicals on the surface. For example, the long lifetime would be expected to decrease in silica gel, dried under vacuum at 600 °C, as water or nitromethane is introduced, because these substances would decrease pore size. Actually the lifetime of *o*-Ps increases at first and only decreases after adding quite a lot of liquid. An interpretation is that the surface does not contain silanol Si—OH groups only and that strained surface groups Si—O*—Si are formed, particularly by heating. These may in turn yield free radicals Si—O· and Si· which quench *o*-Ps by chemical reaction[26], e.g.

$$\text{Si—O}\cdot + o\text{-Ps} \longrightarrow \text{Si—OPs}$$

or conversion quenching, e.g.

$$\text{Si—O}\cdot + o\text{-Ps} \longrightarrow \text{Si—O}\cdot + p\text{-Ps}$$

Introduction of a small amount of water or nitromethane removes these free radical sites[151].

7.5.4.3 Positrons in liquids

Positron annihilation in organic liquids has been extensively studied to attempt to analyse constitutive and substituent effects of the organic molecules on o-Ps formation and annihilation. These studies have been thoroughly discussed[3,7] and the last few years have led to a position where there is a lot of data but less than complete agreement on interpretation[7,8,26,152–154]. This is partly because of experiments[59], which showed very clearly that oxygen had to be very thoroughly removed from many organic liquids to prevent paramagnetic quenching of o-Ps. Extensive earlier results, which were obtained without this precaution, lost much significance. Later results[152,153] from two-component lifetime analysis of annihilation data, obtained with about 200 pure liquids, were presented by the authors as evidence for strong correlations between molecular structure and positronium lifetime. For example, in n-alkanes, normal primary alcohols and 1-chloro substituted n-alkanes, τ_2 decreases as chain length increases and becomes nearly constant for higher members of the series. The cross-section for o-Ps quenching is, however, a linear function of the number of carbon atoms in the molecule. From this observation, partial quenching cross-sections attributable to various substituent groups on molecules were calculated and hence empirical molecular annihilation cross-sections were estimated. These agreed with experiment on a large group of compounds to within two standard deviations. An example of correlations with steric effects is shown in Table 7.3, where the

Table 7.3—Annihilation properties of isomeric pentanes

Compound	τ_2 (10^{-9} s)	$(\sigma v)_{Av}$ (10^{-14} cm^3/s)
Pentane	4.12 ± 0.11	4.69 ± 0.13
2-Methylbutane	4.45 ± 0.07	4.36 ± 0.07
2,2-Dimethylpropane	5.03 ± 0.07	4.08 ± 0.06

straight-chain molecule has the shortest o-Ps lifetime and hence the largest quenching cross-section. The effects of substituents on the behaviour of o-Ps had been studied earlier in halogenated propanes and benzenes[61], where I_2 was found to decrease in both series going from Cl to Br to I. No property of the molecule as a whole was involved. This phenomenon was later ascribed[26] to reactions of the type:

$$\text{Ps} + \text{RX} \longrightarrow \text{PsX} + \text{R}$$

Very recently, a different approach to the data has been taken[154]. First, experimental lifetimes data from three laboratories[41,153,154] are compared giving a degree of agreement, which is very gratifying after so many years of discrepancies. Then the authors set out to calculate the effective number of electrons per atom (Z_{eff}) with which o-Ps annihilates, independent of the structural properties of the molecule. Typical values found are: H (0.028 ± 0.006), C (0.234 ± 0.014), N (0.51 ± 0.02), O (0.304 ± 0.020), F (0.10 ± 0.10), P (0.22 ± 0.10), S (0.72 ± 0.03), Cl (0.45 ± 0.02). Although some of these numbers contain large errors, e.g. for F, the o-Ps lifetimes calculated by means of these quantities show a striking agreement with the experimental data as seen in Table 7.4. It appears that one can attribute to each atom in a molecule an effective number of singlet electrons involved in o-Ps annihilation, which is probably independent of the molecule. However this may not be the complete answer unless extensive temperature variation studies are completed and other data, e.g. for isomeric compounds, are collected.

Solutions of a large number of compounds in organic solvents and water have been extensively studied and their properties, as reflected in their

Table 7.4 Ortho-positronium mean life in some molecular liquids

			$\tau_2 (10^{-9}$ s)			
Entry No.	*Sample*	*Empirical formula*	Bisi *et al.*[154]	Lee and Calitans[41]	Gray, Cook and Sturm[153]	*Back calculated*
1	pentane	C_5H_{12}	4.53 ± 0.12	4.44	4.12	4.27
2	hexane	C_6H_{14}	4.01 ± 0.10	3.94	3.94	4.03
3	heptane	C_7H_{16}	3.73 ± 0.09	3.62	3.70	3.87
4	octane	C_8H_{18}		4.02	3.59	3.78
5	decane	$C_{10}H_{22}$		3.42	3.44	3.65
6	benzene	C_6H_6	3.32 ± 0.10	3.24	3.10	3.13
7	toluene	C_7H_8	3.23 ± 0.08	3.18	3.15	3.16
8	xylene	C_8H_{10}	2.98 ± 0.07		3.12	3.14
9	diethyl ether	$C_4H_{10}O$	4.01 ± 0.10	3.95	3.94	3.78
10	methyl alcohol	CH_4O	3.44 ± 0.09	3.38	3.94	3.43
11	ethyl alcohol	C_2H_6O	3.45 ± 0.09	3.40	3.81	3.44
12	butyl alcohol	$C_4G_{10}O$	3.30 ± 0.09	3.22	3.35	3.33
13	amyl alcohol	$C_5H_{12}O$	3.19 ± 0.08		3.40	3.31
14	octyl alcohol	$C_8H_{18}O$	3.18 ± 0.08	3.15	3.19	3.26
15	acetone	C_3H_6O	3.40 ± 0.09	3.32	3.66	3.45
16	ethylene glycol	$C_2H_6O_2$	2.49 ± 0.06	2.43		2.48
17	glycerol	$C_3H_8O_3$	2.19 ± 0.06	2.14		2.20
18	tetrahydrofuran	C_4H_8O		3.00	2.97	3.07
19	aniline	C_6H_7N	2.36 ± 0.06		2.41	
20	pyridine	C_5H_5N	2.70 ± 0.07	2.48	2.82	
21	chlorobenzene	C_6H_5Cl	2.88 ± 0.07		2.83	
22	carbon disulphide	CS_2	2.25 ± 0.07		2.08	

effects on positron annihilation, have been discussed in extensive reviews[3, 4, 7, 8]. As pointed out earlier in this article, most of the interest probably lies in diamagnetic salts rather than paramagnetic species. Paramagnetic ions of Fe, Co, Ni, Mn and so on cause conversion quenching of *o*-Ps with a consequent shortening of τ_2 and a narrowing in the angular distribution. There is no correlation between the degree of quenching and the number of unpaired spins and, indeed, Sb ions, usually regarded as diamagnetic, quench similarly to paramagnetic ions. (As pointed out some time ago[3], this depends entirely on the structure of the ion in water.) Very recently[155], a new study was made of some very early experiments with solutions of α-α′-diphenyl-8-picryl-hydrazine (DPPH-zine), a diamagnetic derivative of DPPH-zyl, that showed both compounds caused a broadening of the angular distribution. To distinguish between Ps formation inhibition and quenching, the lifetimes and intensities in methanol solutions of these compounds were measured. Both compounds caused no inhibition but strong quenching with rate constants both about 2×10^{10} 1 mol^{-1} s^{-1}. Similarity of rate constants was felt to be due to the three NO_2 groups on an aromatic ring in both molecules and this was confirmed by showing that the quenching rate constant for nitrobenzene in benzene solution was 3.5×10^{10} l mol^{-1} s^{-1}. There was no quenching but inhibition in nitromethane solutions, while neither occurred in nitroglycerine solutions in ethanol. The strong interaction between Ps and these diamagnetic molecules appears to be due to the presence of aromatic nitro groups and might even be the case with paramagnetic DPPH-zyl. This might involve, as suggested for halogenated benzenes[26], a reaction of the type:

$$RNO_2 + Ps \longrightarrow R + PsNO_2$$

which would quench *o*-Ps effectively.

Extensive studies of aqueous solutions of oxyacids have been reported recently[156]. Before the interpretation of the lifetime spectra is discussed it is desirable to outline the basis of a recent semi-quantitative theoretical approach to the chemistry of positronium reactions in liquids proposed by

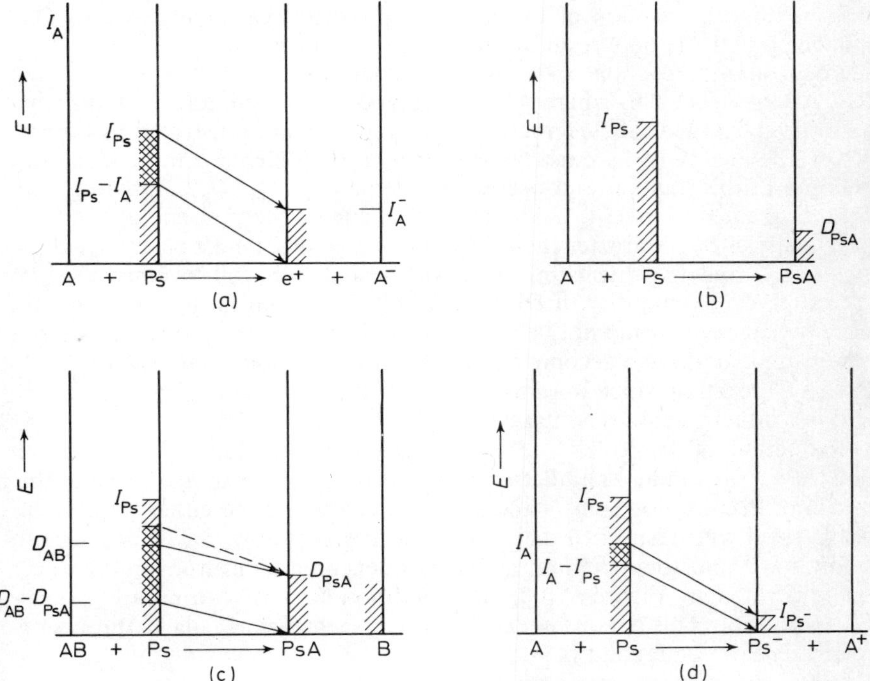

Figure 7.10 Energetics of Ps chemical reactions. *I* is the ionisation energy and *D* is the bond energy. (a) Oxidation by electron transfer, (b) oxidation by compound formation, (c) oxidation by double decomposition, (d) reduction

the same authors[26]. This approach is based primarily on analysis of the shortening of the lifetime of *o*-Ps towards the free annihilation (or *p*-Ps) lifetime. Chemical reaction of Ps may oxidise it to free e^+, bring Ps closer to bound electrons by bond formation, or reduce it to Ps^- ($e^-e^+e^-$) which is stable by only about 0.3 eV. Four types of Ps chemical reactions were considered, by analogy with H reactions, and the energetic conditions for these reactions are shown in Figure 7.10, where the ordinate represents the energy.

Oxidation by electron transfer may be represented by:

$$Ps + A \longrightarrow e^+ + A^-$$

where A is a neutral or charged atom or molecule. As illustrated in Figure 7.10a, if the ionisation potential of A^-, I_{A^-}, is greater than the ionisation potential of Ps, I_{Ps}, the Ps will be oxidised. If I_{A^-} is less than I_{Ps} Ps may be oxidised to e^+ only when it has an energy greater than $I_{Ps^{-1}} - I_{A^{-1}}$. The greater the ionisation potential of A, the greater the probability of reaction and the more *o*-Ps quenching occurs.

Compound formation is represented by: $Ps + A \longrightarrow PsA$. The condition for compound formation is $D_{PsA} > 0$, where D_{PsA} is the Ps—A bond energy. The energetics are depicted in Figure 7.10b. The annihilation of the positron in Ps is greatly enhanced by the proximity of the electron cloud in A. This may cause quenching to a lifetime even shorter than the normal free annihilation lifetime. Interaction of Ps with strong radical scavengers such as O_2, halogen or DPPH may result in this type of reaction.

Positronium reaction by double decomposition is represented by: $Ps + AB \longrightarrow PsA + B$, where AB is a compound or ion, and PsA may be neutral or charged. The energetics of the reaction are illustrated in Figure 7.10c. The energy of Ps must be greater than the difference in dissociation energies of the reactant and product compounds, $D_{AB} - D_{PsA}$. Obviously, if D_{PsA} is greater than D_{AB} oxidation will occur at thermal energies. More practically, if D_{AB} is greater than D_{PsA}, the energy of Ps must be greater than $D_{AB} - D_{PsA}$ and the quenching effect of the reaction will be only to vary the observed Ps intensity. If $D_{AB} - D_{PsA}$ is greater than I_{Ps} no reaction will occur because Ps would not be energetically stable above its binding energy. An example of double decomposition chemical reaction is the $Ps + Cl_2 \longrightarrow PsCl + Cl$ reaction which was indicated in Ps studies in Cl_2– argon mixtures[74] and in synthetic zeolites containing Cl_2 [134].

Reduction of Ps to Ps^- $(e^-e^+e^-)$ which has an ionisation potential of 0.3 eV results in an annihilation rate approaching free annihilation in the medium. Because the spin of the added electron may be either parallel or antiparallel with respect to the positron nearly the same rate of decay considerations hold for reduction chemical quenching as in free annihilation. As illustrated in Figure 7.10d, the condition for Ps^- formation is that $I_A - I_{Ps^{-1}} < I_{Ps}$. This condition exists in many metals, particularly those with relatively low conduction band levels.

When this chemical reaction approach is used to interpret experimental data in liquids from both early and recent research, some interesting and convincing features are observed. Table 7.5 lists values of the ionisation potential, I_{AB}, and bond strength, D_{AB}, for selected molecules and ions and also the bond strength of the hydrogen compounds – the trends of which should be comparable to the trends in the corresponding Ps compounds. The observed values of I_2 for halogenated benzene and propane are as follows: PhF, 24%; PhCl, 14%; PhBr, 6%; PhI, 4%; PrCl, 16%; PrBr, 10% and PrI, 4%. These results cannot be interpreted as due to a variation in the molecule as a whole and point to a direct interaction of Ps at the carbon–halogen bond.

The carbon–halogen bond strength, D_{AB}, and the value of D_{PsA}, depend almost entirely on the halogen, and so the threshold $D_{AB}-D_{PsA}$ depends on the halogen. In the interaction between Ps and the halogenated hydrocarbon the most probable reaction is

$$Ps+RX \longrightarrow PsX+R$$

because the bond between Ps and halogen is stronger than either the Ps—H or the Ps—RX bond. For this reaction D_{PsX} should be generally less than

Table 7.5 Ionisation potentials and bond strengths of selected compounds and ions

Compound or ion	I_{AB} (eV)	*Bond*	D_{AB} (eV)	*Bond*	D_{HA} (eV)	I_2 (%)
H_2O	12.6	H—OH	4.8	—	—	23.0
H_2O_2	12.1	HO—OH	1.5	H—OOH	3.9	4
NH_3	11.0	H—NH_2	4.0	—	—	25
Cl_2	11.3	Cl—Cl	2.5	H—Cl	4.5	0
I_2	9.4	I—I	1.6	H—I	3.1	0
O_2	12.1	O—O	5.2	H—O	4.4	40.0
HF	17.0	H—F	5.9	—	—	17
HCl	12.9	H—Cl	4.7	—	—	23
H_2PO_4	—	O—PO_3H_2—	6.8	H—O	4.4	23
HSO_4	—	O—SO_3H—	4.5	H—O	4.4	18
ClO_4	—	O—ClO_3	—	H—O	4.4	12
NO_3	—	O—NO_2—	2.3	H—O	4.4	1
C_3H_7Cl	10.7	Cl—C_3H_7	3.5	H—Cl	4.7	16
C_3H_7I	9.41	I—C_3H_7	2.4	H—I	3.1	4

D_{RX} and the threshold $D_{RX}-D_{PsX}$ is positive; the quenching consequently will only affect I_2. The quenching effect of the halogen atom in a halogenated hydrocarbon will be in the order: F<Cl<Br<I, because the order of the threshold energies is:

$$(D_{RF}-D_{PsF})>(D_{RCl})-D_{PsCl})>(D_{RBr}-D_{PsBr})>(D_{RI}-D_{PsI})$$

After this digression we may return to consider the four oxyacid–water systems, which were studied over the widest practical range of compositions. Analysis of the lifetime spectra (using dissolved sources of ^{22}NaCl) shows that τ_1 remains constant, τ_2 changes only very slightly, but there are significant changes in I_2 as shown in the experimental points of Figure 7.11 and in the data for the H_2O—H_2SO_4—SO_3 system given in Table 7.6. The first point to be noted is that the curves of I_2 v. mole fraction of oxyacid extrapolate to 17% at zero m.f. instead of the 23% observed in pure water. This is not due to a change in the Ps formation fraction and, at these low acid concentrations, reaction with H^+ may lead to oxidation of Ps by electron transfer:

$$Ps+H^+(aq.) \longrightarrow Ps^+ +H(aq.)$$

As proposed in the theoretical analysis above, and if the excitation energy is negligible, the Ps threshold energy for this reaction is $I_{Ps}-I_H$ where I_H is the ionisation potential of H(aq.) and I_{Ps} is the ionisation potential of Ps. The reduction in I_2 below the value for pure water is 6%, about one quarter,

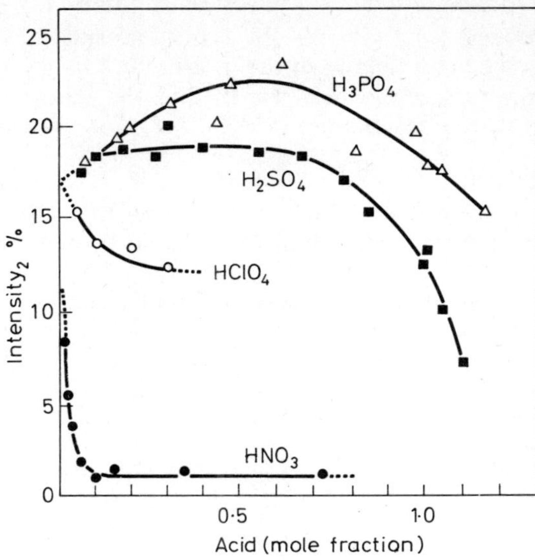

Figure 7.11 Intensities of the second component in lifetime spectra for positron annihilation in some oxyacid-water systems

Table 7.6 The annihilation lifetimes and intensities of the long component for positron annihilation in H_2SO_4–H_2O and H_2SO_4–SO_3 systems

Mole fraction	τ_1(ns)	τ_2(ns)	I_2, %
	H_2SO_4—H_2O *system*		
0.060 H_2SO_4	0.43 ± 0.02	1.82 ± 0.09	17.3 ± 1.0
0.100	0.48 ± 0.02	1.97 ± 0.10	18.2 ± 0.8
0.162	0.41 ± 0.02	1.87 ± 0.09	18.6 ± 0.8
0.216	0.44 ± 0.02	2.09 ± 0.10	18.4 ± 0.8
0.310	0.40 ± 0.02	2.14 ± 0.11	20.2 ± 1.0
0.398	0.42 ± 0.02	2.22 ± 0.11	19.0 ± 0.8
0.545	0.43 ± 0.02	2.38 ± 0.12	18.5 ± 0.8
0.646	0.46 ± 0.02	2.58 ± 0.12	18.5 ± 0.8
0.777	0.42 ± 0.02	2.64 ± 0.12	17.3 ± 0.8
0.85	0.46 ± 0.02	2.90 ± 0.12	15.4 ± 1.0
1.00	0.45 ± 0.02	3.00 ± 0.13	12.6 ± 1.0
	H_2SO_4—SO_3 *system*		
0.01 SO_3	0.46 ± 0.02	2.88 ± 0.13	13.4 ± 1.0
0.04	0.43 ± 0.02	2.69 ± 0.15	9.9 ± 1.0
0.09	0.44 ± 0.02	2.58 ± 0.20	7.2 ± 1.0

indicating that one quarter of the Ps atoms are probably oxidised in this way. Clearly I_{Ps} is much greater than I_H and the rate of the oxidation reaction is very fast.

Figure 7.11 shows that the intensity of τ_2 is nearly constant from acid mole fraction 0.3–0.7. For $HClO_4$, I_2 becomes constant at about 0.5 m.f. The values of I_2 are 22%, 18.5%, 12% and 1% for H_3PO_4, H_2SO_4, $HClO_4$ and HNO_3, respectively. At these concentrations, the acids are in their first dissociated form and it was proposed that they probably react with Ps as follows:

$$Ps + H_2PO_4^- \longrightarrow PsO + H_2PO_3^-$$
$$Ps + HSO_4^- \longrightarrow PsO + HSO_3^-$$
$$Ps + ClO_4^- \longrightarrow PsO + NO_2^-$$

These are double-decomposition reactions which assume the formation of the compound PsO to explain the experimental evidence. (This reaction with NO_3^- differs from the explanation given earlier[157] in that the drop in I_2 is due to Ps formation inhibition because of formation of $e^+NO_3^-$.)

The threshold energy possessed by Ps before these reactions can occur (61) is $D_{AB} - D_{PsO}$ where D_{AB} is the dissociation energy of AB ($H_2PO_3 - O^-$, $HSO_3 - O^-$ etc.) and D_{PsO} is the dissociation energy of the compound PsO. Values of D_{AB} for $H_2PO_4^-$, HSO_4^- and NO_3^- are known (Table 7.5) and the values of D_{PsO} can be obtained from the extrapolation plot of D_{AB} versus the values of I_2 given above. As I_2 approaches zero the extrapolation gives 2.2 eV. This is the bond energy of PsO if the activation energy for the reaction is ignored. From the same graph the unknown bond energy of $(ClO_3—O)^-$ has been estimated to be 3.3 eV.

Similar arguments indicate bond energies of the following Ps compounds, with a max. estimated error of 1 eV:PsOH (1.5 eV), PsF(2.9 eV), PsCl (2.0 eV).

There is a great deal of fascinating chemical information which may be gleaned from positronium studies. However, most of the current studies are directed toward defining Ps reactions rather than using Ps as indicator of the liquid chemical environment and this account can only indicate some of the possibilities. Extensive studies of temperature dependence of I_2 and τ_2 for positrons in various solutions have been made[158–160] and the concept of tunnelling of the electron from Ps to the acceptor has been suggested and explored[161,162]. There are several recent reviews of this subject[4,7,8].

7.6 MESONIC ATOMS

Although there is a high expectation of exciting contributions in the study of the chemistry of mesonic atoms, particularly muonium, the fact is that few people are in a position to carry out such studies. The development of 'meson factories' like those at Zurich ETH and Los Alamos, New Mexico, gives hope for much interesting work with adequate fluxes of mesons. For the present we can discuss the principal features of the transient atoms formed in meson interactions and indicate the nature of chemical processes, which can be explored.

7.6.1 Types and physical properties

Since most studies have involved μ- and π-meson interactions, these mesons will be emphasised. Atomic systems containing mesons were first considered in a classical experiment[163] on the capture of μ-mesons by nuclei. The theoretical basis for the formation of mesonic atoms was established about the same time[164]. Most of the subsequent work has been done with μ^-- and μ^+-mesons which are apparently heavy electrons with masses about 207 times that of the electron. These muons (unlike the π mesons) play no direct part in the strong interactions involved in nuclear binding in complex nuclei and are therefore the most concerned in atomic reactions.

In a mesonic atom the energy levels of the meson are similar to those of an electron in an atom and are described by the same quantum numbers (n,l). The same quantum number, j, for total angular momentum applies to muonic atoms and the individual levels are doublets, because the muon has spin angular momentum, $\frac{1}{2}h$, as for the electron. However in pionic and kaionic atoms, where one electron is replaced by a π^-- or K^--meson (neither possessing spin) the mesonic atom energy levels are singlets and $j = 1$. The energy levels of a mesonic atom lie much deeper than for a normal atom, as a consequence of the large reduced mass ratio μ_m/μ, and are related by

$$E(n,l) = (\mu_m/\mu)E_0(n,l)$$

The mean distance r_0 of a meson from the nucleus in the lowest state is smaller than that of the most tightly bound electron in a normal atom and is given by

$$r_0 = \frac{a_0}{Z} \cdot \frac{\mu}{\mu_m}$$

where a_0 is the Bohr radius (about 5×10^{-11} m) and Z is the atomic number. For muonic silver this means that r_0 is about 5×10^{-15} m and the muon energy states are therefore much influenced by the distribution of electric charge in the nucleus. The muon bond in a muonic atom has therefore been used extensively as a probe of nuclear charge distribution.

The simplest muonic atom is muonic hydrogen where μ^- has replaced the single electron of the hydrogen atom after an intense beam of μ^- (formed by decay of π^--mesons from a high-energy accelerator) is sent into liquid hydrogen. The μp atom in its ground state is a compact structure of radius about 3×10^{-13} m. (It is worth noting that, because of this compactness, nuclear fusion was observed some years ago[165] as a result of the formation of mesonic hydrogen and deuterium. The following reactions of the p and d atoms occurred:

$$\mu\mathrm{p} + \mathrm{D} \longrightarrow \mu\mathrm{d} + \mathrm{H}$$
$$\mu\mathrm{d} + \mathrm{H} \longrightarrow (\mathrm{p}\mu\mathrm{d})^+ + \mathrm{e}$$

Because of the high mass of the muon, the actual separation of p and d in $(\mathrm{p},\mu\mathrm{d})^+$ is only about 2×10^{-13} m and tunnelling of the two nuclei together can occur:

$$\mathrm{p} + \mathrm{d} \longrightarrow {}^3\mathrm{He} + \gamma + 5.5\ \mathrm{MeV}$$

Unfortunately, to obtain a net gain of energy, the μ^- lifetime would have to be about 100 times greater than it is and dreams of fusion power from this process faded.)

A large number of more complex muonic atoms has been studied by bringing μ^- to rest in a suitable target. The muonic atom of the element with atomic number Z will then have, usually, an outer orbital electron replaced by μ^- and $Z-1$ electrons. De-excitation of the atom so formed results in the emission of muonic x-rays, whose energies have been extensively measured to give information on the shape and charge distribution of the nucleus.

Negatively charged π^- and K-mesons, with masses 273 and 966 times that of the electron, also form mesonic atoms and emit x-rays during de-excitation. However these x-rays have much less structure than muonic x-rays because these mesons have no spin.

When a positive meson picks up an electron it forms a structure similar in many ways to a hydrogen atom. Positronium is the best-known example, but muonium has been studied for some years[166, 167]. Muonium can be regarded as a light isotope of hydrogen (positronium being the lightest) and collisions and chemical reactions with regular atoms and molecules can be studied by depolarisation effects. The energy levels for muonium are the same as for hydrogen except for the different reduced mass factor. The Rydberg constant is $R_M = 109{,}209.12 \pm 0.02\ \mathrm{cm}^{-1}$ (for hydrogen $R_H = 109{,}677.58 \pm 0.01\ \mathrm{cm}^{-1}$) and the discrete energy levels are given by $E_n = -R_M/n^2$, where n is the principal quantum number. The ground state of muonium has a hyperfine structure interval which has been measured to a high precision by observing magnetic resonance transitions in strong and weak magnetic fields or by a field-independent Zeemann transition[168, 169] ($\Delta\nu_0 = 4463.317 \pm 0.21$ MHz). The ground-state binding energy is 13.539 eV.

Muonium is unstable because of the decay of the muon, which has a mean life of 2.2×10^{-6} s,

$$\mu^+ \longrightarrow e^+ + \nu_e + \bar{\nu}_\mu$$

The momentum and asymmetry spectrum of the positrons can be calculated with high accuracy; it extends up to a total energy of 52.8 MeV.

7.6.2 Sources and physical measurement techniques

An excellent account of the experimental procedures and data analysis is given in a recent paper on the formation and Larmor precession of muonium[169]. A positive ion decaying at rest ($\pi^+ \longrightarrow \mu^+ + \nu_\mu$) gives a positive muon with spin aligned in a direction opposite to its linear momentum. The decay of the positive muon ($\mu^+ - e^+ \nu_e + \nu_{\bar{\mu}}$) occurs with angular asymmetry so that positron emission in the direction of the muon spin is favoured. Thus polarised muonium can be formed with polarised muons and the effects of collisions and electromagnetic field transitions on the muonium can be studied by observing the changing angular distribution of the positrons resulting from a change in muonium polarisation.

A muon beam was produced from a 380 MeV proton beam falling on an internal target in a synchrocyclotron. The beam entering the gas target in which muonium was formed consisted of 80% muons and 20% pions. The target itself was a high-pressure steel cylinder 9 in diameter and 12 in long containing argon up to 900 lb in^2 and surrounded by Helmholtz coils to provide fields up to ± 17 kG. Detectors were plastic scintillation phosphors with 6810 Å photomultipliers in a coincidence and anticoincidence array.

For details the paper mentioned should be consulted. The positron count is observed as a function of time, when a magnetic field is applied.

7.6.3 Chemical interactions: muonium chemistry

The depolarisation measurement, briefly outlined above, is one method of observing interactions of muonium with atoms and molecules and gives a measure of the depolarisation cross-section for 'impurity' molecules admitted to the argon gas, in which muonium is formed. Another related method depends on measurement of the intensity and width changes in the microwave signal for a magnetic resonance transition between two hfs magnetic substates of muonium. These changes occur as a function of the concentration of various gases added to the argon buffer[170, 171].

In these experiments the effect of adding paramagnetic molecules (O_2, NO) in small amounts is to destroy the resonance signal. Larger concentrations of diamagnetic molecules produce the same effect (C_2H_4, C_2H_6, Cl_2, CH_3Cl, CO_2 and H_2O). Cross-sections for these processes have been obtained[170, 171]. In reaction with paramagnetic molecules, the spin exchange with the unpaired electron is responsible for the depolarisation of muonium observed. (Triplet muonium is converted to unpolarised singlet muonium.) With the other molecules, reactions of muonium (M) of the following types are suggested:

$$\mathrm{M+NO_2 \longrightarrow NO+OM}$$
$$\mathrm{M+C_2H_4 \longrightarrow C_2H_4M}$$

Systematic studies have been made on the amplitude of the μ^+ precession signal in magnetic fields for μ^+ stopped in many different hydrocarbons, solids and liquids[172–174]. Reactions of the following type were proposed:

$$\mathrm{M+RH \longrightarrow R+MH}$$
$$\mathrm{M+RH \longrightarrow RMH}$$

The substitution reaction gives a stable molecule with paired electrons half in the form of ortho MH (parallel spins of μ^+ and proton) and half of para (no spin, no precession). The precession properties of polarised free muonium, MH and RMH vary and the instantaneous concentrations can be found by measuring the amplitude of precession at various frequencies and magnetic field strengths (loss of polarised muonium due to spin exchange and decay of μ^+ must be accounted for). Equations have been derived for these precession amplitudes as a function of concentrations, and hence cross-sections for the above reactions were derived. These cross-sections should be directlv related to those for processes involving H atoms and the kinetic rate constants derivable, with due allowance for the mass difference. The total reaction rate between H and C_6H_6 calculated from the muonium data[174] is 5.5×10^8 $1\ \mathrm{mol^{-1}\ s^{-1}}$ in good agreement with the direct chemical kinetic measurement[175], 6×10^8 $1\ \mathrm{mol^{-1}\ s^{-1}}$. These are clear indications of the value to the study of chemical reactions to be found in muonium studies. Other methods for studying rates of μ-molecule formation have been investigated by determining the shift of μ^+ precession frequencies. The difference in the simple halides between a muon and a proton would be about 2 p.p.m.[176]. Metals which should condense in high-pressure systems containing high muon densities have been shown to be in a highly stable form and should exist in some stellar environments[177]. The probabilities of π^- capture by hydrogen in bases, salts and pseudo-hydrides have been shown to decrease in proportion to the number of hydrogen atoms due to quantum-mechanical resonance[178].

References

1. Burop, E. H. S. (1970). *Contemp. Phys.*, **11,** 335
2. Shearear, J. M. and Deutsch, M. (1949). *Phys. Rev.*, **76,** 462
3. Green, J. and Lee, J. (1964). *Positronium Chemistry,* (New York: Academic Press)
4. Stewart, A. T. and Roellig, L. O. (eds.) (1967). *Positron Annihilation* (New York: Academic Press)
5. Green, J. H. (1965). *Proc. Roy. Aust. Chem. Inst.*, 7
6. Green, J. H. (1966). *Endeavour,* **25,** 16
7. Goldanskii, V. I. (1968). *Atomic Energy Rev.*, **VI(1),** 3; Goldanskii, V. I. (1968). *Physical Chemistry of Positrons and Positronium.* (Moscow: Izadatel 'stoov Nauka)
8. Green, J. H. Merrigan, J. E. and Tao, S. J. (1970). in *Physical Methods of Chemistry. Vol. 1 of Techniques in Chemistry.* (eds. Weissberger and Rossiter) (New York: Interscience)
9. Heymann, F. F. (1961). *Endeavour,* **20,** 225
10. Hughes, V. W., McColm, D. W. and Ziock, K. (1970). *Phys. Rev.*, **1,** 595
11. Johnson, W. R., Buss, D. J. and Carroll, C. O. (1964). *Phys. Rev.*, **135A,** 1232
12. Mazaki, H., Nishi, M. and Shimizu, S. (1968). *Phys. Rev.*, **171,** 408
13. Dirac, P. A. M. (1930). *Proc. Cambridge Phil. Soc.*, **26,** 361
14. Ruark, A. E. (1945). *Phys. Rev.*, **68,** 278
15. Pirenne, J. (1947). *Arch. des Sciences Phys. et Nat.*, **29,** 121
16. Berestetzki, V. B. (1949). *Zh. Exp. Teor. Fiz. SSSR,* **19,** 673
17. Ferrell, R. A. (1951). *Phys. Rev.*, **84,** 858
18. Karplus, R. and Klein, A. (1952). *Phys. Rev.*, **87,** 848
19. Deutsch, M. and Brown, S. C. (1952). *Phys. Rev.*, **85,** 1047
20. Weinstein, R., Deutsch, M. and Brown, S. (1955). *Phys. Rev.*, **98,** 223
21. Heymann, F. F., Osmon, P. E., Veit, J. J. and Williams, W. J. (1961). *Proc. Phys. Soc. (London),* **78,** 1038
22. Paul, D. A. L. Private communication.
23. Tao, S. J., Celitans, G. J. and Green, J. H. (1963). *Proc. Phys. Soc. (London),* **81,** 1091
24. Mohr, C. O. (1955). *Proc. Phys. Soc. (London),* **68,** 342
25. Oré, A. (1949). *Univ. i Bergen Arbok, Natuwitenskapp Rekke,* **Nr. 9**
26. Tao, S. J. and Green, J. H. (1968). *J. Chem. Soc. A,* 408
27. Tao, S. J. Bell, J. and Green, J. H. (1964). *Proc. Phys. Soc. (London),* **83,** 453
28. Paul, D. A. L. and Saint-Pierre, L. (1963). *Phys. Rev. Lett.*, **11,** 493
29. Brandt, W., Berks, S. and Walker, W. W. (1960). *Phys. Rev.*, **120,** 1289.
30. Bell, R. E. and Graham, R. L. (1953). *Phys. Rev.*, **90,** 644
31. Ferrell, R. A. (1958). *Phys. Rev.*, **110,** 1355
32. Dixon, W. R. and Trainor, L. E. H. (1955). *Phys. Rev.*, **97,** 733
33. Deutsch, M. and Dulit, E. (1951). *Phys. Rev.*, **84,** 601
34. Bisi, A., Fasana, A. and Zappa, L. (1961). *Phys. Rev.*, **124,** 1487
35. Berko, S. Private communication
36. Tao, S. J. and Green, J. H. (1966). *J. Chem. Phys.*, **44,** 4007
37. Takhar, P. S. (1967). *Phys. Lett.*, **24A,** 169
38. Celitans, G. J. and Green, J. H. (1963). *Proc. Phys. Soc.*, **82,** 1002
39. Brandt, W. and Paulin, R. (1968). *Phys. Rev. Lett.*, **21,** 193
40. Celitans, G. J., Tao, S. J. and Green, J. H. (1964). *Proc. Phys. Soc.*, **83,** 833
41. Lee, J. and Celitans, G. J. (1966). J. Chem. Phys., **44,** 2506
42. Majumdar, C. K. and Rhide, M. G. (1968). *Phys. Rev.*, **169,** 295
43. Tao, S. J. (1968). *Trans. I.E.E.E., Nuclear Science,* **15,** 431
44. Ogata, A., Tao, S. J. and Green, J. H. (1968) *Nucl. Instr. Methods,* **60,** 141; Tao, S. J. (1969). *AEC-NYO-3661-27*

45. Schwarzchild, A. Z. and Warburton, E. K. (1968). *Ann. Rev. Nucl. Sci.*, **18,** 265
46. Pond, T. A. (1954). *Phys. Rev.*, **93,** 478
47. Celitans, G. J. and Green, J. H. (1964). *Proc. Phys. Soc.*, **83,** 823
48. Green, J. H. (1967). Ref. 4, p95
49. Bertolaccini, M., Bussolati, C. and Zappa, L. (1965). *Phys. Rev.*, **139A,** 696
50. Greenberger, A., Mills, A. P., Thompson, A. and Berko, S., (1970). *Phys. Lett,* **32A,** 72
51. Mokrushin, A. D. and Goldanskii, V. I. (1968). *Soviet Phys. JETP.*, **26,** 314
52. Hsu, F. H. H. and Wu, C. S. (1967). *Phys. Rev. Lett.*, **18,** 889
53. Tao, S. J. (1965). *Ph.D. Dissertation,* Univ. of New South Wales
54. Roellig, L. O. (1967). Ref. 4; Roellig, L. O. and Kelly, T. M. (1967). *Phys, Rev. Lett.*, **18,** 387
55. Tao, S. J., Bell, J. and Green, J. H. (1964). *Proc. Phys. Soc.*, **83,** 453
56. Green, J. H. and Tao, S. J. (1963). *J. Chem. Phys.*, **39,** 3160
57. Paul, D. A. L. and Saint-Pierre, L. (1963). *Phys. Rev. Lett.*, **11,** 493
58. Leung, C. Y. and Paul, D. A. L. (1967) in Ref. 4
59. Lee, J. and Celitans, G. J. (1965). *J. Chem. Phys.*, **42,** 437
60. Hatcher, C. R. (1961).*J. Chem. Phys.*, **35,** 2266
61. Kerr, D. P. and Hogg, B. G. (1962). *J. Chem. Phys.*, **36,** 2109
62. Tao, S. J. (1970). *J. Chem. Phys.*, **52,** 752
63. Ormrod, J. H. and Hogg, B. G. (1961). *J. Chem. Phys.*, **34,** 624
64. Green, R. E. and Bell, R. E. (1957). *Can. J. Phys.*, **35,** 398
65. Tao, S. J., Celitans, G. J. and Green, J. H. (1963). *Proc. Phys. Soc.*, **81,** 1091
66. Tao, S. J., Bell, J. and Green, J. H. (1964). *Proc. Phys. Soc.*, **83,** 453
67. Paul, D. A. L. (1964). *Proc. Phys. Soc.*, **84,** 563
68. Falk, W. A. and Jones, G. (1964) *Can. J. Phys.*, **42,** 1751
69. Falk, W. A., Orth, P. H. R. and Jones, G. (1965) *Phys. Rev. Lett.*, **14,** 447
70. Osmon, P. E. (1965). *Phys. Rev.*, **140A,** 8
71. Osmon, P. E. (1965). *Phys. Rev.*, **138B,** 216
72. Tao, S. J. (1969). *Phys. Rev.*, 1970, 1, A, 1257
73. Goldanskii, V. I. and Sayasov, Yu. S. (1964). *Phys. Rev. Lett.*, **13,** 300
74. Tao, S. J. (1965). *Phys. Rev. Lett.*, **14,** 935
75. Massey, H. S. W., Lawson, J. and Thompson, G. (1966). *Quantum Theory of Atoms, Molecules, Solid State,* 203 (New York: Academic Press)
76. Drachman, R. J. (1966). *Phys. Rev.*, **138A,** 1582
77. Drachman, R. J. (1966). *Phys. Rev.*, **144,** 25
78. Drachman, R. J. (1966). *Phys. Rev.*, **150,** 10
79. Drachman, R. J. (1968). *Phys. Rev.*, **173,** 190
80. Hogg, B. G., Laidlaw, G. M., Goldanskii, V. I. and Shantarovich, V. P. (1968). *At. Energy Rev.*, **6,** 149
81. Tao, S. J. and Kelly, T. M. (1969). *Phys. Rev.*, **185,** 135
82. Leung, C. Y. and Paul, D. A. L. (1969), *J. Phys. B (London)*, **2,** 1278
83. Lee, G. F., Orth, P. H. R. and Jones, G. (1969). *Phys. Rev. Lett.*, **A28,** 674
84. Orth, P. H. R. and Jones, G. (1969). *Phys. Rev.*, **183,** 16
85. Tao, S. J. (1970). *Phys. Rev. A,* **1,** 1257
86. Urbanovich, S. I. (1969). *JETP (USSR) Engl. Transl.*, **9,** 425
87. Leventhal, M. (1970). *Proc. Nat. Acad. Sci. U.S.*, **66,** 6
88. Lang, G., DeBenedetti, S. and Smoluchowski, R. (1955). *Phys. Rev.*, **99,** 596
89. Stewart, A. T. (1957). *Can. J. Phys.*, **35,** 168
90. Berko, S. and Plaskett, J. B. (1958). *Phys. Rev.*, **112,** 1877
91. Ferrell, R. A. (1956). *Rev. Mod. Phys.*, **28,** 308
92. Williams, D. L., Becker, E. H., Petijevich, P. and Jones, G. (1968). *Phys. Rev. Lett.*, **20,** 448
93. Sueoka, O. (1967). *J. Phys. Soc. Jap.*, **23,** 1246
94. Fujiwara, K. and Sueoka, O. (1966). *J. Phys. Soc. Jap.*, **21,** 1947
95. Berko, S., Cushner, S. and Erskine, J. C. (1968). *Phys. Lett.*, **27A,** 668
96. Mijnarends, P. E. (1969). *Phys. Rev.*, **178,** 622
97. Melngailis, J. (1970) *Phys. Rev. B,* **2,** 563
98. Sedov, V. L. (1967). *Fiz. Tverd. Tela.*, **9,** 1957
99. Carbotte, J. P. and Arora, H. L. (1967). *Can. J. Phys.*, **45,** 387
100. Stewart, A. T., Shand, J. B. and Kim, S. M. (1966). *Proc. Phys. Soc.*, **88,** 1001

101. Stewart, A. T., Kasmiss, J. H. and March, R. H. (1963). *Phys. Rev.*, **132**, 495
102. Kusmiss, J. H., Esseltine, C. D., Snead, C. D. and Goland, A. N. (1970). *Phys. Lett. A*, **32**, 175
103. Gustafson, D. R. and Barnes, G. T. (1967). *Phys. Rev. Lett.*, **18**, 3
104. Berko, S. and Zuckerman, J. (1964). *Phys. Rev. Lett.*, **13**, 339a
105. Dekhtyar, I. Ya., Mikhalenkov, V. S. and Sakharova, S. G. (1967). *Dokl. Akad. Nauk SSSR*, **174**, 803
106. Faraci, G. and Spadoni, M. (1969). *Phys. Rev. Lett.*, **22**, 928
107. Kugel, H. W., Funk, E. G. and Mihelich, J. W. (1966). *Phys. Lett.*, **20**, 364
108. Weisberg, H. and Berko, S. (1967). *Phys. Rev.*, **154**, 249
109. Garg, J. C. and Saraf, B. L. (1969). *J. Phys. Soc. Jap.*, **27**, 1697
110. Holt, W. H., Rose, M. F. and Chuang, S. Y. (1970). *Bull. Amer. Phys. Soc.*, **15**, 576
111. Hautojaervi, P., Tamminen, A. and Jauho, P. (1970). *Phys. Rev. Lett.*, **24**, 459
112. Connors, D. C. and West, R. N. (1969). *Phys. Lett.*, **30A**, 24
113. Cole, G. D. and Walker, W. W. (1965). *J. Chem. Phys.*, **62**, 1692
114. Clark, H. C. and Hogg, B. G. (1962). *J. Chem. Phys.* **37**, 1898
115. Tao, S. J. and Lee, J. (1967). *Bull. Amer. Phys. Soc.*, **12**, 534
116. Cottini, C., Fabri, G., Gatti, E. and Garmagnoli, E. (1960). *J. Phys. Chem. Solids*, **17**, 65
117. Chuang, S. Y. and Tao, S. J. (1970). *Phys. Lett.*, **33A**, 56
118. Petersen, K., Eldrup, M. and Trumpy, G. (1970). *Phys. Lett.*, **31A**, 109
119. Wang, C. and Ache, H. J. (1970). *J. Chem. Phys.*, **52**, 5492
120. Bisi, A., Fiorentini, A. and Zappa, L. (1964). *Phys. Rev.*, **134A**, 328
121. Stewart, A. T. and Pope, N. K. (1960). *Phys. Rev.*, **120**, 2033
122. Bisi, A., Bussolati, C., Cova, S. and Zappa, L. (1966). *Phys. Rev.*, **141**, 348
123. Williams, T. and Ache, H. J. (1969). *J. Chem. Phys.*, **51**, 3536
124. Garg, J. C. and Saraf, B. L. (1970). *Chem. Phys. Lett.*, **5**, 591
125. Paulin, R. and Ambrosini, G. (1966). *C. R. Acad. Sci. Paris*, **263**, 207
126. Brandt, W. and Paulin, R. (1968). *Phys. Rev. Lett.*, **21**, 193
127. Brandt, W., Coressot, G. and Paulin, R. (1969). *Phys. Rev. Lett.*, **23**, 522
128. Stewart, A. T., Hodges, C. H., McKee, B. T. A. and Triftshauser, W. (1970). *Bull. Amer. Phys. Soc.*, **15**, 811
129. Greenberger, A., Mills, A. P., Thompson, A and Berko, S. (1970). *Phys. Lett.*, **32A**, 72
130. Paulin, R. and Ambrosini, G. (1968). *J. de Phys.*, **26**, 253
131. Bartenev, G. M., Varisov, A. Z., Goldanskii, V. I., Levin, B. M., Mokrushin, A. D. and Tsyganov, A. D. (1970). *Sov. Phys. Solid State (Engl. Transl.)*, **11**, 2575
132. Goldanskii, V. I., Levin, B. M. and Mokrushin, A. D. (1970). *JETP. Lett. (USSR) (Engl. Transl.)*, **11**, 23
133. Perkal, M. B. and Walters, W. B. (1970). *J. Chem. Phys.*, **53**, 190
134. Brandt, W. and Fahs, J. H. (1970). *Phys. Rev.*, **B2**, 1425; (ref. also 1960 *Phys. Rev.*, **120**, 1289 and 1966 *Phys. Rev.*, **142**, 231)
135. Khan, M. N. G. A. (1969). *Proc. Roy. Aust. Chem. Inst.*, **36**, 246
136. Tao, S. J. and Green, J. H. (1965). *Proc. Phys. Soc.*, **85**, 463
137. Green, J. H. and Tao, S. J. (1965). *Brit. J. Appl. Phys.*, **16**, 981
138. Chandra, G., Kulkarni, V. G., Lagu, R. G., Patankar, A. V. G. and Thosar, B. V. (1965). *Phys. Lett.*, **16**, 40
139. MacKenzie, I. K. and McKee, B. T. A. (1966). *Can. J. Phys.*, **44**, 435
140. Brandt, W. and Spirn, I. (1966). *Phys. Rev.*, **142**, 231
141. Khan, M. N. G. A. (1968). *Ph. D. Thesis*, University of New South Wales
142. Khan, M. N. G. A., Carswell, D. J. and Bell, J. (1969). *Phys. Lett.*, **29A**, 237
143. Bisi, A., Gatti, A. and Zappa, L. (1961). *Nuovo Cimento*, **22**, 266
144. Khan, M. N. G. A. (1969). *J. Chem. Phys.*, **50**, 3639
145. Khan, M. N. G. A. (1970). *J. Phys. D. Appl. Phys.*, **3**, 663
146. Hock, C. W. (1966). *J. Polymer Sci.*, **A4**, 227
147. Ogata, A. and Tao, S. J. (1970). *J. Appl. Phys.*, **41**, 4261
148. Chuang, S. Y. and Tao, S. J. (1970). *J. Chem. Phys.*, **52**, 749
149. Acher, E. G. (1970). *J. Coll. Interface Sci.*, **32**, 41
150. Chuang, S. Y. and Tao, S. J. (1969). *Bull. Amer. Phys. Soc.*, **14**, 89
151. Chuang, S. Y. and Tao, S. J. (1971). To be published.
152. Gray, P. R., Sturm, G. P. and Cook, C. F. (1967). *J. Chem. Phys.*, **46**, 3487

153. Gray, P, R., Cook, C. F. and Sturm, G. P. (1968). *J. Chem. Phys.,* **48,** 1145
154. Bisi, A., Gambarini, G. and Zappa, L. (1970). *Nuovo Cimento,* **67B,** 75
155. Goldanskii, V. I., Mogensen, O. E. and Shantarovich, V. P. (1970). *Phys. Lett.,* **32A,** 98
156. Tao, S. J. and Green, J. H. (1969). *J. Phys. Chem.,* **73,** 882
157. Green, R. E. and Bell, R. E. (1957). *Can. J. Phys.,* **35,** 398
158. McGervey, J. D., Horstman, H. and DeBenedetti, S. (1961). *Phys. Rev.,* **124,** 1113
159. Jackson, J. E. and McGervey, J. D. (1963). *J. Chem. Phys.,* **38,** 300
160. Horstman, H. (1965). *J. Inorg, Nucl. Chem.,* **27,** 1191
161. Goldanskii, V. I., Firsov, V. G. and Shantarovich, V. P. (1964). *Dokl. Akad. Nauk SSSR,* **155,** 636
162. Bertolaccini, M., Bisi, A. and Zappa, L. (1966). *Nuovo Cimento,* **46,** 237
163. Conversi, M., Pancini, E. and Piccioni, O. (1947). *Phys. Rev.,* **71,** 209
164. Fermi, E. and Teller, E. (1947). *Phys. Rev.,* **72,** 399
165. Alvarez, L. W., Braidner, H., Crawford, F. S., Crawford, J. A., Falk-Vairant, P., Good, N. L., Gow, J. D., Rosenfeld, A. H., Solmitz, F. T., Stevenson, M. L., Ticho, H. K. and Tripp, R. D., (1957). *Phys. Rev.,* **105,** 1127
166. Hughes, V. W. (1966). *Ann. Revs. Nucl. Sci.,* **16,** 445
167. Mobley, R. M., Bailey, J. M., Cleland, W. E., Hughes, V. W. and Rothberg, J. E. (1965). *Proc. Fourth Intl. Conf. Physics of Electronic and Atomic Collisions,* 194, Science Bookcrafters: (Ed. Kerwin, L. and Fite, W.) (Hastings-on-Hudson, New York)
168. Ehrlich, R. D., Hofer, H., Magnon, A., Stowel, D., Swanson, R. A. and Telegdi, V. L. (1969). *Phys. Rev. Lett.,* **23,** 513
169. Hughes, V. W., McColm, D. W., Ziock, K. and Prepost, R. (1970). *Phys. Rev.,* **1,** 595
170. Mobley, R. M., Bailey, J. M., Cleland, W. E., Hughes, V. W., and Rothberg, J. E. (1966). *J. Chem. Phys.,* **44,** 4354
171. Mobley, R. M., Amato, J. J., Hughes, V. W., Rothberg, J. E. and Thompson, P.A. (1967). *J. Chem. Phys.,* **47,** 3074
172. Firsov, V. G. and Byakov, V. M. (1965) *Soviet Phys. JETP,* **20,** 719
173. Myasishcheva, G. G., Obukhov, Yu. V., Roganov, V. S. and Firsov, V. G. (1968). *Soviet Phys. JETP,* **26,** 298
174. Babaev, A. J., Balats, M. Ya., Myasishcheva, G. G., Obukhov, Yu. V., Roganov, V. S. and Firsov, V. G. (1966). *Soviet Phys. JETP,* **23,** 583
175. Melville, H. W. and Robb, J. C. (1950) *Proc. Roy. Soc. A,* **202,** 181
176. Breskman, D. and Kanofsky, A. (1970). *Nuovo Cimento,* **68B,** 147
177. McKenna, P. and Pastine, D. J. (1969). *Phys. Rev. Lett.,* **23,** 1508
178. Krumshtein, Z. V., Petrukhin, V. I., Ponomarev, V. L. and Prokoshkin, Yu. D. (1968). *Zhur. Eksp. Teor. Fiz.,* **55,** 1640

8
High-Energy Nuclear Chemistry

A. C. PAPPAS, J. ALSTAD and E. HAGEBØ
University of Oslo

The Reader should take the attitude of General Banquo in the Witches Scene in Shakespeare's *Macbeth*[1] when reading this review:

'If you can look into the seeds of time,
And say which grain will grow, and which will not,
Speak, then, to me, who neither beg nor fear
Your favours nor your hate.'

8.1 INTRODUCTION

8.1.1 Nuclear chemistry

Our view is that Wheeler[2] has clearly expressed the 'policy' of modern nuclear chemistry:

'The analysis of regularities from nucleus to nucleus, like the analysis of the regularities from molecule to molecule, often provides a better answer and deeper understanding than any attempt at a calculation from first principles.'

The field of nuclear chemistry embraces a broad region not only of nuclear charge (Z) and mass (A), but also of energy scale. Thus today's studies explore many steps on 'the quantum ladder'[3], i.e. they are concerned with energy concentrations reaching from fractions of eV/nucleon to about 100 GeV/nucleon the upper limit depending only on the availability of high energy accelerators. It should be recalled, that at super-high energies, nuclear matter is under the most 'unusual' conditions as seen from the solar system. Such conditions, however, occur frequently as 'catastrophic' happenings in the universe.

Since the early days of nuclear chemistry the climbing of 'the quantum ladder' has always taken place with a simultaneous expansion within a given level of energy concentration. The latter, which from the point of view of 'the triangle of the natural sciences' should be regarded as a 'one-dimensional' trend, is characteristic and well known from research in chemistry and most other parts of the natural sciences. It is only nuclear science which follows the 'two-dimensional' expansion. Nuclear chemistry has therefore become a large and far-reaching discipline which influences many studies in other fields of science, ranging from those of the smallest microcosmos to those of the largest macrocosmos.

8.1.2 Aims of the review

Both the main basic concepts and the main trends in experimental approaches and methods are essentially the same in low- and high-energy nuclear chemistry and are reviewed by Lefort in Chapter 4 of this volume. Their discussion is therefore omitted from the present chapter, which will treat papers concerned with nuclear reactions and the behaviour of nuclear matter at high energies, i.e. above 100 MeV per incident nucleon.

Nuclear reactions above 100 MeV include fission, spallation and fragmentation, with incident protons, photons and mesons. In this connection the production of nuclides far off the stability line and attempts to reach super-heavy elements by direct or indirect 'use' of high-energy induced reactions will be mentioned.

High-energy nuclear chemistry has introduced new ideas, insight and tools in many other fields of science. Perhaps the largest impact is found in the 'chemistry and physics of the Universe'. Therefore these derivatives, or 'arms of the nuclear chemistry octopus' are included.

It is not possible to do justice to all the papers published in the period under review; nevertheless we hope to give glimpses of the beauty and richness of our field and the present state of our problems.

Since 1958 many reviews of high-energy induced reactions have appeared[4−10]. Spallation and fragmentation, and fission are reviewed in references (11–14) and (15–18), respectively. In addition to these published reviews, fission and spallation, and their application to cosmic ray-phenomena are regularly discussed during the biennial 'Informal European Conferences on the Interactions of High-Energy Particles and Complex Nuclei' and during the 'Gordon Conferences in Nuclear Chemistry', which discuss these phenomena every second year. Proceedings are not published from any of these conferences.

8.2 MODEL OF HIGH-ENERGY NUCLEAR REACTIONS

During the forties, a two-step model for high-energy nuclear reactions in medium heavy nuclei was suggested by Serber[19] and further developed by Goldberger[20]. According to this model, the incident particle collides with a nucleon within the nucleus. The two collision partners may escape or each undergo further collisions and so on. The energy of the struck nucleons, which are unable to escape, will remain as nuclear excitation. This will eventually give rise to evaporation. Thus it is assumed that the reaction starts with a fast nucleonic cascade and is terminated by nuclear de-excitation through slow evaporation of particles. This basic model has remained almost unchanged as a 'code' for our 'understanding' of high-energy induced reactions. The term *spallation*, originally suggested by Sullivan and Seaborg, for *complex* reactions is now also used to include *simple* reactions in which only very few nucleons are emitted as a result of insufficient excitation*.

8.2.1 The cascade step

In the reactions under consideration the incident particle energy is very much higher than the interaction energy between individual nucleons in nuclear matter. Furthermore, the reduced de Broglie wavelengths of energetic nucleons are less than the average nucleon–nucleon distance of about 2.5 F in nuclear matter. (For a 250 MeV nucleon this wavelength is about 0.3 F.) It is therefore assumed that elastic collisions take place within the nuclear boundary as two-body interactions. Except for the limitation due to the Pauli exclusion principle, and that the struck nucleon has an initial momentum, the collisions will be the same as between two free nucleons.

This somewhat over-simplified picture allows the use of the 'impulse approximation'[21], i.e. elementary particle scattering cross-sections and angular distributions at the appropriate energy. (Cross-sections for like nucleons are much smaller than for unlike nucleons[22].)

8.2.1.1 The intra-nuclear cascade

The intra-nuclear cascade has a time-scale, determined by the time it takes a fast nucleon to traverse the target nucleus, of the order of magnitude 10^{-22} s.

*Thus the word *spallation* does not define a class of nuclear reactions in terms of reaction mechanism as, for instance, the word *fission* does.

The nature and development of this fast cascade will depend upon the probabilities and kinematics of a large number of two-body collisions occurring over a wide energy range. The process is therefore well suited to treatment using Monte Carlo calculations. At incident energies above 400 MeV, meson production becomes significant and the propagation of the cascade is influenced by inelastic nucleon–nucleon collisions. Thus consideration of the production and re-absorption of mesons in nuclear matter is necessary. The meson influence becomes more severe the higher the energy and causes more complex reactions.

8.2.1.2 *The energy equilibration*

The nucleus at the end of the cascade will contain cascade particles (nucleons and mesons) trapped within its boundary, and holes originating from ejected nucleons. An equilibration will take place, and ultimately result in a random distribution of deposited energy among all available degrees of freedom. Previously, this was assumed to take place within a time similar to that of the cascade and was therefore included with it. Today equilibration is expected to be slower with an estimated time-scale of $10^{-20}-10^{-21}$ s[23]. This time should increase with energy because of the increasing number of available quantum states.

According to Harp *et al.*[23] the number of nucleons that will evaporate before energy equilibrium is attained is low, about 5% of the total in a medium heavy nucleus, when the excitation energy is of the same order of magnitude as its total binding energy. This seems to support the 'classical' way of treating a high-energy reaction as two (independent) steps, a cascade and an evaporation.

As a result the cascade step will lead to nuclei each having a certain amount of excitation energy—*cascade residuals*. Depending on the development of the cascade, these nuclei will have a range of Z and A values and a broad spectrum of excitation energy. The latter may rarely extend all the way up to that of the compound nucleus and is in general considerably lower[22, 24].

8.2.2 Calculations of the cascade step

The Monte Carlo calculations of the cascade step give information on the identity, energy and angular direction of the escaping particles and their multiplicities; A, Z, excitation energy (E^*) and linear and angular momentum of each cascade residual.

8.2.2.1 *Constant density nucleus*

These calculations use a square well potential and differ as to the choice of basic parameters such as depth of potential well, mean free path, cut-off (escape) energy, etc. The most detailed calculations along these lines are those by Metropolis *et al.*[22, 24] which give results up to 1.8 GeV protons on a number of targets.

The results have been extensively compared with experiments and show good agreement in many cases. Serious discrepancies are, however, also found and attributed by many mainly to the nuclear model used.

8.2.2.2 Core-skin nucleus

A model with a high density core surrounded by a skin in which the density gradually decreases from that of the core to zero has been used by Bertini[25], Chen *et al.*[26], Cohen[27] and most recently Bertini and Guthrie[28]. The latter considered incident particles up to 3 GeV on representative targets*.

All authors assume a density distribution following the shape of the charge distribution in the nucleus as given by Hofstadter[29] from electron-scattering experiments. In the Monte Carlo calculations this is approximated by step functions with several steps. The use of a degenerate Fermi gas momentum distribution within each concentric region results in a composite momentum distribution for the entire nucleus which differs from that of a degenerate Fermi distribution. A further consequence of the use of step functions is that the nuclear potential differs from one concentric region to the next. Energy conservation requires that the nucleons must change their kinetic energy when crossing the border between regions of different density. This is usually taken into account by changing the kinetic energy of the struck nucleon without changing its direction of motion (STEP (NO)).

Assuming that the variation of the nuclear potential is entirely due to central forces, the trajectory of a particle when entering a denser region will be more radial. This will increase the path length of the particle in question and thus the probability of interaction. In[41] (Refs. 26, 27) the effect of refraction and reflection due to the spatial non-uniformity potential has been investigated (STEP (YES)).

Extensive comparisons both with a great variety of experimental results[26] and with only (p, xn) and (p, pxn) cross-sections[30] seem to indicate that as a whole the simplest of the two approaches, STEP (NO), gives best agreement. This is rather surprising since from a nuclear model point of view STEP (NO) is less satisfactory than STEP (YES).

In addition, it should be borne in mind that the total reaction cross-section decreases when the skin thickness is reduced[25] and this explains why reactions in the skin, i.e. *peripheral* reactions, may be observed to have 'anomalously' low cross-sections.

8.2.3 The evaporation step

According to the usual assumptions a relatively long time will elapse before the second step occurs in the Serber–Goldberger, and more recent, models. This is the de-excitation of the cascade residuals which takes place by successive evaporations of nucleons (and smaller combinations of these), each with a few MeV of kinetic energy and assumed to be isotropically distributed in

*Computer programs for the Monte Carlo calculations[26, 28] are available from their authors.

the moving system. Each evaporation is assumed to be independent of the preceding one, and similar to the decay of a compound nucleus. The final nuclei are the (*primary*) *spallation products.*

8.2.3.1 Lifetimes of cascade residuals

The lifetimes of the cascade residuals depend on E^*. For heavy nuclei with E^* ten to a few tens of MeV lifetimes of 10^{-16}–10^{-17} s are reported[31,32]. At E^* hundreds of MeV, as is likely for cascade residuals resulting from GeV interactions, lifetimes are estimated to be rather short, 10^{-20}–10^{-21} s[33]. The narrow overlap between the cascade step and the de-excitation step[26] introduces a complexity which has not yet been taken into consideration in calculations.

From time to time the validity of the two-step model for certain reactions has been questioned. Most recently Poskanzer *et al.*[34,35], studying energy spectra and formation cross-sections of *cis*-Na elements from U and Ag + 5.5 GeV protons, arrive at the conclusions that the 'classical' model cannot explain their results. A reaction model, however, where a pre-equilibration evaporation is considered with a smooth transition from the cascade step to the evaporation step may be preferable[36].

8.2.3.2 Evaporation-fission competition

Depending on Z, A and E^* of the cascade residuals fission may compete with evaporation during each step of the de-excitation process. The probability for fission is highest in *trans*-Pb nuclei, but this process is also observed in lighter nuclei provided enough energy is available for nuclear deformation, cf. Ref. 37. The new-born *fission fragments* will de-excite through neutron evaporation forming (*primary*) fission products.

8.2.4 Calculations of the de-excitation step

8.2.4.1 The evaporation model

As the de-excitation typical of nuclei with low excitation energy is assumed applicable also to the highly-excited cascade residuals, the final step of high-energy reactions is usually described by statistical mechanics. In the calculations the evaporation formalism developed by Weisskopf[38] is used.

Monte Carlo calculations are again very powerful and are used to follow the fate of the cascade residuals. Extensive calculations are those by Dostrovsky *et al.*[39–43] using the results from the cascade calculations in Ref. 22 as input data. Calculations for low-mass nuclei were recently made by Bertini[44] and Dostrovsky *et al.*[45] with input data from Ref. 25 and Ref. 26, respectively. Due to limitations on computer time, all Monte Carlo calculations are today limited to events with reasonably high probabilities. Thus the analytical approach by Rudstam[46] which covers both low- and high-probability events is a valuable contribution for cross-section calculations.

Angular momentum effects and γ-competition are ignored in all these calculations. According to Grover[47, 48] angular momentum effects can alter the competition between neutron and γ-emission. The possibility of severe γ-competition should therefore be taken into consideration at high E^* and high spins. This would result in increased probabilities for high mass isotopic spallation products.

Competition from fission is taken into account by adding fission as an additional mode of de-excitation during each evaporation step. In doing so the neutron to fission probabilities, expressed through the respective widths, have to be given some type of energy dependence[49–51]. The approach is, however, somewhat arbitrary as the dependence of the fission widths with Z, A and E^* is not known at high excitation energies. Effects of angular momentum on this competition have been considered[52].

8.2.4.2 The break-up model

Here de-excitation takes place as a one step radioactive decay of the cascade residual into nucleons (and smaller combinations of these) and a daughter product (the primary spallation product). Thus the quantum mechanical[53] 'golden rule 2' is applied between the initial state (cascade residual) and the final state (all decay products). The expectations from the evaporation and the 'break-up' model have been compared with experimental results[54] and both seem to work equally well. This is explained by the fact that in the reactions studied where the excitation energy of the cascade residuals approached their total binding energy the dominant parameter is the density of the final states. This enters in the same manner in both approaches.

Epherre *et al.*[55] have studied the relative merits of these two models. They conclude: 'both are based on the fact that, when sufficiently averaged over, the nuclear parameters depend very little upon anything and may just as well be replaced by constants'.

Further calculations of spallation cross-sections are considered in Section 8.5, 'numerology', but first nuclear structure effects in spallation reactions (Section 8.3) and the products from high-energy reactions (Section 8.4) must be discussed.

8.3 NUCLEAR STRUCTURE EFFECTS AND SIMPLE HIGH-ENERGY REACTIONS

The contribution to the production of cascade particles from secondary and higher order collisions is substantial in heavy nuclei. In light and medium heavy nuclei, however, the effect of such collisions is much reduced due to smaller amount of nuclear matter traversed.

As the skin in light nuclei makes up a substantial part of the entire nucleus, a particle incident in the skin with a large impact parameter will only traverse low density matter. Hence, the probability for multiple interactions becomes still lower and structural factors may be much more apparent than when the particle enters the depths of a heavy nucleus. The effect of the skin and the core is clearly reflected in the angular distributions found with emulsion work[56, 57]. For low multiplicity events these distributions closely resemble similar events in free-particle collisions due to peripheral mechanisms, while high multiplicity events are somewhat modified by interactions taking place in the core of the nucleus.

When the contribution from cascades involving more than one collision is very small, simple reactions such as (p, n), (p, 2p), (p, pn) and (p, 2n) occur*; because there are a limited number of reaction paths resulting in cascade residuals with Z and A close to those of the target. Further, a single collision gives insufficient E^* to evaporate particles, not only are semi-classical calculations applicable, but the reactions can be used as crucial tests for Monte Carlo calculations. It turns out that for simple reactions the majority of Monte Carlo calculations fail because the reactions are composed of two types of events, core and peripheral, which according to Shapiro[58] should be treated independently, i.e. either as complex reactions which are dominated by phase-space factors or as simple reactions which are related to the energy variation of the nuclear probability.

Mechanisms assumed to be of importance in simple reactions are direct interactions with low momentum transfer to the struck nucleus and can be divided into[13] knock-out (P,PN) and (P,2P), inelastic scattering with subsequent nucleonic evaporation (P,Pn), (P,Pp), and charge exchange scattering (P,N), eventually with subsequent evaporation (P,Nn), (P,Np). At energies above the meson threshold emission of pions must be included.

8.3.1 Nuclear levels

Apparently Porile and Tanaka[59] were among the first to correlate structures in (p,pn) cross-sections with the availability of the uppermost neutron level for the (P,N) mechanism. As (p,pn) cross-sections are not very sensitive to fluctuations in single-particle neutron energies with neutron number, studies of (p,n) reactions are preferable even if the cross-sections are one to two orders of magnitude smaller than (p,pn) cross-sections.

Assuming direct ejection of the least-bound and next least-bound neutron, it was possible to calculate[60] the relative contribution from single particle neutron states to (p,n) cross-sections and to estimate reasonable values and trends of binding energies of the next least-bound neutron. Depending on the neutron states and the hole energy left by the knocked-out neutron, 50–100% of the (p,n) cross-section is due to neutrons emitted from the uppermost level, i.e. (P,N) mechanism. Thus trends of cross-sections in proton induced reactions involving single and two nucleon interactions should be predictable on the basis of (P,N) cross-sections and the spatial distribution of neutron states within the struck nucleus and the stability of these. The latter depends on whether the hole energy is smaller or larger than the binding energy of the

*In the following, overall reactions will be written in the standard way (i.e. in small letters), while when it is useful to distinguish between different reaction mechanisms, capital letters are used for incident particle and cascade nucleons, i.e. P, N, and small letters for evaporated nucleons, i.e. p, n.

most loosely bound neutron. The same conclusions are drawn from (γ,n) and (γ,p) reactions on light nuclei[61].

8.3.1.1 *Isomeric yield ratios*

Isomeric yield ratios from simple reactions are closely related to the level structures of specific nuclei. Caretto[62] measured the isomeric yield ratio of ^{117}In in ^{118}Sn (p,2p). Considering this as a (P,2P) reaction, a proton has to be knocked out of one of the four uppermost proton levels[63] in the target nucleus. If other protons were knocked out, the excitation energy of the In nucleus would be so high that a neutron evaporates.

The theoretical isomeric yield ratios were calculated[62] by taking into account the statistical populations of these four levels, the proton availabilities for successful (P,2P) events of each proton level[64] and the decay of the excited states to the isomeric states considering the minimum possible number of dipole γ-emissions. The results were in excellent agreement with experiment, which in addition shows energy independence (0.3–18 GeV).

8.3.2 Mechanisms in simple reactions

The validity of range and angular distribution studies of recoiling nuclei from simple (high-energy) nuclear reactions in singling out different reaction mechanisms and evaluation of their relative contribution to the total cross-section has most clearly been demonstrated by Remsberg *et al.*[65–67].

Angular distribution properties depend on relativistic kinematics only and not on details of the reaction model[67]. From studies[66] of ^{65}Cu (p,pn) experimental support for the correctness of the assumption[68] of independent contributions from the various mechanisms for (nucleon, 2 nucleon) reactions to the total cross-sections seems established. Thus in (p,pn) reactions the (P,Pn) mechanism is responsible for 20–30% of the total cross-section, while the (P,Np) mechanism contributes less than 2–3%. This leaves 70–80% to the knock-out mechanism (P,PN), including cascades accompanied by emission of pions. In the ^{63}Cu (p,n) and ^{63}Cu (p,2n) reactions, about two-thirds of the total cross-sections[67] seems to be due, respectively, to (P,N) and to (P,Nn).

Cross-sections are, however, rather sensitive to small changes in nucleonic configuration of the light target nuclei, both for simple and more complicated reactions[69]. A warning against generalising too much from the above results might therefore be justified.

Measured cross-sections of (p,xp) reactions in light nuclei with 154 MeV protons when compared[45] with calculated cross-sections[26, 40] confirm that the (p,xp) reaction products arise almost entirely from (P,2P), (P,3P), etc., and are but little influenced by evaporation.

8.3.3 Climbing the quantum ladder

When mesons are produced by proton interaction with nuclei it is very likely that a pion–nucleon isobar is formed within the nucleus. If low momentum (less than twice the Fermi momentum, i.e. below 0.5 GeV/c) transfers are involved in the intranuclear collision and this occurs in the skin, the decay products of the isobar might escape. Such reactions are:

(a) $p+p_{nucl} \rightarrow n_{nucl}+p+\pi^+$ with $Z_{target} \rightarrow (Z-1)_{product}$

(b) $\left\{\begin{array}{l} p+n_{nucl} \rightarrow p_{nucl}+p+\pi^- \\ p+n_{nucl} \rightarrow p_{nucl}+n+\pi^0 \end{array}\right\}$ with $Z_{target} \rightarrow (Z+1)_{product}$

Some very exciting studies involving such reactions have been reported[65, 70–72].

Remsberg[70] has studied trends in the excitation function for $^{65}Cu\ (p,p\pi^+)\ ^{65}Ni$ in view of calculations[73, 74] based on the one-pion exchange theory. The latter is known to describe pion production cross-sections well in free proton–proton collisions with low momentum transfer. Using for the energy dependence of the π^+–p cross-sections, experimental values and a constant value respectively, he is able to show the importance of the $(\frac{3}{2},\frac{3}{2})$ pion-nucleon resonance in the excitation function. In addition, this isobar, $\frac{3}{2}\Delta(1236)$ is 'observed'[71]. Because the kinetic-energy spectra of recoiling nuclei are directly related to the effective mass of all outgoing particles the 'observation' suggests that isobar–nucleon cross-sections are similar to nucleon–nucleon cross-sections. Only then could the isobar have a significant probability to leave the nucleus before it decays[70].

Similar studies of reaction type (b) also involve pion–nucleon resonances, i.e. the isobars $\frac{1}{2}N'(1520)$, $\frac{1}{2}N(1670)$ or $\frac{1}{2}N(1688)$ in addition to $\frac{3}{2}\Delta(1236)$. Calculated recoil kinetic-energy spectra at 15 degrees (2.5 GeV protons)[71] show that distinct peaks are expected. The (p,n) reaction which gives the same product as (b) will not interfere at these energies because the products will recoil at about 90 degrees[67].

Cross-sections for photo production of charged pions as a function of bremsstrahlung energy[75, 76] give results in good agreement with the above studies.[65, 70].

Thus such studies as reported in References 65, 70 and 71 have demonstrated the possibility of detection and of production of pion–nucleon (and strange particle) isobars by radiochemical means.

8.3.4 Skin thickness effects

Cross-sections for (p,pn) reactions just below the meson production threshold show a remarkable independence of A for medium heavy and heavy nuclei. These cross-sections are scattered within a narrow band (60 ± 5 mb). However, ^{58}Ni is a clear cut exception. Karol and Miller[77] investigated the effect of the thickness of the nuclear skin in this reaction and found that a skin thickness for ^{58}Ni slightly smaller than that for neighbouring nuclei could explain the low (p,pn) cross-section for this nucleus.

Neither absolute nor relative constancy of skin thickness is supported by electron scattering experiments. On the contrary, these show that for isotopic nuclei the A dependence of the charge radius is lower than the usual $A^{\frac{1}{3}}$ and that the skin thickness, generally assumed to be about 2.4 F, decreases with increasing A[78–80].

8.3.5 Pairing energy effects

Odd-even effects in formation cross-sections of products from high-energy reactions due to differences in nucleonic separation energies (i.e. pairing) and other details of the nuclear energy surface were suggested[81] long ago.

In radiochemical studies, however, such effects were not observed. This was explained[82] by nucleonic effects arising from level densities. Differences in separation energies which would favour even configurations should be cancelled by differences in level densities which would favour odd configurations. Marginally significant odd-even effects, however, have from time to time been tempting suggestions. Only quite recently were such effects proved experimentally.

Using on-line mass spectrometry Thibault[83] found (to her surprise) a 10–20% effect (favouring even N species) in the yields of isotopes of Na produced from a variety of targets with multi GeV protons. Tracy *et al.*[84] using 'a method of differences' was able to show that not only are these results significant, but also that similar studies[85] of isotopes of K, Rb and Cs produced in the same reactions show odd-even effects. The effect is about 5% for K, but much lower (insignificant?) for Rb and Cs.

Odd-even effects in the production cross-sections of light isotopes of *cis*-Na elements in U+5.5 GeV protons have been reported by Poskanzer *et al.*[34]. Even-even nuclides are favoured more than odd–odd ones.

Thus it is clear that odd-even effects on the nuclear energy surface are reflected in formation cross-sections. The results and their trends are not surprising as with increasing A, neglecting nuclear personalities, the general tendency is for a much slower decrease in pairing energies than increase in level densities.

8.3.6 Cluster effects

The most energetic particles (D, He, Li, Be, B nuclei) produced in the high energy interactions of protons with emulsion nuclei show a tendency for forward peaking in their angular distribution. This cannot be described by statistical processes involving only nucleons, but by the assumption of clusters or sub-structures in the skin.

The hypothesis is that there exist in the skin of the nuclei temporary groupings of nucleons which may give nucleons, deuterium, α and heavier clusters. Accepting the possibility that incident particles interact with such clusters, an analysis[86] is made on the basis of elastic or inelastic nucleon-cluster collisions[87] assuming appropriate pre-formation probabilities for the clusters in question and 'emission' of these early in the process, i.e. during the cascade step. Emulsion studies[27, 88–91] seem to support the cluster picture.

Such studies, however, do not answer the obvious question whether such structures are not also present in the core of medium heavy and heavy nuclei. If so, these clusters would hardly be observable due to the difficulty for a complex particle produced in a nucleon-cluster collision to escape from the nuclear core without being lost in secondary collisions.

That the experimental distributions are explicable assuming a finite probability of preformed clusters must not be taken to mean that these really exist in the skin of nuclei, but merely to indicate that nucleons in the skin may correlate in a manner similar to that in the assumed clusters, i.e. a temporary presence of quasi-clusters.

The same conclusions seem valid for the interactions of antiprotons with emulsion nuclei. Antiprotons deposit more energy than protons, but the angular distribution of emitted particles seems to be similar in the two cases[92–94] supporting the view that these particles are emitted on a fast time-scale, i.e. prior to the evaporation step.

Studies of the properties and emission of these and heavier clusters, or, better named, fragments, have quite recently taken a new line, and will be further discussed in relation to *fragmentation* in high-energy nuclear reactions (Section 8.7).

8.4 PRODUCTS FROM HIGH-ENERGY NUCLEAR REACTIONS

In high-energy particle-induced nuclear reactions the products become more complex and numerous the higher the incident energy and the heavier the target nucleus. This 'mess' finally consists of cascade and evaporated particles, and fragmentation, spallation and fission products. Most products give rise to radioactive decay chains and the half-lives involved range from infinity (stability) to fractions of seconds. Many species in the short-lived region are unidentified.

Our knowledge, however, is continuously expanding in this respect thanks to fast 'rabbit' and on-line techniques linked to accelerators or reactors. For information on ISOL (Isotope Separator On-Line) projects, the reader is referred to a recent review[95]. Studies of nuclear decay schemes made using such facilities are of the greatest value for nuclear reaction studies, since radiochemical cross-section measurements depend on pertinent decay schemes.

Complex reactions at high energies have been studied for target nuclei ranging all over the Periodic System and with incident particle energies up to about 70 GeV.

Recently, as high energy (GeV) electrons have become available, nuclear reactions induced by bremsstrahlung have again become popular after many years of 'magic sleep'. Such studies are, however, hampered by the continuous energy distribution of the photons, from threshold up to the maximum electron energy. Only energies above the photo meson threshold (150 MeV) are of interest for the present review. At these energies production of pions and cascade nucleons takes place – the latter giving rise to the well-known spallation and fission reactions. If the pions are adsorbed about 70 MeV is given to each of two nucleons which in turn may escape or produce nuclear excitation. The results will therefore be similar to medium-energy particle-induced reactions.

8.4.1 Presentation of formation cross-sections

Formation cross-sections $\sigma(Z, A)$ or yields of products from spallation, fission and fragmentation are often called primary or independent yields. These can be correlated in different ways depending on the information sought:

(1) Mass yields or isobaric yields $\sigma(A) = \Sigma\sigma(Z, A)$. These are related to the respective mass number in mass distributions, i.e. $\sigma(A)$ v. A (Figure 8.1).

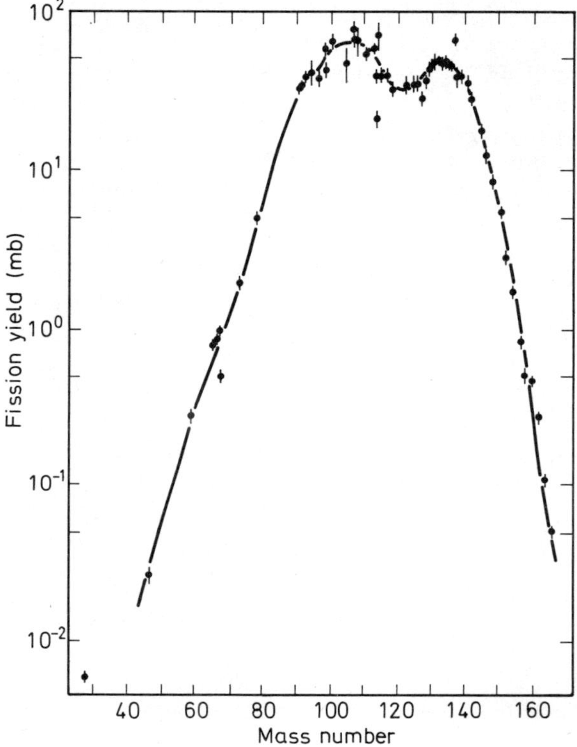

Figure 8.1 Mass distribution curve from the interaction of 170 MeV protons with uranium[172, 181, 182] (Section 8.6.2.1)

(2) Yields $\sigma(Z, A)$ or single products are used as such or as fractional yields, i.e. fraction of the corresponding isobaric yield: $f = \sigma(Z, A)/\sigma(A)$. Three different correlations are of interest here:

(a) Charge dispersion is $\sigma(Z, A)$ or f as a function of Z for constant A. The Z value corresponding to the maximum in the dispersion is the most probable primary charge Z_p. The variation of Z_p with A gives the charge distribution[17]. The N value corresponding to Z_p is N_p $(= A–Z_p)$, (Figure 8.2).

(b) Isotopic dispersion $\sigma(Z. A)$ as a function of A for constant Z, (Figure 8.3).

(c) N/Z dispersion (often confusingly called charge dispersion) or FFGY plots after Friedlander *et al.*[96] who were the first to plot $\sigma(Z, A)$ as a function of N/Z of the product (Z, A). The N/Z ratio corresponding to the maximum in

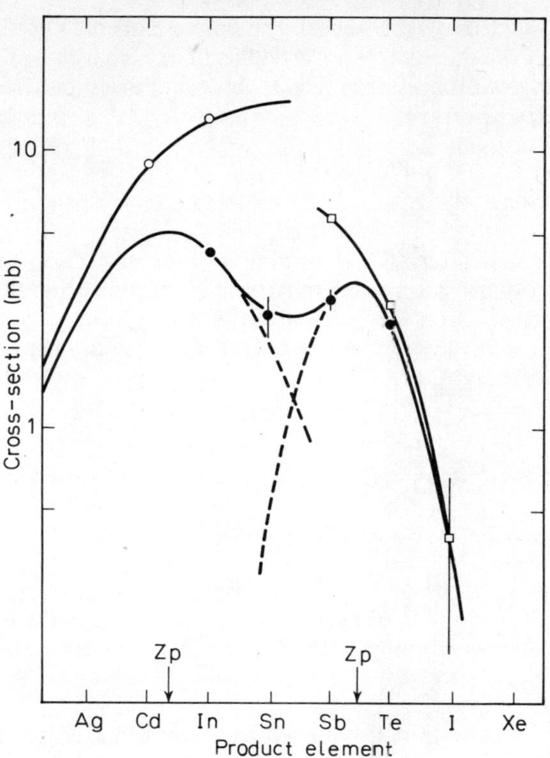

Figure 8.2 Charge dispersion for mass number 117 resulting from the interaction of 18 GeV protons with uranium[204, 234] (Section 8.6.2.2)

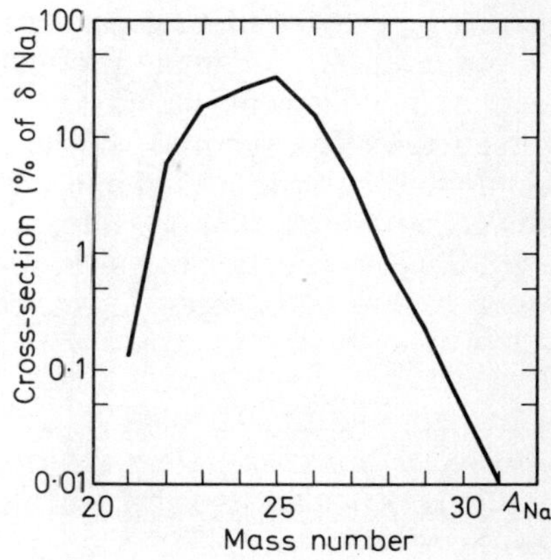

Figure 8.3 Isotopic dispersion for Na produced by the reaction of 25 GeV protons with uranium[83]. Odd-even effects in the cross-sections are demonstrated (Section 8.3.5)

this dispersion is $(N/Z)_m$. In general, the conversion of N/Z dispersions to charge dispersions requires[97,98] knowledge of the charge and mass distribution otherwise N/Z dispersions lead to apparent discrepancies with results from charge dispersions.

8.5 SPALLATION

Much of the present interest and emphasis on cross-section measurements of spallation products is directed towards testing the general applicability of spallation cross-section formulae and towards problems concerned with the origin of the solar system, space chemistry* (radiation ages, meteoritic studies) and nucleo-synthesis.

8.5.1 Cross-sections of interest to astrophysical problems

8.5.1.1 Isotopes of Li, Be *and* B *from light elements*†

Formation cross-sections of $^{6,7}Li$, $^{7,9,10}Be$ and $^{10,11}B$ from spallation of C, N and O with up to tens of GeV protons and up to multi-hundred MeV α-particles are now available[100–108]. The majority of the cross-sections is determined by mass spectrometry which correctly gives higher values than emulsion studies.

The isotope ratios obtained have led to revised models for the reactions responsible for nuclidic abundances. Substantial support is now given to the hypothesis of Gradsztajn *et al.*[86,109] for the spallation origin of Li, Be and B in the solar system and in the galactic cosmic radiation. The initial formation ratios have not been significantly altered by subsequent processes, such as (n,γ) reactions which were previously postulated.

8.5.1.2 Heavier products from 'meteoritic' elements

Formation cross-sections of spallation products from 'meteoritic' elements are today known for many products from the majority of the lighter and medium light elements including those of the first transition series. Protons with energies (few hundred MeV to about 30 GeV) representative of galactic cosmic rays have been used and the studies also include short-lived species thus giving information on isotopic and charge dispersions[104,110,111].

Studies of production cross-sections of isotopes of K are of importance for deducing radiation ages in meteorites. Many results are now avail-

*The chemical composition of galactic cosmic rays is mainly hydrogen, but helium is not negligible; about 10% compared to hydrogen. Higher elements are much less abundant.

†For a survey of available cross-sections for proton-induced reactions in He, C, N, O as of 1967, see Ref. 99.

able[83, 85, 111–113] which can be used to estimate production cross-sections of the stable isotopes $^{39,41}K$ in artificially bombarded 'meteoritic' elements (Fe,Ni). Combined with direct measurements in meteorites these give the original K content. Further reference is given elsewhere[114].

8.5.2 Cross-section formulae

The general pattern of formation cross-sections (yields) of spallation products from particle- and photon-induced reactions can advantageously be represented as a smooth spallation yield surface[115]. This is a three-dimensional representation with the nuclear chart as base, i.e. Z and N as horizontal axes and yield as the vertical axis. For products lighter than a few A and Z units below those of the target and ignoring the very light products, this surface has the form of a long and narrow ridge rising towards the target. The charge dispersion is approximately Gaussian with a full width at half maximum (FWHM) approximately independent of A and apparently independent of the target, incident particle and its energy. The peak values of the charge dispersions (i.e. the top of the ridge) and the mass distribution decrease exponentially with distance from the target. The Z_p follows the trend of the β stability line (Z_A), slightly on the neutron excess side, but at very high particle energies, Z_p may cross Z_A and move slightly over to the neutron deficient side. The slope of the ridge depends strongly on the incident energy and decreases for proton induced spallation up to a few GeV whereafter it levels off and seems to remain constant[116]. For photon-induced spallation the decrease is not as fast and does not seem to level off below the maximum energies (16 GeV) used[117].

8.5.2.1 Empirical formulae

These have been the success of numerology in its good sense.

The regular variation of spallation yields with Z and A of the products led Rudstam[82] to approximate the spallation yield surface by a simple four parameter formula. By proper choice of these parameters as deduced from a fit to measured yields, the final formula was good within a factor of two in predicting yields of products from 3 to 30 A units below that of the target mass ($A_t = 51$–75).

Results using this and other formulae cannot claim the same reliability as directly determined values. Its practical application, however, required a generalisation in order to cover most spallation conditions, and a fifth parameter was introduced[118]. These five ($P,R,S,T,\hat{\sigma}$) are regarded as free parameters determined by comparison with experimental results. Functional constants are introduced to take care of the variation of these parameters with such experimental conditions as the energy of the incident particle (in P), Z and A of the spallation product (R), Z_p and A_p of the products on the top of the spallation ridge (S,T), Z_t and A_t of the target, and the total inelastic cross-section $\hat{\sigma}(\sigma_i,A_t,E)$. The selection of the final form of the formula is based on best fit with about 1200 measured yields. A symmetric modified Gaussian

Charge Distribution together with an exponential Mass Distribution (CDMD) turned out to be good within an average factor of 3.

$$\sigma(Z,A) = f(\hat{\sigma},P,A_t)\cdot\exp[PA - R\,|\,Z - SA + TA^2\,|^{3/2}]$$

Considering the large range of cross-sections covered, a ratio of more than 10^6 between the highest and lowest yield and not least the uncertainties in the experimental values, this agreement, which is very good, strongly suggests that the yields of spallation products are really the sum of contributions from a very large number of reaction paths.

It was not possible for Rudstam to establish if the parameter P defining the slope of the ridge depends on the target for Z_t above 50, but some support[119] has later been found for a mass dependence above this value.

The CDMD (analytical) equation has been approximated by an integral equation[120]. Even if not as generally applicable as the original equation, this integral equation is very useful in helping to elucidate the spallation contribution in complex nuclear reactions (see for instance Reference 121).

During studies of spallation yields a strong correlation between yields and $(N/Z)_t$ of the target has been observed[112, 122]. This led to the introduction of two new parameters[113] in the CDMD formula. According to tests[111] this improved CDMD formula seems to fit experimental data only slightly better than the original version. There are, however, only few data available from enriched target isotopes with large mass differences.

The CDMD[118] formula has also been found applicable to photon-induced reactions[123].

Some authors have expressed the hope that new empirical formulae with even more parameters can be developed. It is well known that by having enough variable parameters, 'everything can be fitted to everything', however, with increasing loss of reality. It is to be hoped that such a 'science of numerology' will not flourish in future spallation studies.

8.5.2.2 Semi-empirical formulae

The recent years have seen some successful and interesting approaches in predicting spallation yields on a more physical basis. One of the first was probably that of Audouze *et al.*[124] as a result of studying complex reactions in light nuclei on the basis of an analysis made by Shapiro[58].

Experimental yields are governed by parameters related to the phase space available, and especially by the third component of the iso-spin, i.e. $T_Z = \frac{1}{2}(N-Z)$[109]. Spallation yields for all products corresponding to a given $\Delta T_Z = T_{Z_p} - T_{Z_t}$ (product–target) or one half the difference between the number of protons and of neutrons emitted, plotted v. ΔT_Z, give a Gaussian distribution as a function of ΔT_Z with a FWHM increasing with energy[124]. The position of the maximum in this distribution T_{Z_m} depends on T_{Z_t}. This is in accordance with results from studies of heavier targets, i.e. that $(N/Z)_t$ is strongly reflected in the most probable product in a memory type of effect (cf. Section 8.5.3) and is also borne out by Monte Carlo calculations for the cascade step[22, 24] giving the ratios of emitted neutrons to protons.

The total spallation yield for a product (A_p, T_{Z_p}) from a target (A_t, T_{Z_t}) is given by a CDMD-like formula, but which inherently (as in Ref. 113) allows for the Z and N composition of the target:

$$\sigma(A_t, T_{Z_t}; A_p, T_{Z_p}; E) = W\cdot\exp[-P\Delta A - R(T_{Z_p} - T_{Z_m})^2]$$

Here $\Delta A = A_t - A_p$ = total number of nucleons emitted, and W, P, R, T_{Zm} are phenomenological parameters. The estimates are claimed to be good within a factor of three for $T_{Zp} = 0$ and 1 as for the CDMD[118] formula, but for larger values of T_{Zp} not better than five.

The analytical approach to the evaporation step by Rudstam[46] (see Section 8.2.4) is based on a recursion formula for obtaining the probability for formation of a product Z,A within a given excitation energy interval by starting from an initial target nucleus with a given excitation energy. The Dostrovsky *et al.*[40] formulation of the evaporation theory[38] is used and different forms of the potential barrier tried (independent of, as well as dependent, on energy). Level density parameters, the radius constant and barrier penetration factors are varied in order to obtain the best fit to observed spallation yield distributions as represented by the CDMD formula[118]. The parameter values thus obtained are in accordance with those obtained by other means, but, as stressed by the author, the parameters involved are interrelated. Furthermore, the description used for each evaporation step and the level density formulae are theoretical or semi-theoretical and not satisfactorily proved to be valid at very high excitation energies, although they are at low energies.

8.5.3 Reaction mechanisms

8.5.3.1 Particle induced spallation

In high energy interactions in *cis*-Hg nuclei fission is expected to contribute only slightly to the total cross-section[37]. Studies of reactions in light and medium heavy nuclei should therefore give information mainly concerning spallation (and fragmentation).

Such studies show that whereas fission products have a most probable N/Z ratio (i.e. N_p/Z_p) which does not differ too much from that of the target nucleus, spallation products from 3 to 29 GeV proton interactions, e.g. from Ag[125] have a much lower N_p/Z_p (*c.* 1.17) than that of the target nucleus (*c.* 1.30) with the difference slightly increasing with increasing mass distance from a given target species. For a fixed isobaric chain the N_p/Z_p increases with target mass or when $(N/Z)_t$ increases[122, 125–128].

These experiments show that, not only for light targets[109, 124], products formed with small changes in the third component of the iso-spin seem to be favoured.

Katcoff *et al.*[125] have shown that the shapes of the curves describing the independent yield v. N/Z of the product are nearly identical (FWHM about 1.4 N/Z units) in different product mass regions (except close to or very far from the target) and also that the N/Z corresponding to the peaks of these distributions, $(N/Z)_m$ do not change from 3 to 29 GeV. This must be ascribed to an extended constancy of nuclear transparency beyond 2 GeV.

For isotopes of Sb and I from Ta and Au + 18.2 GeV protons[36,127] the isotopic dispersions are somewhat broader (FWHM *c.* 4 A units) than for light products like Sc (FWHM *c.* 3.5 A units). The peaks of the dispersions are respectively on the neutron deficient and neutron excess side of Z_A. As σ_f/σ_{Tot} is small for these targets and one may exclude products with $Z = 51$ and 53 arising from fission, the isotopic dispersions observed must be due to spallation alone. The importance of this for separating spallation and fission contributions in GeV interactions in U, as suggested by Rudstam and Sørensen[36], will be taken up in Section 8.6.

Another interesting observation in GeV interactions concerns the slopes of the different parts of the mass distributions. For Ag the total isobaric yields for masses above A *c.* 60 are higher at 3 GeV than at 29 GeV and vice versa for masses below A *c.* 60, where the change is at its largest below A *c.* 35[125]. For Au and Pb[121,129] the energy effect is such that at 29 GeV spallation plays a larger role than at 3 GeV in producing nuclides around A *c.* 80. where the total cross-section decreases between 3 and 29 GeV while the yields of neutron-deficient products increase slightly (Au) or remain constant (Pb). Thus as the incident particle energy increases, events requiring very high excitation energy become more probable at the expense of events requiring more moderate excitation energy. The effect which is barely noticeable in Cu, but rather pronounced in medium heavy (Ag) and heavy nuclei (Au, Pb, U)[17,121,125,130] may be ascribed to an increase in the pion multiplicities and a subsequent absorption of some of the pions produced. The absorption must be greater in heavy than in light nuclei, but with increasing energy it should reach a saturation due to increasing pion escape probability with increasing particle energy in the 10 GeV region.

8.5.3.2 Photon-induced spallation

The interest in photo-nuclear reactions is because quantum mechanical interference phenomena can be studied by (γ,xn) and other simple reactions.

The(γ,n) cross-sections above the photo meson threshold (150 MeV) turn out to be much greater[131–133] in Cs and I than expected from estimates of pion and nucleon transparencies. This is, however, most likely due to large contributions from photons in the giant resonance region as (γ,xn) cross-sections with x less than 10, from threshold up to 900 MeV, are well explained on the basis of free nucleon photo-pion cross-sections. The experimental (γ,xn) cross-sections[133] are in good agreement with results of cascade and evaporation calculations[134]. The measured (γ,xn) cross-sections are also well reproduced in calculations based on the quasi-deuteron model coupled with cascade and evaporation calculations[135].

8.5.3.3 Conclusion

A large number of formation cross-sections of spallation products from incident particles with energies up to 30 GeV on a great variety of targets are now available.

As a general conclusion concerning spallation studies one is tempted to propose that even if our theoretical knowledge of the reaction mechanisms is of a more general nature and that these mechanisms quite often are not too well understood, rather little remains to be done on cross-section measurements themselves, except for heavy target nuclei and high (GeV) incident-particle energies where the situation is considerably more complicated as result of overlap of all 'known' reactions (i.e. spallation, fission, fragmentation). The spallation of heavy elements is therefore discussed in connection with fission in the following section.

8.6 NUCLEAR FISSION

8.6.1 Fission cross-sections and fissilities

8.6.1.1 Cross-sections

Determination of fission cross-sections, σ_f, in the GeV region is seriously hampered by the difficulty of unambiguously distinguishing fission from other types of reactions. Radiochemical methods give integrated mass distribution curves, but only upper and lower cross-section limits in the GeV region[17]. This is because at these energies the distributions are very broad, with poorly defined limits, and the peaks well established at lower energies are more or less smoothed out. Thus track detector techniques based on direct identification or detection of the two complementary fission fragments should be a superior technique.

(a) *Proton-induced fission* – Fission cross-sections have been collected between 150 and 400 MeV for a large number of elements[136] and above 400 MeV for fewer elements[137, 138]. While at energies below 400 MeV there is in general a good agreement between results using different methods; above this energy results from different methods scatter within a factor of about two.

The trend of σ_f, as observed from loaded emulsion studies[139], that σ_f increases with energy, passes a broad maximum and then decreases monotonically, is borne out by the new data. For U, σ_f rises steeply, seems to be fairly constant at about 1.4 b from 150 MeV to about 1 GeV, then decreases to about 0.7 b at 29 GeV. Bi and Pb show similar trends, but rise slower reaching the flat maximum at 0.5–1.0 GeV and then decreasing. The relative cross-section values are for U:Bi:Pb are roughly 10:2:1 and 17:1:1 at the maxima and at 29 GeV, respectively.

By extrapolating results from Monte Carlo calculations of the cascade step[24], it can be deduced that with increasing energy about the 1 GeV region (1) a lowering of the fissionability parameter Z^2/A for *trans*-Pb elements occurs as result of knocked out neutrons and protons in the ratio of about $(N/Z)_t$, and (2) a substantial increase in the mean excitation energy of the cascade residuals will take place. According to Hudis and Katcoff[140] the σ_f v energy curves show that the effect on Z^2/A is more important than the increase in excitation energy. This is in contrast with the probability for fragmentation which is a weak function of Z^2/A, but more strongly dependent on the mean excitation energy[141].

The trends of σ_f with energy also seem to indicate that if the cascade residuals involved de-excite prior to fission, this de-excitation will most likely take place by balanced evaporation of protons and neutrons as a result of the high mean excitation energies concerned. If only neutrons were evaporated, Z^2/A would rise rather quickly down the evaporation chain.

From data on Au, Pt, Re and Ta it appears that σ_f for Au may reach a plateau at *c.* 1 GeV, while the data available for the other three elements are only good up to *c.* 0.6 GeV. Levelling off of σ_f for these elements only occurs at far higher energies[142].

Determinations of the energy dependence of σ_f for elements in the middle of the periodic system are extremely crude and fragmentary due to the techniques used, which do not register lighter products than *A c.* 30.

From the information given on paired and unpaired tracks in mica and the mass number limit for identification[140] it looks as if the asymmetry in fission of U increases with bombarding energy. Identification of such fission events is, however, extremely difficult, and it may turn out that these events may not be due to real fission, but partly to spallation, fragmentation or some 'fission-like' process. 'Deep' spallation is a term introduced by Crespo *et al.*[143] describing reactions resulting in strongly neutron-deficient products far removed from the target and having a threshold above 0.5 GeV. Fragmentation products are observed also below this energy and should have heavy partners. It may be possible to solve these questions by angular distribution studies as fragmentation products (*A* less than *c.* 30) should give abnormal distributions and not those corresponding to normal spallation or fission (Section 8.6.2.4).

(b) *Photon-induced fission* – Photo fission seems to be due to absorption of dipole E1 or M1 photons[144, 145]. The reported energy dependence of photo fission cross-section $\sigma_{\gamma,f}$ shows many contradictory results and interpretations, moreover some authors use bremsstrahlung[146–150] to induce fission while others use electrons[145] and deduce $\sigma_{\gamma,f}$ by different 'unfolding procedures'. The photo fission cross-sections for U and Th have rather sharp maximum at *c.* 400 MeV then decrease and start to level off at *c.* 800 MeV. This behaviour seems to support the suggestion that the photo mesons are most important in these processes[151]. Moretto *et al.*[145] using electron induced fission reject this view.

(c) *Antiproton-, pion-, and kaon-induced fission* – The $\bar{p}$–p and $\bar{p}$–n total interaction cross-sections decrease strongly with energy. For 2–3 GeV antiprotons these are about 70–80 mb, of which somewhat less than one-half is due to annihilation. For 30 GeV antiprotons the annihilation accounts for less than 10% of a total of about 45 mb[152].

Antiprotons will survive only a few collisions in complex nuclei due to annihilation, and a large fraction of the annihilation energy will appear in the kinetic energy of pion and cascade nucleons. In AgBr + 3.7 GeV antiprotons the average excitation energy is estimated to about 1.5 times that from protons with the same kinetic energy[92].

The fission cross-sections of U, Bi, Au + 2.5 GeV antiprotons are a factor of about 2 higher than for protons of the same energy[153]. The absorption cross-section for antiprotons in U is only about 30% higher than for protons[153].

The high antiproton fission cross-section may therefore be partly explained as due to greater angular momenta resulting from antiproton than from proton interactions, as the former has on the average a larger impact parameter[153].

This is due to the shorter mean free path of antiprotons which should also increase the mean deposition energy after the initial shorter cascade. The final result will most likely be a slightly higher average Z^2/A for antiproton than for proton interactions which, together with higher angular momenta, should result in higher cross-sections for antiproton-induced fission.

Pion-induced fission is an old story – observed by Perfilov *et al.* in 1951[154]. For information on this and other early works, see Ref. 9.

The first studies reported with GeV π^- are those by Husain and Katcoff[153]. The (2.5 GeV) π^- fission cross-sections for U, Bi and Au turn out to be the same as those for protons. A preliminary study of kaon (K^-) induced fission in AgBr has been reported[155].

The limiting factors in such studies are beam intensities and purities. With the meson factories under construction possibilities will be opened to study product distributions from meson induced reactions and thereby help clarify the interactions of these particles in nuclear matter.

8.6.1.2 *Nuclear fissilities*

The empirical relation between nuclear fissility at maximum fission cross-section, $\sigma_{f_{max}}/\sigma_{Tot}$ for proton-induced fission and Z^2/A is given by Perfilov[37]:

$$f = \sigma_{f_{max}}/\sigma_{Tot} \propto \exp[0.682(Z^2/A - 36.25)]$$

with $\sigma_{Tot} = \pi(1.26\ A^{1/3} - 0.41)^2 \cdot 10^-2$ b. When compared with the available data this equation is valid from U down to Z^2/A just below Ta, but experimental values for lower Z^2/A deviate increasingly with decreasing Z^2/A down to the lowest available data for Ag [156, 157].

The liquid drop model predicts that the fission barrier increases from a few MeV in the U region to a maximum of several tens of MeV in the Br region and decreases for still lighter elements[159, 160]. On this basis Nix and Sassi[161] examined nuclear fissilities. The trends obtained for σ_f/σ_{Tot} versus Z^2/A agree reasonably well with the Perfilov relation down to Z^2/A about 25. For lighter nuclei the Nix–Sassi dependence deviates significantly, showing a minimum in the Ag region (Z^2/A about 20), whereafter the fissility increases rapidly for still lower Z^2/A values.

It has been shown by Mitrofanova *et al.*[147] that in photo-fission fissilities of *trans*-Lu elements follow the Perfilov relation ($\sigma_{Tot_\gamma} = 335 \cdot A \cdot 10^{-6}$ b [162]). ‘exciting’ results, however, are due to Methasiri and Johansson[149] who studied photo-fission cross-sections of Pb, Au, Ta, Yb, Ho, Gd, Nd, La, Sn, Ag, Mo, Cu and Ni with 300 to 900 MeV bremsstrahlung using glass detectors. They found a well defined minimum in the fissility at Z^2/A about 24 (the La region) with $f = \sigma_{\gamma, f}/\sigma_{Tot}$ c. 2–3 $\times 10^{-4}$ and a steep increase with further decrease in Z^2/A reaching 3×10^{-2} for Ni, this element having a photo-fission cross-section above the threshold of about 0.6 mb, or about 0.1 that of Pb and a fissility like Au. Thus the trends indicated by the Nix–Sassi approach have found substantial support even if they suggest a minimum at Z^2/A about 20 and the predicted fissility for Ni is a factor 10–100 lower than observed.

Using the Perfilov[37] and the Methasari–Johansson results[149], proton fission cross-sections of about 20 mb can be predicted in the Ni region as compared to a σ_{Tot} of about 0.6 b or only about 1/3 that of Au for 600 MeV protons.

A new field for fission studies seems to have been opened where insight into new and probably exciting aspects of fission might result. The light nuclei have a fissionability parameter $x = Z^2/A/(Z^2/A)_{Crit.} \approx 0.3$, i.e. in the region where pear shape (odd) deformations are expected to become unstable[159, 163].

8.6.1.3 *Competition between neutron evaporation and fission*

An important question in high-energy fission studies is at what step in the competitive de-excitation of the cascade residuals fission occurs – early or late. There are indications that in heavy elements fission occurs late in the de-excitation when the cascade residuals are the result of interactions with protons above 100 MeV.

The fission probability at each evaporation step can be expressed through the respective widths (Γ), for fission, neutron evaporation and, if enough excitation energy is available, for proton evaporation. Different assumptions can be tried, e.g. that Γ_f/Γ_n is independent of excitation energy, shows a sharp break at a given excitation energy or follows the statistical model. The results are rather contradictory; some authors report that fission occurs late in the de-excitation of the cascade residuals while others are in favour of early fission.

Radiochemical studies have been made by measuring yields of products formed close to the cascade residual in complex reactions or to the target in simple reactions. These products cannot be formed as a result of fission, but their yields are decreased by fission. Thus, by comparing measured yields with yields calculated on the basis of the assumptions mentioned above, information concerning this competition can be obtained[30, 50, 150, 164, 165].

A more difficult approach is the direct measurement of post- and pre-fission neutrons. Most recently Cheifetz *et al.*[166] found that of a total of 11 neutrons 5.8 ± 1.0 and 6.9 ± 1.0 are pre-fission neutrons in U + 155 MeV protons and Bi + 155 MeV protons, respectively, as compared to 10.5 ± 1 in an earlier experiment[167, 168].

The experiments suggest that fission *on the average* occurs during the last half of the evaporation chain in proton induced fission. In photon-induced reactions (Au + 300 to 900 MeV bremsstrahlung) fission is estimated to take place close to the end of the de-excitation, i.e. after emission of 11 ± 2 neutrons out of a total of about 15[150].

According to Cheifetz *et al.*[166] their neutron results, radiochemical yields of spallation products in the vicinity of the target and fission cross-sections are most compatible if the relative fission probability, Γ_f/Γ_n, at least for Bi, is assumed to decrease at higher excitation energies.

For Bi, this is further supported by the necessity of a low fission barrier[169] in order to explain the yields of Po isotopes in Bi+156 MeV protons[30] and by the far greater level density for fission than for neutron emission needed to explain the total fission cross-sections in Pb, Bi+150 MeV protons[170]. As will be shown in Section 8.6.2.1, Γ_f/Γ_n decreasing with increasing excitation energy for U+155, 170 MeV protons is consistent with results from semiconductor studies[171] and from radiochemical studies[172] when examined in the light of the treatment of U+170 MeV protons by Rudstam and Pappas[173].

8.6.2 Distribution functions

8.6.2.1 Mass distribution

In drawing fission product distributions, it should be noted that the idea of reflection symmetry, which unfortunately evolved from crude studies, may be only qualitative, but many authors still assume it is justified and consequently bias their interpretations. Such reflection symmetry is not compatible with the broad spectrum of deposited energy in the cascade residuals. As fission at high excitation energies is dominated by the symmetric component and at low excitation energies by the asymmetric component, the composite mass distribution cannot be expected to be reflection symmetric. In high-energy interactions a fissioning nucleus cannot therefore be defined, only an average one. Thus even if yields of two mass numbers, i.e. a light and a heavy, are the same, it cannot be taken to signify that the sum of these mass numbers represents the average fissioning nucleus.

(a) *Symmetric and asymmetric fission components*—An analytical treatment by Rudstam and Pappas[173] on the basis of the Turkevich–Niday[174] two-component hypothesis and assuming that the probability of the symmetric and asymmetric component depends only upon excitation energy of the fissioning nucleus, disclosed that the mass distribution for fission of U+170 MeV protons could be expected to be neither single peaked nor reflection symmetric. Remsberg *et al.*[175] have studied the kinetic energy and momenta in the fission of U+2.9 GeV protons and shown that these properties are also not symmetric about some mean mass.

That the asymmetric fission component contributes substantially in the fission of heavy elements up to very high energies is shown experimentally both by the mass distribution resulting from very detailed radiochemical studies of U+170 MeV protons[96, 97, 172, 176–182] and by kinetic energy distribution studies of U+155 MeV protons using semiconductor detectors[171]. In the latter study the authors use the erroneous assumption that the mass distribution has an axis of symmetry at one-half of the estimated mass number of the fissioning nucleus. They are, however, aware that the mass distributions as calculated from the kinetic energy data hide valuable information which is disclosed in radiochemical studies.

(b) *Few hundred MeV fission*—The mass distribution in U+170 MeV protons is shown in Figure 8.1, based on collected data[172, 181, 182]. The light and heavy peaks are centred around mass 105 and 130, respectively, with apparent FWHM of about 23 and 18*A* units.

An analysis of the mechanisms behind the different parts of the mass distribution curves cannot be presented before the results on charge dispersion and charge distribution and on the recoil properties of the fission products have been reviewed.

Galin *et al.*[171] obtain about a 30% asymmetric fission component in the fission of U+155 MeV protons.

The excitation energy above which the symmetric fission component

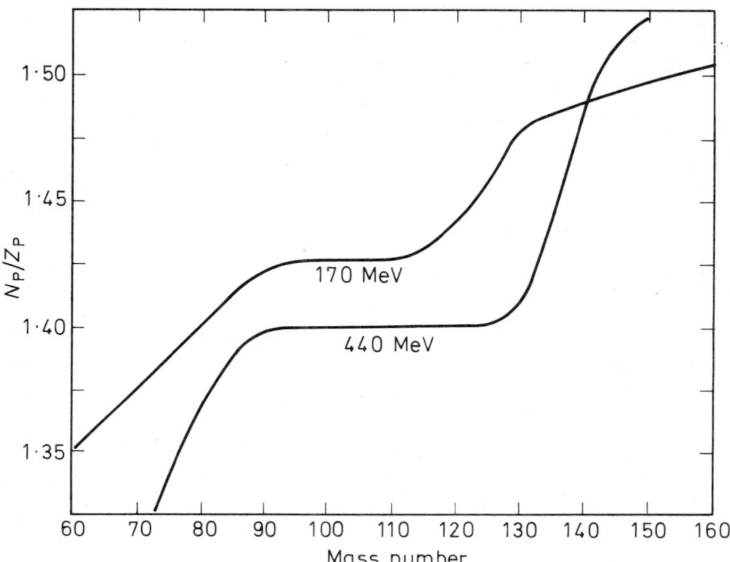

Figure 8.4 Neutron to proton ratios for most probable products N_p/Z_p as a function of product mass number. This is one way of representing the charge distribution. The examples are taken from the reactions of 170 MeV [172] and 440 MeV [98] protons with uranium

could be expected to dominate depends on both Z and A of the target, but for $Z = 89$ and 93, it is about 20 MeV [183] and 50 MeV respectively[173–184].

Substantial contributions from asymmetric fission are also confirmed by studies of the ranges in loaded nuclear emulsions[185]. Fission of U is most symmetric at a deposition energy 60–100 MeV. At higher energies asymmetric fission increases again as corroborated by Bychenkov and Perfilov[186] from studies of U, Bi, W+660 MeV protons. These authors also report increasing probability for range ratios from 1.5 to 2.5 with lower Z values of the target element, i.e. the mass distribution apparently broadens.

In W ($\sigma_f = 21$ mb), the mass distribution is a broad smooth hump over the mass region 22 to 140 (FWHM about 43 A units) If this is due to fission–as 'deep' spallation[143] and fragmentation should not be included in the correlated tracks used for fission identification–and off-hand explanation might be that the average deposition energy increases. The average deposition energy reported, 200–300 MeV, is substantially higher than expectations from Monte Carlo calculations[24], which give less than 200 MeV. It might therefore be that the energy distribution for events leading to fission is different from the total energy distribution of the cascade residuals.

(c) *GeV fission* – Increasing the incident proton energy to 3 GeV results, according to Remsberg *et al.*[175], in a much broader mass spectrum of fissioning nuclei than at lower energies, with an overall average mass number corresponding to the removal of roughly 36 nucleons from the target and an average deposition energy of 250 to 400 MeV. The fission mechanism, however, neither alters for U nor for Bi[175,187].

Studies of angular correlations between pairs of fission products show that the lowest excitation energies result predominantly in asymmetric fission with the total kinetic energy of the pairs averaging around 170 MeV, their masses being distributed around 135 and 95. The estimated contribution is less than 10% of the total fission events[175]. The asymmetric fission component seems to be correlated with low momentum transfer in the cascade step and with only a few evaporation steps.

As in fission of U + 170 MeV protons[172], light products seem to be associated with low average total mass (i.e. high deposition energy events) and heavy products with high average total mass (i.e. low or moderate deposition energy events).

Higher excitation energies lead predominantly to a symmetric fission component with total kinetic energy 155–160 MeV and average mass *c.* 105. Thus it looks as if the average separation between the charge centres of the products at scission is larger for the symmetric than for the asymmetric fission component[175].

A conjecture of the trend of symmetric fission probability using the data in Ref. 173 suggests that the asymmetric fission component should vanish in the multi GeV region.

With steadily increasing bombarding energies not only will the mass distribution become broader and flatter, but the most probable mass (if such an expression is still valid) will move towards lower mass numbers, i.e. from about 105 at 170 MeV (Figure 8.1) to *c.* 90 at 28 GeV[188] (Figure 8.6).

A phenomenological division of the different parts of the mass distribution at multi GeV energies seems still possible however, as pointed out by Crespo *et al.*[189]. These authors suggest that products heavier than 190 are spallation products due to the de-excitation, without fission, of cascade residuals. Products lighter than 45 are emitted particles (fragments, clusters) ejected in some way or other during the cascade or evaporation step. The mass region in between is reserved for fission products which, however, must be understood as superimposed on a base line due to 'deep' spallation.

In addition to the broadening of the mass distribution, products are formed in the fission product mass region with properties unknown for fission products. These 'new species' have N/Z ratios lower than those of fission products[17,36,96,127,190] and shorter recoil ranges[191–194], properties which are commonly ascribed to spallation products. In Section 8.5.2.1 it is stressed that the overall shape of the spallation surface is very little influenced by the target element and the bombarding energy; the main parameter is the separation in mass number between the target and the product.

Rudstam and Sørensen[36] compared isotopic dispersion curves for I formed in U + 590 MeV protons with those from U + 18 GeV protons.

While the yields increase in a smooth manner in the first process from about 0.08 mb for ^{118}I to a plateau at around 10 mb for ^{126}I and higher A isotopes, the 18 GeV dispersion shows substantially higher yields (0.5 mb for ^{118}I) of isotopes lighter than 123, at the expense of the heavier species which are lowered by a factor about 2. In addition a small peak at $A = 121$ is indicated.

The light wing of the dispersion at 18 GeV is almost identical to that of the dispersion following spallation of Au, Ta + 18 GeV protons, which has

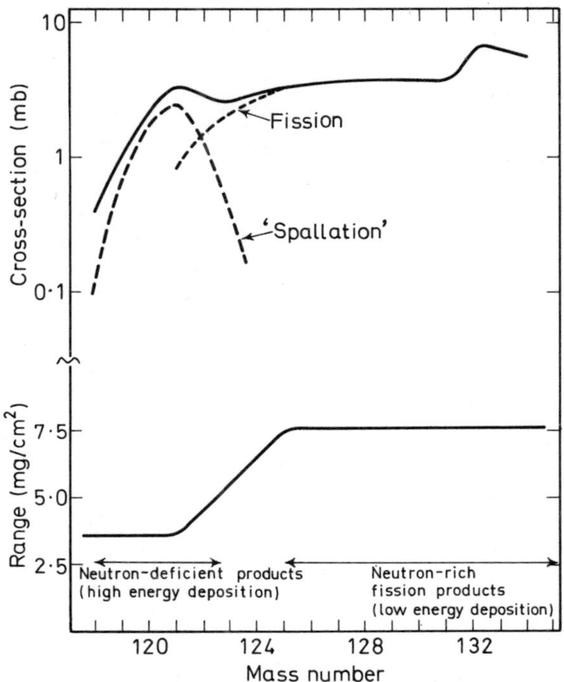

Figure 8.5 Isotopic dispersion of iodine from U + 18.2 GeV p decomposed in a spallation and fission part[36]. The interrelation to recoil ranges in uranium is shown[192].

its peak value at 121. Subtraction of the Au (Ta) curve from the 18 GeV U curve removes the spallation contribution in the latter and gives an isotopic dispersion having the same shape as that from U + 560 MeV. Hagebø and Ravn[127] find general support for this approach when isotopes of Sb are considered. For Br and other light products[193], however, the same authors[194] are more doubtful concerning the validity of this subtraction method (Figure 8.5).

The highest energy used in studies of mass distributions from proton interactions with U is, as of today, 28 GeV. Studies by Chu *et al.*[188] give an overall mass distribution and a few charge dispersions based on their own and other data[85, 121, 129]. The products in the mass region 50–160 have an integrated cross-section of about 2.3 b. Hudis and Katcoff[140] have measured by means of track detectors the binary fission cross-section for U at this energy and obtained a value of 0.67 b. Thus of the yields in this mass region 1.0 b may be due to processes other than fission. Chu *et al.* ascribe products beyond 160 to spallation (Figure 8.6).

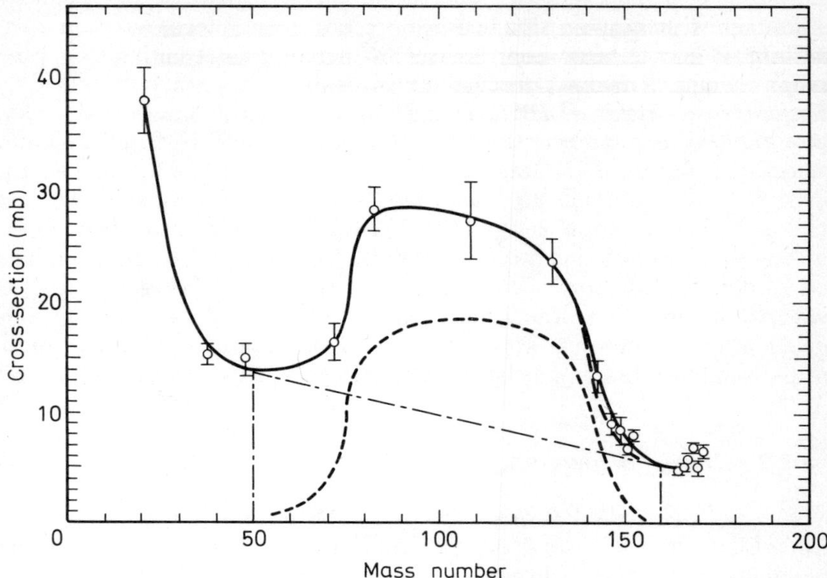

Figure 8.6 Mass yields versus A of some products of the interaction of U + 28 GeV p reproduced from the work of Chu *et al.*[188]. The dashed curve shows the fission contribution. (From Chu *et al.*[188], by courtesy of the American Institute of Physics)

It should be noted, no doubt, that neutron excess products in the mass region 50–160 result from binary fission. But as shown in the charge and isotopic dispersion studies by many authors, neutron deficient products are also formed within the same region (see Section 8.6.2.2).

If the spallation yield formula (Section 8.5.2) can be applied to 28 GeV interactions with U, a maximum of one-half of the 1.0 b non-fission events in the mass region 50–160 can be ascribed to spallation products[188]. The assumption of a fragmentation contribution of about 7 mb/mass number[195] results in a total of 0.7 b to be compared with a lower value of 0.5 b. In view of the crudeness of this whole approach a fragmentation contribution seems plausible and is further supported by recoil studies (Section 8.6.2.4).

(d) *Photo-fission* – Fission induced by bremsstrahlung energies ($E_{\gamma_{max}}$) up to 3 GeV has been studied in U, Th, Bi and Au[196–199]. The resulting mass distributions are double peaked, i.e. similar to low energy fission. The large amount of asymmetric fission is due to the low excitation energy (12–18 MeV) caused by the dominating giant resonance. The symmetric fission component is due to that part of the bremsstrahlung spectrum which is above the photo meson threshold. The peak to valley ratios v. $E_{\gamma_{max}}$ for U follow the trend already found by Schmitt and Sugarman[200] reaching values of about 4 and 2 at 650 MeV[196], and 3–4 GeV[197], respectively. Photo-fission[198, 199] of Bi and Au gives a single-peaked mass distribution slightly broader for Bi than for Au.

An interesting photo fission mass distribution study is the successful attempt by Schrøder *et al.* [199] to separate the cross-section of the symmetric and the asymmetric component in Th, U + 0.3 to 1.1 GeV $E_{\gamma_{max}}$. These authors show that both low and high deposition energies give rise to asymmetric fission. Thus there is no significant difference in the mechanisms involved.

Photo-fission of heavy elements seems therefore less exciting than photo fission of light elements as discussed previously in this review.

(e) *Mesonic fission* – Budick *et al.*[201] have recently shown that in slow pion-induced fission (excitation energy about 80 MeV) of ^{235}U the kinetic energy distribution of the fragments is centred around 57 MeV and that the distribution corresponds to the symmetric fission component. Fission of muonic ^{235}U (excitation energy 15 to 20 MeV), however, shows two peaks in the energy distribution centred around 55 and 82 MeV, i.e. the asymmetric fission component. For a determination of the relative contribution of the two fission types and other characteristics of these types of fission much higher intensity meson beams are required. A mesa-like mass distribution for pion-induced fission has been found earlier by Russell and Turkevich[202].

8.6.2.2 Charge dispersion

(a) *Few hundred MeV fission* – In U + 170 MeV protons the charge dispersion is Gaussian with a mass dependent FWHM[97, 172, 180]. This is narrower in the light and heavy than in the 'symmetric' region, i.e. in Z units: about 2.2 for A below 95 and above 140, and about 2.8 in the mass region 100–140. The results on the FWHM from the energy dependence studies[96] as obtained from N/Z dispersions at 100 MeV (3.3) and 200 MeV (3.8) in the mass region 127–134, when corrected for the observation[172] of a strong increase in N_p/Z_p with A in this mass region, give a FWHM 2.9 Z units in full agreement with the direct measurements. Recent mass spectrometric studies of Rb and Cs in U + 156 MeV protons[85] when based on the N_p/Z_p dependence in Ref. 172 give FWHM 2.1 and 3.0 Z units at mass around 90 and 135, respectively, in agreement with previous results from U + 170 MeV protons.

Charge dispersion studies of U + 450 to 600 MeV protons are known for limited mass regions only and seem to indicate that the increase in energy from 170 MeV does not change the FWHM in the region 100–130, it remains at 2.8–2.9 Z units[203, 204], but at around mass 140, however, the FWHM increases considerably with energy from 2.2 to 3.2 Z units. At high mass ($A = 153$) a value of 2.6 Z units is found[205].

(b) *GeV fission* – Above 1 GeV interactions with U a double-peaked charge dispersion starts to develop. This seems to have been discovered independently by Friedlander *et al.*[96], Alexander *et al.*[191] and by Alstad[206]. This discovery catalysed a veritable avalanche of charge dispersion studies[17, 36, 85, 121, 129, 188, 204, 207–210].

It appears that charge dispersions between 10 and 30 GeV show a more or less pronounced asymmetry over the mass region studied, i.e. *c.* 100–160 (cf. Figure 8.2). At some masses even distinct double peaks are observed with Z_p's on either side of Z_A.

The FWHM of the neutron excess part follows the trends from lower energies, i.e. constant in the medium mass region and increases at heavier masses. The neutron-deficient part with a FWHM about 2 Z units shows a resemblance to spallation.

8.6.2.3 *Charge distribution*

Studies of the variation of Z_p and of N_p/Z_p with A of the product will give information concerning average properties of fission. Here some very interesting results have been found[97, 98, 172, 180, 211]. The extensive data measured in U + 170 MeV proton fission show[172] (Figure 8.4) that N_p/Z_p is only constant over a small mass region 90–115, corresponding to the symmetric peak in the mass distribution which is centred around mass 105 with a FWHM about 23 A units. (Figure 8.1.) An adjustment of the data from 170 MeV to 450 MeV and including results from U + 450 MeV proton fission leads to a similar curve[98], i.e. N_p/Z_p is only constant in the mass region 90–125 (Figure 8.4). The flat plateau seems to increase in length with increasing energy, however, at a lower absolute N_p/Z_p value.

At high excitation energy fission will, as seen in Section 8.6.2.1, be predominantly symmetric and the products will have N_p/Z_p ratios not far from that of the fissioning nucleus and its excitation energy will be divided between the fragments according to the ratio of their masses. The N_p/Z_p ratio is therefore a measure also of the excitation energy after the cascade step (the neutron binding energies in the cascade residuals and in the neutron-excess fission fragments are about equal). On this basis the average number of neutrons evaporated prior to fission and/or from both fission fragments as function of fission fragment mass can be calculated[172].

The neutron emission curve can be transformed to a deposition energy curve by assuming that energy is dissipated at a rate of about 10 MeV per neutron emitted, and the result compared with those from recoil studies[98] which give the variation of average cascade deposition energy leading to (Z_p, A) as a function of A_p. The shapes of the curves turn out to be very similar and reflect the N_p/Z_p dependence, having a steep fall to A about 90, a long flat plateau to about 110 (170 MeV) or 130 (450 MeV), then a fall and levelling off towards A 150. It is very encouraging that two such different approaches lead to the same picture.

The previous discussion shows that one cannot draw conclusions of a general nature from measurements of charge dispersions at only a very few mass numbers. Therefore not only more, but also more detailed, charge dispersion studies are a necessity in order to reach a better understanding of fission reactions at GeV energies, where charge distributions are completely unknown.

8.6.2.4 *Kinetic energies and angular distributions*

The high kinetic energy of fission fragments is a consequence of the mutual coulomb repulsion at scission. The manner in which spallation products are generally formed, i.e. a cascade followed by an evaporation step, should result in a much lower average momentum than if the products arose from fission.

(a) *Models*—The analysis of experimental data is based on the two-step model, which in angular distribution and kinetic energy studies is expressed using the vector model, i.e. the velocity of a reaction product is given by the vector sum of the velocity at the first step and that at the second step. It is a fundamental assumption that these two velocities are independent of each other. In this analysis Monte Carlo calculations are used[26].

With this model a measurement of the angular distribution and average kinetic energy (T) for a given primary product gives information on the angular anisotropy of the products and the average E^* of the cascade residual.

As mentioned earlier in this review, the STEP (NO) approach[26] seems, surprisingly. to be in best agreement with the experimental results concerning the properties of cascade particles, cascade residuals, etc. Among others *Crespo et al.*[189] have tested the two-step vector model and conclude that their data (U+2.2 GeV protons) are consistent with this model. These authors deduce deposition energy values for cascade residuals leading to different products (neutron excess as well as deficient) and conclude that here also STEP (NO) is superior to STEP (YES).

Some authors, however, find a breakdown of the assumed relation between the deposition energy and the projection of the momentum parallel to the beam. This relation, when expressed relative to E^* of the compound nucleus and to the momentum of the incident particle, should not be strongly dependent on the nuclear model, target nucleus or incident particle energy[26].

Others go further and doubt the validity of the two-step model itself in explaining some of the data[34, 35, 191, 195, 209] and use this doubt as an argument for fragmentation (Section 8.7).

(b) *Angular distribution*—The angular distribution is of the form $a+b\cos^2\theta$, where θ is the angle between the direction of the moving fragment and the beam. Sugarman *et al.*[211] in studies of the products from U+450 MeV protons found a significant anisotropy which was further studied by Hogan and Sugarman[98], showing an increase in the anisotropy parameter b/a from about -0.1 at low deposition energies, reaching 0 at 120 MeV, and rising to about 0.3 at 280 MeV. This result seems to be supported by data for neutron excess species at low E^* in the fission of U+2.2 GeV protons[189]; however, the correlation[98, 211] with E^* does not seem to hold at 2.2 GeV.

Crespo *et al.*[189] used the Chen *et al.*[26] calculations to estimated spatial distribution of the angular momentum vectors for cascade residuals. The results here do not depend on using STEP (YES) or STEP (NO) and show, that at deposition energies E^* below 50 MeV, an orientation parallel to the beam is preferred. Above 250 MeV the angular momentum vector is preferentially perpendicular to the beam. The latter is the case for all events involving compound nuclei. The calculations of Chen *et al.* extend to 378 MeV only. However, it should be expected that cascades leading to low E^* will not change at higher incident energies from the anisotropy found for neutron-excess species in U+2.2 GeV protons[189]. Thus the negative anisotropy parameter found for heavy neutron-excess fission species[98, 189, 211] suggests that these probably arise from asymmetric fission of residual nuclei with low average excitation energy and a configuration close to that of the target.

Light neutron excess fission products do not seem to give a negative anisotropy parameter and therefore arise from a higher average deposition energy.

In U+2.2 GeV protons, the angular distribution of neutron-deficient products is strikingly different from those of neutron excess. The former are preferentially emitted forward and the latter sideways to the beam. Among the neutron deficient products a small amount of low velocity products is found and some spallation-like process seems necessary to account for these.

(c) *Total kinetic energy*—In general, the average total kinetic energy, for fission fragments from U, is only slightly lower at high energies than in thermal fission due to the lowering of Z and A of the fissioning nucleus (Table 8.1).

Table 8.1 Average total kinetic energy for fission fragments

Target element	*Proton energy*	*Total T* MeV	FWHM* MeV	*Reference*
U	156 MeV	168	31	212
	450	163		211
	2.9 GeV	160	40	175
Bi	2.9 GeV	135	30	187
Ta	156 MeV	118	24	212

*Approximate values.

The total kinetic-energy spectrum broadens as expected due to the increase in the number of fission channels available with the incident particle energy. The average total kinetic energy decreases for less fissile nuclei.

The recoil ranges of the neutron-deficient species drop by a factor about 2 between 0.4–0.7 GeV and 18 GeV protons on U, while the ranges of the neutron excess species change only a few per cent over the entire energy region.

This was probably first realised for isotopes of I from U+0.72 and 6.2 GeV protons by Alexander *et al*[191] and later shown for many other products[192, 193, 194, 209]. In the mass region 100–125 the drop in the isobaric ranges occurs over about 2 Z units and in the isotopic ranges over about 4 A units (Figure 8.5).

The short-range products have been described earlier in this review (Section 8.6.2.1) as spallation products. We will return to these once more in Section 8.7.

8.6.2.5 Isomeric yield ratios

Isomeric yield ratios (high- to low-spin isomers) of products from high energy fission are scanty and poorly understood. A few ratios from proton-induced fission in Bi and U have been shown to increase with energy up to about 200 MeV [204, 207]. Above this energy all the measured ratios stay constant with energy and are mostly characterised by a strong preference for the high-spin isomer. It may seem natural to ascribe this increase to an increase in angular momentum of the fissioning nucleus as the bombarding energy is increased.

Measurements on the anisotropy of fission products[98] show that the average orbital angular momenta of the fission fragments are in good agreement with the Monte Carlo calculations[26]. When it comes to the intrinsic angular momenta of the fragments, however, the situation becomes more complex because of the angular momentum produced during the fission act. This angular momentum, which is shown to be about 5 ħ (e.g.[213]) in low-energy fission, is ascribed to the wriggling and bending modes of excitation of the fissioning nucleus at scission[163]. This angular momentum is excitation energy dependent. As will be discussed in Section 8.6.3, recoil studies[98, 194] seem to indicate that the elongation of the fissioning nucleus increases with excitation energy. Thus the angular momenta given to the fragments at scisson may increase above the values predicted by the liquid-drop model.

Simple calculations[214] of fission fragment spins based on observed isomeric yield ratios for Sb products in fission (U + 570 MeV protons) indicate that the root-mean-square angular momentum of the fragment leading to these products is of the order of 20 ħ. The average cascade deposition energies giving these products are between 120 and 200 MeV[194].

8.6.3 Fission mechanisms

Asymmetric fission in heavy nuclei predominantly occurs in nuclei with low excitation energy associated with low momentum transfer and a short evaporation sequence. Symmetric fission results predominantly from fissioning nuclei with high excitation energy.

This was first pointed out by Sugarman *et al.*[211] from studies of U + 450 MeV protons. Remsberg *et al.*[175] have shown by measuring kinetic energies, velocities and angular correlations between complementary fission fragments in the mass region 85–110 from U + 2.9 GeV protons that the mass spectrum perpendicular to the beam shows a distinct contribution from asymmetric fission. The total kinetic energy about 170 MeV is centred around masses 135 and 95. Symmetric fission centred around mass 105 is predominantly a result of high deposition energy and shows a total kinetic energy 155 to 160 MeV.

The mean fragment velocity agrees with estimates based on the liquid drop model[163] and is only slightly sensitive to either charge or mass of the fissioning nucleus as well as[189] to a wide range of deposition energies.

Studies of mean fission fragment velocities in U + 2.9 GeV protons[175] gave 1.19 $(\mathrm{MeV/amu})^{\frac{1}{2}}$, and is close to the value for the low-energy fission of Bi 1.18 $(\mathrm{MeV/amu})^{\frac{1}{2}}$, which is representative for symmetric fission. For thermal fission of U, representative of asymmetric fission, the velocity is 1.22 $(\mathrm{MeV/amu})^{\frac{1}{2}}$.

The postulate by Sugarman *et al.*[211] that products in the mass region 47–140 from U + 450 MeV protons result from 'conventional' fission is corroborated by Crespo *et al.*[189] in U + 2.2 GeV protons from angular distribution and differential range studies of neutron-excess products in the mass region 90–140.

Recoil ranges of neutron excess nuclides in the mass region 100–115[209] show a deposition energy of the cascade residuals leading to these products of 146 ± 13 MeV at 11.5 GeV as compared to 123 ± 3 MeV at 450 MeV[211]. This agreement has been taken to suggest that the selectivity in the fission process at GeV and multi-hundred MeV incident energies is similar for cascade events leading to the most probable products.

The cascade deposition energy increases linearly with increasing Z along an isobaric chain with 25–35 Me/V unit[98, 203, 211]. This linearity is rather extensive both in Z (more than 5 units) and in deposition energy (a factor of about 10).

These studies also confirm that neutron deficient products are formed from higher deposition energy events rather than neutron-excess species. According to studies of U + 0.57 and 18.2 GeV protons[194] the slopes of deposition energy v. A for isotopic products seem to be independent of the target, incident particle energy and product type (Pd to Cs) and to average about 12 MeV/amu.

The study of isobaric chains at A about 140 by Hogan and Sugarman[98] also disclosed that the total kinetic energy decreases linearly with increasing Z of the isobar with about 3.0 MeV/Z unit.

It is then possible to deduce a relation between the total kinetic energy released and the cascade deposition energy:

$$E^* = \text{const.} - \frac{35}{3} T$$

These findings by Hogan and Sugarman[98] are explained as due to three different effects, each resulting in about a 10% reduction in the total kinetic energy release for neutron-deficient products relative to the neutron excess ones:

(1) Much of the high deposition energy which results in neutron-deficient products appears as excitation energy in the corresponding new-born fission fragments. This gives rise to a larger neutron emission and therefore larger loss in kinetic energy from neutron deficient products than for neutron excess ones, the latter arising from low deposition energy events.

(2) By conservation of momentum, which determines the kinetic energy at scisson, a heavy fission fragment which is highly excited*, receives a smaller fraction of the initial kinetic energy than a less heavy fragment (not complementary), leading to the same product mass.

(3) The effect of greatest interest is of increasing deformation with increasing excitation energy of the fissioning nucleus.

Crespo *et al.*[189] in U + 2.2 GeV protons also find a decrease in the total kinetic energy released, but using the liquid-drop model[163] they claim that most of the decrease is due to post-fission neutron emission (effect 1). Thus they get longer evaporation chains for neutron-deficient products and shorter ones for neutron excess products than suggested by Hogan and Sugarman[98].

Remsberg *et al.*[175], also from studies of U + 2.9 GeV protons, do not, however, observe the decreasing kinetic energy with increasing excitation energy of the fragment. These authors claim that the difference in total kinetic energy of

*It is assumed that fission occurs before appreciable neutron evaporation and that the total excitation (deposition) energy is divided in proportion to the masses.

isobaric neutron excess and neutron-deficient fission products is due to symmetric fission having the lower kinetic energy and asymmetric fission the higher.

New details on deformation studies have been obtained from U + 600 MeV protons[194]. The relation between the deformation at scission given by Hogan and Sugarman[98], expressed as the ratio of total kinetic energy to the total coulomb energy of touching spheres, and the total excitation energy, does not seem to be exactly the same[18] for the additional nuclides studied[194].

Instead of a smooth decrease with increasing total excitation energy, Hagebø and Ravn[194] find a rather sharp transition region between high kinetic energy fragments from asymmetric fission and low kinetic energy (more deformed) fragments in the symmetric region. Thus it looks as if there is an increasing deformation at scission with increasing excitation energy.

The exciting results concerning cascade deposition energies, kinetic energies and their relation to the final product as found by recoil and other techniques certainly require more detailed attention.

8.7 FRAGMENTATION

In previous sections we have occasionally mentioned fragmentation without however, giving a clear definition of this process. This is quite explicable, as the process or processes by which products in the mass region below 30 are produced in multi GeV proton interactions with heavy nuclei have been a puzzle ever since they were first observed.

The term fragmentation was first used by Kruger and Sugarman for a phenomenon which they could not explain as spallation in the classical sense. Using incident particle energies in the GeV region on heavy targets, Wolfgang *et al.*[215] observed the formation of what came to be known as fragmentation products. These are light nuclides, such as ^{18}F, ^{22}Na, ^{24}Na, ^{28}Mg and ^{32}P, i.e. 'pieces' or 'lumps' of nuclear matter which in some way or other are ejected from the nucleus. This 'process' has been discussed by many authors[69,191,216,218] and most recently by Beg and Porile[195] and by Poskanzer *et al.*[34,35]. If a fragment is emitted during the cascade, one would expect that a highly-excited residual nucleus is formed, which due to de-excitation through evaporation of many nucleons will tend to be situated very close to stability, or even slightly on the neutron deficient side of Z_A. The angular distribution of these 'heavy counterparts' will, owing to momentum conservation, be different from that of spallation products in the same mass region and their recoil ranges should be larger.

Studies to clarify the fragmentation mechanism have followed three lines: (1) energy and target dependences of cross-sections of 'classical' fragmentation products such as ^{22,24}Na, ^{32,33}P, (2) energy dependence of recoil properties and excitation functions of nuclides which might be expected to be in the mass region of possible 'counterparts' of the fragments, and (3) energy specta and angular distribution of 'lighter' fragmentation products. In each of these three categories interesting results have been obtained quite recently. More details of the background for such studies are given in the introduction to references[35, 195].

8.7.1 Fragmentation products

8.7.1.1 'Classical' products

Among the first which seemed to disobey the two-step model were ^{22}Na and ^{24}Na. Cross-sections of these and ^{32,33}P[83, 124, 182, 217, 219–225] are now available for a large variety of targets covering the region up to U and with incident proton energies 0.1–70 GeV.

The ratio of ^{24}Na (^{33}P) cumulative yield to ^{22}Na (^{32}P) primary yield from a given target is independent of energy[221, 223], while the absolute yields increase rapidly with energy. For a given energy the cross-sections show an exponential decrease with increasing target mass number A_t in agreement with expectations from spallation. However, this dependence holds only to A_t about 60, where the cross-sections go through a minimum in the mass region 100–200 and rise again for A_t above 200[216, 217, 219–221]. Superimposed on this general distribution are 'fission-like' peaks for targets with Z beyond 30[219, 221]. These peaks are most pronounced if cross-sections are plotted v. Z or N of the target[219, 220].

These maxima, when compared with the results of recoil studies, seem primarily due to low deposition energy events. Furthermore it is agreed that when produced from high-energy incident protons on heavy targets these fragments are the result of a 'break-up'-like process. Additional support for such a process is found in studies of Ne and Ar isotopes from Au, U + 3.3 GeV protons[121] and of ^{8}Li from Cu, Ag, Au[226].

Whether this 'break up'-like process is an extremely asymmetric fission[189], ternary fission[34, 227] or a very rapid process occurring during the energy equilibration in the cascade step[218] is not yet really settled.

8.7.1.2 'Counterpart' products

Beg and Porile[195] have studied the recoil properties of neutron-deficient isotopes of Sr and Ba from U as a function of proton energy (0.45–11.5 GeV). A constant recoil range corresponding to fission is observed up to *c.* 1 GeV, whereafter the ranges drop to about one-half over the region 1–5 GeV. Using a Rudstam–Sørensen[36] approach for decomposing the data into a spallation and fission part and taking into account excitation functions for neutron-deficient species, the authors compare their spallation part ranges with ranges from fission of targets (A_f) lighter than U and proton energies sufficiently low that fission is the main process. An assumption of a unique relation between the range of a product and A_f, irrespective of A_t and the proton energy, is inherent in this approach. Thus Beg and Porile arrive at an apparent fissioning nucleus with mass about 234 for energies below 1 GeV, then a fast decrease in A_f down to about 165 at energies around 5 GeV where it remains unchanged. A low A can also be estimated for the average nucleus resulting from a 5 GeV initial cascade followed by nucleonic evaporation.

Assuming that the average cascade deposition energy corresponds to about 10 MeV per nucleon emitted and using extrapolated values of average deposition energies from Monte Carlo calculations[24], a total of 70–75 nucleons is expected to be lost in an overall (spallation) process.

The apparently low A_f necessary for production of the neutron-deficient Sr and Ba isotopes would require an unreasonably large increase at about 5 GeV relative to about 1 GeV in probabilities of interactions involving deposition energies above 500 MeV at the expense of events leading to deposition energies around 200 MeV [195].

Experimental fission cross-sections in this target mass region increase with energy by a factor of at most 2 [140]. This would indicate a fission cross-section for the production of the neutron-deficient nuclides involved a factor about 300 lower than the measured cross-section[195].

These and additional arguments lead to the conclusion[195] that, provided all products in mass region 20–60 are the result of the same process, only fragmentation can explain the occurrence of the neutron-deficient Sr and Ba isotopes at about 5 GeV incident proton energies.

8.7.1.3 'Light' products

Studies of some of the light fragments, i.e. A below 20, gave strong support for including the nuclear energy surface when describing the fast process responsible for their formation[69]. Thus in the evaporation formalism used the binding energy of these products in the 'parent' nucleus is involved.

Studies of light fragments have recently been enjoying a new 'boom'. Thus Poskanzer *et al.*[34, 35] using physical methods, i.e. dE/dx, E measurements with semiconductor detector telescopes, have studied energy spectra, angular distribution and yields of fragments with Z up to 14 and 18 from 5.5 GeV proton interactions in U and Ag, respectively. However, because of threshold effects, individual isotopes can be studied only up to Z around 7. Again the nuclear energy surface is important and the yield surface is roughly similar to that known from spallation. However, it falls off more steeply on the neutron deficient than on the neutron excess side. The kinetic energy spectra also show distinct differences, as the high-energy tail is more pronounced and flatter for neutron deficient than for neutron-excess products. When resolved the neutron-deficient products are more forward peaked than the neutron excess ones.

In order to fit the energy spectra using the two-step model and assuming these products evaporated, high apparent nuclear temperatures, i.e. 10–13 and even up to 20 MeV are required in addition to a low Coulomb barrier (about one half the nominal one). This together with the forward peaking seems incompatible with the two-step model.

The simplest application of the two-step model to Monte Carlo calculations give results which we have seen in previous sections, seem to explain satisfactorily most phenomena occurring below GeV energies.

The growing experimental information which has become available in recent years particularly on the production of lighter fragments requires a more detailed and less prejudiced approach.

As stated by Poskanzer *et al.* the evaporation of fragments can be treated much better than hitherto by taking into account, for instance, pre-equilibrium evaporation (from the fragments), liquid-drop distortions, fission competition and angular momentum. This would exploit the full capabilities of the two-step model with less limiting assumptions and would be necessary for a realistic comparison with experimental data. Only when the model predictions are known in detail will it be possible to find the experimental aspects which require the introduction of a completely new reaction mechanism.

8.8 SUPER-HEAVY ELEMENTS

The search for super-heavy elements is of interest because theory[228–230] predicts the existence of nuclides with atomic numbers around 114 and neutron numbers around 184 with half-lives long enough to be isolated chemically once produced.

The studies discussed in Section 8.7.1.3 are of great interest in this respect as they clearly indicate the possibility of secondary reactions, i.e. reactions between the fragments produced and original target nuclei, because a significant fraction of the fragments has energies in excess of those required to induce reactions in U. Access to energetic neutron excess projectiles can thus be obtained, in addition to possible accelerated fission products[231]. Thus a new possibility for the production of super-heavy elements seems to have been opened.

In large angle scattering of multi GeV protons on heavy target nuclei, these might receive recoil energies high enough to pass the coulomb barrier in the target[232].

An attempt along this line is reported by Marinov *et al.*[232] using 24 GeV proton bombardment of W. However, great uncertainties exist in our understanding of both the production of the fast recoils and the subsequent reaction between the two heavy nuclei and demand more extensive experimental tests. In the scattering of 76 GeV neutrons on Ta, Flerov *et al.*[233] estimate the cross-section for production of 1 GeV Ta nuclides to be less than $2 \times 10^{-}3$ μb. Many nuclear chemists with access to heavy targets irradiated with multi GeV protons for long periods are presently looking for eka-elements, and new results will hopefully be available in the near future. The more relaxed ones look for similar reactions in lower Z targets in order to study the above mentioned reaction types by looking for heavier, but known, nuclides.

Acknowledgements

We are indebted to many colleagues who kindly made available results from their groups for this review, especially to Drs B. Forkman, G. Friedlander, E. K. Hyde and R. Klapisch.

We are very thankful to the following young members of the nuclear chemistry group at the University of Oslo for collecting the large amount of published information on high-energy nuclear chemistry and for additional help during the manuscript preparation: Mr T. Danielssen, Miss A. M. Habbestad, Mrs I. Haldorsen, Mr T. Lund and Mr M. Skarestad. We are deeply thankful to Mrs A. Birch-Aune for her never failing helpfulness which has contributed in no small measure to the preparation of the final manuscript.

One of the authors (ACP) expresses his gratitude to the University of Oslo for leave granted and to the Norwegian Research Council for Sciences and Humanities for additional support.

During the finalisation of such a review it is our experience that the probability that any of us will be found with his family decreases in a dramatic way. We wish to express our sincere thanks to our wives, who, as always, have shown admirable patience.

References

1. Shakespeare, W. *Macbeth,* Act 1, Sc. III, in The Works of W. Shakespeare, Odhams Press Ltd. (1944).
2. Wheeler, J. A. (1968). *Maria Sklodowska-Curie Centenary Lectures* (Proc. Symp. Warsaw 1967), 25, IAEA, Vienna
3. Weisskopf, V. F. (1963). *Int. Sci. Techn.*, June, 61
4. Templeton, D. H. (1953). *Ann. Rev. Nucl. Sci.,* **2,** 93
5. Spence, R. W. and Ford, G. P. (1953). *Ann. Rev. Nucl. Sci.,* **2,** 399
6. Rudstam, G. (1957). *Svensk Kem. Tidskr.,* **69,** 378
7. Miller, J. M. and Hudis, J. (1959). *Ann. Rev. Nucl. Sci.,* **9,** 159
8. Perfilov, N. A., Lozhkin, O. V. and Shamov, V. P. (1960). *Soviet Phys. Usp.* (English Transl.), **3,** 1
9. Hyde, E. (1964). *The Nuclear Properties of the Heavy Elements,* Vol. III (Englewood Cliffs: Prentice-Hall)
10. Hudis, J. (1968). *Nuclear Chemistry,* Vol. I, Ch. 3 (New York: Academic Press)
11. Harvey, B. H. (1959). *Progr. Nucl. Phys.,* **7,** 89
12. Cumming, J. B. (1963). *Ann. Rev. Nucl. Sci.,* **13,** 261
13. Grover, J. R. and Caretto, A. A. (1964). *Ann. Rev. Nucl. Sci.,* **14,** 51
14. Lefort, M. (1964). *Ann. Phys.,* **9,** 249
15. Perfilov, N. A. (1958). *Physics of Nuclear Fission,* Ch. 7 (London: Pergamon Press)
16. Ivanova, N. S. (1958). *Physics of Nuclear Fission,* Ch. 8 (London: Pergamon Press)
17. Friedlander, G. (1965). 'Physics and Chemistry of Fission' (*Proc. Symp. Salzburg* 1965), **2,** 265. IAEA, Vienna
18. Pappas, A. C., Alstad, J. and Hagebø, E. (1969). 'Physics and Chemistry of Fission' (*Proc.* II. *Symp. Vienna* 1969), 669. IAEA, Vienna
19. Serber, R. (1947). *Phys. Rev.,* **72,** 1114
20. Goldberger, M. L. (1948), *Phys. Rev.,* **74,** 1269
21. Chew, G. and Goldberger, M. (1952). *Phys. Rev.,* **87,** 778
22. Metropolis, N., Bivins, R., Storm, M., Turkevich, A., Miller, J. M. and Friedlander, G. (1958). *Phys. Rev.,* **110,** 185
23. Harp, G. D., Miller, J. M. and Berne, B. J. (1968), *Phys. Rev.,* **165,** 1166
24. Metropolis, N., Bivins, R., Storm, M., Miller, J. M., Friedlander, G. and Turkevich, A. (1958). *Phys. Rev.,* **110,** 204
25. Bertini, H. W. (1963). USA Report ORNL-3383, 1963; *Phys. Rev.,* **131,** 1801
26. Chen., K., Fraenkel, Z., Friedlander, G., Grover, J. R., Miller, J. M. and Shimamoto, Y. (1968). *Phys. Rev.,* **166,** 949
27. Cohen, J. P. (1966). *Nucl. Phys.,* **84,** 316
28. Bertini, H. W. and Guthrie, M. P. (1971). *Nucl. Phys.,* **A169,** 670
29. Hofstadter, R. (1956). *Revs. Modern Phys.,* **28,** 214
30. Beyec, Y. Le and Lefort, M. (1967). *Nucl. Phys.,* **A99,** 131

31. Brown, F., Marsden, D. A. and Werner, R. D. (1968). *Phys. Rev. Lett.*, **20,** 1445
32. Nielsen, K. O. and Gibson, W. M. (1969). *Bull. Amer. Phys. Soc.* **14,** 629
33. Ericson, T. (1960). *Advan. Phys.*, **9,** 425
34. Poskanzer, A. M., Butler, G. W. and Hyde, E. K. (1971). *Phys. Rev.*, **C3,** 882
35. Hyde, E. K., Butler, G. W. and Poskanzer, A. M. (1971). *USA-report UCRL*-19991, Berkeley
36. Rudstam, G. and Sørensen, G. (1966). *J. Inorg. Nucl. Chem.*, **28,** 771
37. Perfilov, N. A. (1962), *Soviet Phys. JETP,* **14,** 623
38. Weisskopf, V. F. (1937). *Phys. Rev.*, **52,** 295
39. Dostrovsky, I., Rabinowitz, P. and Bivins, R. (1958). *Phys. Rev.*, **111,** 1659
40. Dostrovsky, I., Fraenkel, Z. and Friedlander, G. (1959). *Phys. Rev.*, **116,** 683
41. Dostrovsky, I., Fraenkel, Z. and Winsberg, L. (1960). *Phys. Rev.*, **118,** 781
42. Dostrovsky, I., Fraenkel, Z. and Rabinowitz, P. (1960). *Phys. Rev.*, **118,** 791
43. Dostrovsky, I., Fraenkel, Z. and Hudis, J. (1961). *Phys. Rev.*, **123,** 1452
44. Bertini, H. W. (1968). *Phys. Rev.*, **171,** 1261
45. Dostrovsky, I., Gauvin, H. and Lefort, M. (1968). *Phys. Rev.*, **169** 836
46. Rudstam, G. (1969). *Nucl. Phys.*, **A126** 401
47. Grover, J. R. (1962). *Phys. Rev.*, **127,** 2142
48. Grover, J. R. and Gilat, J. (1967). *Phys. Rev.*, **157,** 814
49. Dostrovsky, I., Fraenkel, Z. and Rabinowitz, P. (1958). '2nd Int. Conf. Peaceful Uses of Atomic Energy' (*Proc. Conf. Geneva* 1958) **15,** 301, UN, Geneva
50. Lindner, M. and Turkevich, A. (1960). *Phys. Rev.*, **119,** 1632
51. Huizenga, J. R. and Vandenbosch, R. (1962). *Nuclear Reactions* (ed. P. M. Endt and P. B. Smith), **2,** 42. (Amsterdam: North Holland Publ. Co.)
52. Plasil, F. (1963). *USA-report UCRL-11193*, Berkeley
53. Fermi, E. (1950). *Nucl. Phys., Rev. Ed.*, 142, University of Chicago Press
54. Epherre, M. and Gradsztajn, E. (1967). *J. Phys.*, **28,** 48
55. Epherre, M., Gradsztajn, E., Klapisch, R. and Reeves, H. (1969). *Nucl. Phys.*, **A139,** 545
56. Bhowmik, B. and Shivpuri, R. K. (1968). *Can. J. Phys.*, **46,** 2527
57. Bhowmik, B. and Shivpuri, R. K. (1969). *Can. J. Phys.*, **47,** 195
58. Shapiero, I. S. (1967). 'Interactions of High Energy Particles with Nuclei' *(Proc. Int. School of Physics* 'Enrico Fermi' Varenna 1966), 210 (New York: Academic Press)
59. Porile, N. T. and Tanaka, S. (1963). *Phys. Rev.*, **130,** 1541
60. Read, J. B. J. and Miller, J. M. (1965). *Phys. Rev.*, **140B,** 623
61. Adler, J. O., Andersson, G., Forkman, B., Jonsson, G. G. and Lindgren, K. (1971). *Swedisch report NPR-LUND- 7104*
62. Caretto, A. A. (1967). *Nucl. Phys.*, **A92,** 133
63. Ross, A. A., Mark, H. and Lawson, R. D. (1956). *Phys. Rev.*, **102,** 1613
64. Benioff, P. A. (1960). *Phys. Rev.*, **119,** 324
65. Poskanzer, A. M., Cumming, J. B. and Remsberg, L. P. (1968). *Phys. Rev.*, **168,** 1331
66. Remsberg, L. P. (1968). *Phys. Rev.*, **174,** 1338
67. Remsberg, L. P. (1969). *Phys. Rev* **188,** 1703
68. Remsberg, L. P. and Miller, J. M. (1963). *Phys. Rev.*, **130,** 2069
69. Dostrovsky, I., Davis Jr., R., Poskanzer, A. M. and Reeder, P. L. (1965). *Phys. Rev.*, **139B,** 1513
70. Remsberg, L. P. (1965). *Phys. Rev.*, **138B,** 572
71. Remsberg, L. P. (1969). *Phys. Rev.*, **188,** 1698
72. Nikitjuk, L., Pokrovsky, V. and Ribakov, V. (1970). Preprint P6–5248, JINR, Dubna
73. Ericson, T., Selleri, F. and Van de Walle, R. T. (1962). *Nucl. Phys.*, **36,** 353
74. Selleri, F. (1967). *Phys. Rev.*, **164,** 1475
75. Nydahl, G. and Forkman B. (1968). *Nucl. Phys.*, **B7,** 97
76. Blomqvist, I., Nydahl, G. and Forkman, B. (1971). *Nucl. Phys.*, **A162,** 193
77. Karol, P. J. and Miller, J. M. (1968). *Phys. Rev.*, **166,** 1089
78. Hofstadter, R., Nöldeke, G. K., Oostrum, K. J. van, Suelzle, L., Yearian, M. R., Clark, B. C., Herman, R. and Ravenhall, D. G. (1965). *Phys. Rev. Lett.*, **15,** 758
79. Oostrum, K. J. van, Hofstadter, R., Nöldeke, G. K., Yearian, M. R., Clark, B. C., Herman, R. and Ravenhall, D. G. (1966). *Phys. Rev. Lett.*, **16,** 528
80. Ehrlich, R. D., Fryberger, D., Jensen, D. A., Nissim-Sabat, C., Powers, R. J. and Telegdi, V. L. (1967). *Phys. Rev. Lett.*, **18,** 959
81. Pappas, A. C. (1954). *Fra Fysikkens Verden,* **16,** 161 (Quoted in Ref. 82)

82. Rudstam, G. (1956). 'Spallation of Medium Light Elements'.' *Thesis*. Univ. of Uppsala, Uppsala
83. Thibault, C. (1971). 'Etude par spectrométrie de masse en ligne de noyaux légeres exotiques produits dans les réactions à haute énergie.' *Thesis*. Univ. of Paris, Orsay
84. Tracy, B. L., Chaumont, J., Klapisch, R., Nitschke, J. M., Poskanzer, A. M., Roeckl, E. and Thibault, C. (1971). *Phys. Rev.* (in press)
85. Chaumont, J. (1970). 'Contribution à l'étude de la fission par spectrométrie de mass en ligne.' *Thesis*. Univ. of Paris, Orsay
86. Gradsztajn, E. (1965). *Ann. Phys. (Paris)*, **10**, 791
87. Lozhkin, O. V. and Perfilov, N. A. (1965). *Jadernaya Khimiya, Izd. Nauka*, Moscow
88. Chakkalakal, D. A. and Barkow, A. G. (1966). *Nuovo Cimento*, **41**, 249
89. Grigorev, E. L., Lozhkin, O. V., Maltsev, V. M. and Yakovlev, Y. P. (1968). *Soviet J. Nucl. Phys.*, **6**, 507
90. Azimov, S. A., Aripov, R., Beter, E. V., Gulyamov, U. G., Igamberdiev, K. and Lozhkin, O. V. (1970). *Soviet J. Nucl. Phys.*, **10**, 652
91. Balashov, V. V. and Markov, V. I. (1971). *Nucl. Phys.*, **A163**, 465
92. Katcoff, S. (1967). *Phys. Rev.*, **157**, 1126
93. Baumann, G., Henny, D. and Cüer, P. (1967). *C. R. Acad. Sci. Paris*, **B264**, 1832
94. Daetwyler, J. J., Czapak, G. and Jeannet, E. (1968). *Helv. Phys. Acta*, **41**, 251
95. Talbert, Jr., W. L. (1970). *Proc. Int. Conf. Properties of Nuclei Far From the Region of Beta-Stability, Leysin.* CERN 70–30, **1**, 109, Geneva
96. Friedlander, G., Freedmann, L., Gordon, B. and Yaffe, L. (1963). *Phys. Rev.*, **129**, 1809
97. Hagebø, E., Pappas, A. C. and Aagaard, P. (1964). *J. Inorg. Nucl. Chem.*, **26**, 1639
98. Hogan, J. J. and Sugarman, N. (1969). *Phys. Rev.*, **182**, 1210
99. Audouze, J. Epherre, M. and Reeves, H. (1967). *French report IPN- 91*, Orsay
100. Bernas, R., Epherre, M., Gradsztajn, E., Klapisch, R. and Yiou, F. (1965). *Phys. Lett.*, **15**, 147
101. Epherre, M., Gradsztajn, E. (1967). *J. Physiol.*, **28**, 48
102. Yiou, F. (1968). *Ann. Phys. (Paris)*, **3**, 169
103. Yiou, F., Baril, M., Citres, J. D. de, Fontes, P., Gradsztajn, E. and Bernas, R. (1968). *Phys. Rev.*, **166**, 968
104. Rayudu, G. V. S. (1968). *J. Inorg. Nucl. Chem.*, **30**, 2311
105. Yiou, F., Seide, C. and Bernas, R. (1969). *J. Geophys. R.* **74**, 2447
106. Fontes, P., Perron, C., Lestringuez, J., Yiou, F. and Bernas, R. (1971). *Nucl. Phys.*, **A165**, 405
107. Jung, M., Jacquot, C., Baixeras-Aiguabella, C., Schmitt, R. and Braun, H. (1968). *C. R. Acad. Sci. Paris*, **266**, 1154
108. Jung, M., Jacquot, C., Baixeras-Aiguabella, C., Schmitt, R., Braun, H. and Girardin, L. (1969). *Phys. Rev.*, **188**, 1517
109. Bernas, R., Gradsztajn, E., Reeves, H. and Schatzmann, E. (1967). *Ann. Phys.* **44**, 426
110. Klapisch, R. (1970). *Proc. Int. Conf. Properties of Nuclei Far From the Region of Beta-Stability Leysin*, CERN 70–30, **1**, 21, Geneva
111. Cline, J. E. and Nieschmidt, E. B. (1971). *Nucl. Phys.*, **A169**, 437
112. Chackett, K. F. (1965). *J. Inorg. Nucl. Chem.*, **27**, 2493
113. Chackett, K. F. and Chackett, G. A. (1967). *Nucl. Phys.*, **A100**, 633
114. Kohman, T. P. and Bender, M. L. (1967). *High Energy Nuclear Reactions in Astrophysics*, Ch. 7, 169 (New York: Benjamin)
115. Halpern, I., Debs, R. J., Eisinger, J. T., Fairhall, A. W. and Richter, H. G. (1955). *Phys. Rev.*, **97**, 1327
116. Rudstam, G., Brunninx, E. and Pappas, A. C. (1962). *Phys. Rev.*, **126**, 1852
117. Fulmer, C. B., Toth, K. S., Williams, I. R., Handley, T. M., Dell, G. F., Callis, E. L., Jenkins, T. M. and Wyckoff, J. M. (1970). *Phys. Rev.*, **C2**, 1371
118. Rudstam, G. (1966). *Z. Naturforsch.*, **21a**, 1027
119. Neidhart, B. and Bächmann, K. (1971). *J. Inorg. Nucl. Chem.*, **33**, 2751
120. Schwarz, U. and Oeschger, H. (1967). *Z. Naturforsch.*, **22a**, 972
121. Hudis, J., Kirsten, T., Stoenner, R. W. and Schaeffer, O. A. (1970). *Phys. Rev.*, **1C**, 2019
122. Porile, N. T. and Church, L. B. (1964). *Phys. Rev.*, **133B**, 310
123. Jonsson, G. G. and Person, B. (1970). *Nucl. Phys.*, **A153**, 32
124. Audouze, J., Epherre, M. and Reeves, H. (1967). *Nucl. Phys.*, **A97**, 144
125. Katcoff, S., Fickel, H. B. and Wyttenbach, A. (1968). *Phys. Rev.*, **166**, 1147

126. Kaufman, S. (1963). *Phys. Rev.*, **129,** 1866
127. Hagebø, E. and Ravn, H. (1969). *J. Inorg. Nucl. Chem.*, **31,** 897
128. Trabitzsch, U. and Bäckmann, K. (1970). Personal communication to E. Hagebø
129. Porile, N. T. (1966). *Phys. Rev.*, **148,** 1235
130. Hudis, J., Dostrovsky, I., Friedlander, G., Grover, J. R., Porile, N. T., Remsberg, L. P., Stoenner, R. W. and Tanaka, S. (1963). *Phys. Rev.*, **129,** 434
131. Napoli, V. de, Dobici, F., Salvetti, F. and Carvalho, H. G. de (1967). *Nuovo Cimento,* **48B,** 1
132. Kato, T., Tsai, H. T. and Oka, Y. (1970). *Bull. Chem. Soc. Jap.*, **43,** 576
133. Jonsson, G. G. and Forkman, B. (1968). *Nucl. Phys.*, **A107,** 52
134. Jonsson, G. G. and Lindgren, K. (1970). *Nucl. Phys.*, **A141,** 355
135. Gabriel, T. A. and Alsnieller, Jr., R. G. (1969). *Phys. Rev.*, **182,** 1035
136. Konshin, V. A., Matusevich, E. S. and Regushevskii, V. N. (1965). 'Physics and Chemistry of Fission' (*Proc. Symp. Salzburg,* 1965), **2,** 349, IAEA, Vienna
137. Brandt, R., Carbonara, F., Cieslak, E., Piekarz, H., Piekarz, J. and Zakrzewski, J. (1971). *European report CERN-71-2,* Geneva
138. Rémy, G., Ralarosy, J., Stein, R., Debeauvais, M. and Tripier, J. (1971). *Nucl. Phys.*, **A163,** 583
139. Carvalho, H. G. de, Cortini, G., Muchnik, M., Potenza, G., Rinzivillo, R. and Loek, W. O. (1963). *Nuovo Cimento,* **27,** 468
140. Hudis, J. and Katcoff, S. (1969). *Phys. Rev.*, **180,** 1122
141. Gorichev, P. A., Lozhkin, O. V. and Perfilov, N. A. (1967). *Soviet J. Nucl. Phys.*, **5,** 19
142. Konshin, V. A., Matusevich, E. S. and Regushevskii, V. N. (1966). *Soviet J. Nucl. Phys.*, **2,** 489
143. Grespo, V. P. Cumming, J. B. and Alexander, J. M. (1970). *Phys. Rev.*, **C2,** 1777
144. Ranyuk, Y. N. and Sorokin, P. V. (1967). *Soviet J. Nucl. Phys.*, **5,** 377
145. Moretto, L. G., Gatti, R. C., Thompson, S. G., Routti, J. T., Heisenberg, J. H., Middleman, L. M., Yearian, M. R. and Hofstadter, R. (1969). *Phys. Rev.*, **179,** 1176
146. Carbonara, F., Carvalho, H. G. de, Rinzivillo, R., Sassi, E. and Murtas, G. P. (1965). *Nucl. Phys.* **73,** 385
147. Mitrofanova, A. V., Ranyuk, Y. N. and Sorokin, P. V. (1968). *Soviet J. Nucl. Phys.*, **6,** 512
148. Methasiri, T. (1970). 'Studies of High-Energy Photofission and Ternary Fission', Paper 1, *Thesis,* University of Lund
149. Methasiri, T. and Johansson, S. A. E. (1971). *Nucl. Phys.*, **A167,** 97
150. Lindgren, K. and Jonsson, G. G. (1971). *Nucl. Phys.*, **A166,** 643
151. Bernardini, G., Reitz, R. and Segré, E. (1953). *Phys. Rev.*, **90,** 573
152. Allaby, J. C., Bushnin, Y. B., Gorin, Y. P., Denisov, S. P., Giacomelli, G., Diddens, A. N., Dobinson, R. W., Donskov, S. V., Klovning, A., Petrukhin, A. I., Prokoshkin, Y. D., Stahlbrandt, C. A., Stoyaneva, D. A. and Shuvalov, R. S. (1971). *Soviet J. Nucl. Phys.* **13,** 295
153. Husain, L. and Katcoff, S. (1971). *Phys. Rev.*, **C4,** 263
154. Perfilov, N. A., Ivanova, N. S., Lozhkin, O. V., Ostroumov, V. I. and Shamov, V. P. (1955). 'USSR Peaceful Uses of Atomic Energy.' *Chemical Sciences,* 79, Moscow
155. Juric, M., Popov, S. and Todorovic, Z. (1970). *Nucl. Phys.*, **A140,** 154
156. Perfilov, N. A. (1965). 'Physics and Chemistry of Fission.' *(Proc. Symp. Salzburg,1965),* **2,** 283. IAEA, Vienna
157. Pappas, A. C. (1963). *2nd Informal European Conf. on the Interactions of High Energy Particles and Complex Nuclei,* Leysin, Quoted in ref. (158)
158. Pappas, A. C. (1966). *Z. Naturforsch,* **21a,** 995
159. Cohen, S. and Swiatecki, W. J. (1963). *Ann. Phys. (N.Y.),* **22,** 406
160. Myers, W. D. and Swiatecki, W. J. USA report UCRL-11980, 1965 and UCRL-17070, 1966, Berkeley
161. Nix, J. R. and Sassi, E. (1966). *Nucl. Phys.*, **81,** 61
162. Roos, C. E. and Peterson, U. Z. (1961). *Phys. Rev.*, **124,** 1610
163. Nix, J. R. and Swiatecki, W. J. (1965). *Nucl. Phys.*, **71,** 1
164. Lefort, M., Simonoff, G. N. and Tarrago, X. (1961). *Nucl. Phys.*, **25,** 216
165. Pate, B. D. and Poskanzer, A. M. (1961). *Phys. Rev.*, **123,** 647
166. Cheifetz, E., Fraenkel, Z., Galin, J., Lefort, M., Peter, J. and Tarrago, X. (1970). *Phys. Rev.*, **C2,** 256
167. Harding, G. N. (1956). *Proc. Phys. Soc. (London),* **A69,** 330

168. Harding, G. N. and Farley, F. J. M. (1956). *Proc. Phys. Soc. (London)*, **A69** 853
169. Beyec, Y. Le, Lefort, M. and Peter J. (1966). *Nucl. Phys.*, **88,** 215
170. Joopari, A. K. (1966). *USA report UCRL-16489,* Berkeley
171. Galin, J., Lefort, M., Peter, J., Tarrago, X., Chiefetz, E. and Fraenkel, Z. (1969). *Nucl. Phys.*, **A134,** 513
172. Pappas, A. C. and Hagebø, E. (1966). *J. Inorg. Nucl. Chem.*, **28,** 1769.
173. Rudstam, G. and Pappas, A. C. (1961). *Nucl. Phys.*, **22,** 468.
174. Turkevich, A. and Niday, J. B. (1951). *Phys. Rev.*, **84,** 52
175. Remsberg, L. P., Plasil, F., Cumming, J. B. and Perlman, M. L. (1969). *Phys. Rev.*, **187,** 1597
176. Lindner, M. and Osborne, R. M. (1954). *Phys. Rev.*, **94,** 1323
177. Hicks, H. G. and Gilbert, R. S. (1955). *Phys. Rev.*, **100,** 1286
178. Kjelberg, A. and Pappas, A. C. (1956). *Nucl. Phys.*, **1,** 322
179. Stevenson, P. C., Hicks, H. G., Nervik, W. E. and Nethaway, D. R. (1958). *Phys. Rev.*, **111,** 886
180. Pappas, A. C. and Alstad, J. (1961). *J. Inorg. Nucl. Chem.*, **17,** 195
181. Skarestad, M. (1971). 'Radiokjemiske studier av fisjon i Uran indusert av 170 MeV protoner', *Thesis,* University of Oslo
182. Haldorsen, I. R. (1971). Personal Communication to A. G. Pappas
183. Britt, H. C., Wegner, H. E. and Gursky, J. C. (1963). *Phys. Rev.*, **129,** 2239
184. Konecny, E. and Schmitt, H. W. (1968). *Phys. Rev.*, **172,** 1213
185. Shamov, V. P. and Lozhkin, O. V. (1956). *Soviet Phys. JETP,* **2,** 111
186. Bychenkov, V. S. and Perfilov, N. A. (1967). *Soviet J. Nucl. Phys.*, **5,** 186
187. Remsberg, L. P., Plasil, F., Cumming, J. B. and Perlman, M. L. (1970). *Phys. Rev.*, **C1,** 265
188. Chu, Y. Y., Franz, E. M. Friedlander, G. and Karol, P. J. (1971). *Phys. Rev. C.*, in press
189. Grespo, V. P., Cumming, J. B. and Poskanzer, A. M. (1968). *Phys. Rev.*, **174,** 1455
190. Rudstam, G. (1965). 'Physics and Chemistry of Fission' *(Proc. Symp. Salzburg, 1965),* **2,** 323. IAEA, Vienna
191. Alexander, J. M., Baltzinger, C. and Gazdik, M. F. (1963). *Phys. Rev.*, **129,** 1826
192. Brandt. (1965). 'Physics and Chemistry of Fission' *(Proc. Symp. Salzburg, 1965),* **2,** 239. IEAE, Vienna
193. Brandt, R. (1968). *Habilitationsschrift,* Philipps-University, Marburg
194. Hagebø, E. and Ravn, H. (1969). *J. Inorg. Nucl. Chem.*, **31,** 2649
195. Beg, K. and Porile, N. T. (1971). *Phys. Rev.*, **C3,** 1631
196. Sakamtoto, K. and Kuroda, P. K. (1966). *J. Inorg. Nucl. Chem.*, **28,** 679
197. Williams, I. R., Fulmer, C. B., Dell, G. F. and Engebretson, M. (1968). *J. Phys. Lett.*, **26B,** 140
198. Komar, A. P., Bochagov, B. A., Kotov, A.A., Ranyuk, Y. N., Semenchuk, G. G., Solyakin, G. E. and Sorokin, P. V. (1970). *Soviet J. Nucl. Phys.*, **10,** 30
199. Schrøder, B., Nydahl, G. and Forkman, B. (1970). *Nucl. Phys.*, **A143,** 449
200. Schmitt, R. A. and Sugarman, N. (1954). *Phys. Rev.*, **95,** 1260
201. Budick, B., Cheng, S. C., Macagno, E. R., Rushton, A. M. and Wu, C. S. (1970). *Phys. Rev. Lett.*, **24,** 604
202. Russel, I. J. and Turkevich, A. Unpublished. See Ref. 9, Ch. 12
203. Panontin, J. A. and Porile, N. T. (1968). *J. Inorg. Nucl. Chem.*, **30,** 2017
204. Hagebø, E. (1970). *J. Inorg. Nucl. Chem.*, **32,** 2489
205. Alstad, J. (1966). *3rd Informal European Conf. on the Interactions of High Energy Particles and Complex Nuclei,* Geneva.
206. Alstad, J. (1961). *1st Informal European Conf. on the Interactions of High Energy Particles and Complex Nuclei, Geneva*
207. Hagebø, E. (1967). *J. Inorg. Nucl. Chem.*, **29,** 2515
208. Ravn, H. (1969). *J. Inorg. Nucl. Chem.*, **31,** 1883
209. Panontin, J. A. and Porile, N. T. (1970). *J. Inorg. Nucl. Chem.*, **32,** 1775
210. Panontin, J. A. and Porile, N. T. (1971). *J. Inorg. Nucl. Chem.*, in press
211. Sugarman, S., Münzel, H., Panontin, J. A., Wielgoz, K., Ramaniah, M. V., Lange, G. and Menchero, E. L. (1966). *Phys. Rev.*, **143,** 952
212. Stephan, C. J. and Perlman, M. L. (1967). *Phys. Rev.*, **164,** 1528
213. Sarantites, D. G., Gordon, G. E. and Corvell, C. D. (1965). *Phys. Rev.*, **108B,** 353
214. Hagebø, E. (1971). *Studies of Proton-Induced Nuclear Fission, Inaugural Dissertation,* Universitetsforlaget, Oslo

215. Wolfgang, R., Baker, E. W., Caretto, A. A., Cumming, J. B., Friedlander, G. and Hudis, J. (1956). *Phys. Rev.*, **103**, 394
216. Crespo, V. P., Alexander, J. M. and Hyde, E. K. (1963). *Phys. Rev.*, **131**, 1763
217. Korteling, R. and Kiefer, R. (1970). *Phys. Rev.*, **C2**, 957
218. Cumming, J. B., Cross, R. J., Jr., Hudis, J. and Poskanzer, A. M. (1964). *Phys. Rev.*, **134B**, 167
219. Korteling, R. G. and Caretto, A. A. (1967). *J. Inorg. Nucl. Chem.*, **29**, 2863
220. Korteling, R. G. and Caretto, A. A. (1970). *Phys. Rev.*, **C1**, 1960
221. Hudis, J. (1968). *Phys. Rev.*, **171**, 1303
222. Juliano, A. and Porile, N. T. (1967). *J. Inorg. Nucl. Chem.*, **29**, 2859
223. Moskaleva, L. P., Fedoseev, G. A. and Khalemskii, A. N. (1971). *Soviet J. Nucl. Phys.*, **12**, 472
224. Simonoff, G. N. and Vidal, C. (1966). *Phys. Lett.*, **20**, 30
225. Lagarde, M. (1971). Personal Communication to A. G. Pappas
226. Katcoff, S. Baker, E. W. and Porile, N. T. (1965). *Phys. Rev.*, **140B**, 1549
227. Iyer, R. H. and Cobble, J. W. (1968). *Phys. Rev.*, **172**, 1186
228. Strutinsky, V. M. (1967). *Nucl. Phys.*, **A95**, 420
229. Strutinsky, V. M. (1968). *Nucl. Phys.*, **A122**, 1
230. Nilsson, S. G., Tsang, C. F., Sobiczewski, A., Szymanski, Z., Wyceck, S., Gustafson, C., Lamm, I. L., Möller, P. and Nilsson, B. (1969). *Nucl. Phys.*, **A131**, 1
231. Cheifetz, E., Gatti, R. C., Jared, R. C. and Thompson, S. G. (1970). *Phys. Rev. Lett.*, **24**, 148
232. Marinov, A., Batty, C. J., Kilvington, A. I., Newton, G. W. A., Robinson, V. J. and Hemingway, J. D. (1971). *Nature (London)*, **229**, 464
233. Flerov, G. N. Gangrsky, Y. P. and Orlova, O. A. (1971). *Soviet Report E7-5887*, JINR, Dubna
234. Hagebø, E. (1970). Unpublished results